ZELDOVICH
REMINISCENCES

ZELDOVICH
REMINISCENCES

Edited by

R. A. Sunyaev

CHAPMAN & HALL/CRC

A CRC Press Company

Boca Raton London New York Washington, D.C.

Library of Congress Cataloging-in-Publication Data

Zeldovich : reminiscences / edited by R.A. Sunyaev.
p. cm.
ISBN 0-415-28790-1 (alk. paper)
1. Zel§'ovich, ëì. B. (ëìkov Borisovich)—Contributions in nuclear physics. 2. Zel§'ovich, ëì. B. (ëìkov Borisovich)—Contributions in chemical physics. 3. Zel§'ovich, ëì. B. (ëìkov Borisovich)—Contributions in astrophysics. 4. Zel§'ovich, ëì. B. (ëìkov Borisovich)—Contributions in cosmology. I. Siuniaev, R. A. (rashid Alievich)

QC16.Z45Z45 2004
530′.092—dc22 2004043572

Direct all inquiries to CRC Press LLC, 2000 N.W. Corporate Blvd., Boca Raton, Florida 33431.

Visit the CRC Press Web site at www.crcpress.com

International Standard Book Number 0-415-28790-1
Library of Congress Card Number 2004043572
Printed in the United States of America 1 2 3 4 5 6 7 8 9 0
Printed on acid-free paper

CONTENTS

Introduction

It has now been 15 years since Yakov Borisovich Zeldovich left us. He mentioned to me once that all the wrong ideas and concepts die with their creators. Now it seems that we are able to say that he had many brilliant ideas which outlive him. His ideas and his name are still in the air during many astrophysical and cosmological conferences.

Zeldovich worked successfully in many fields of science, from chemistry to the theory of elementary particles. He was successful everywhere, and he addressed the widest variety of problems with great interest and admiration for what Mother Nature was hiding in these problems.

Cosmology was his last love and I remember that after 1970 he had lost practically all interest in any other branch of science (nuclear physics, elementary particles, physical chemistry, mechanics, applied mathematics, theory of shock waves and combustion, relativistic astrophysics, etc.) which were of great interest to him and where he got great results before the age of 55. He continued to write books on different topics with his colleagues and participated actively in conferences on different subjects, but we all knew that he devoted the last 20 years of his life to cosmology. And he was happy with his last love. He worked every day up to the last day of his life. Many years ago he told me how nice and interesting it is to enter a new field of science. It is necessary to make an effort to learn 10% of all general information about a subject. Then it is very easy to begin to work. Working hard very quickly you will reach the level where you understand 90% of the whole field. Then it is time to leave and to look for a new subject because to understand the last 10% it is necessary to spend a lot of time and effort. He did so, working in all his previous fields of interest but he was never able to think of leaving cosmology.

In recent years we have returned to the same stage where mankind was during the time of Magellan. For the majority of people Earth was infinite, full of mysteries before Magellan, and after his expedition anyone interested recognized that Earth is finite and has a rather simple geometry. After Magellan a lot of interesting problems still existed in geography but those were problems without global consequences relative to our understanding of our position in the Universe.

Now we are confronted with a similar situation with the whole Universe. Detailed studies of cosmic microwave background radiation, distant supernovae, clusters of galaxies, $Ly-\alpha$ clouds, etc. are providing us with all the main parameters and constants defining our Universe. In several years we may be able to recognize that our Big Bang Model is completely right and that the Universe is a very simple physical object or maybe we will find new unknown factors influencing its behavior. Dark energy is the first example.

The last 20 years of Zeldovich's life were unbelievably successful. He introduced the spectrum of primordial perturbations, which today is confirmed by all existing observations and is well known as the Harrison–Zeldovich spectrum. He predicted the acoustic peaks in the angular distribution of the microwave background radiation, proven three years ago by BOOMERANG and MAXIMA balloon experiments, and reconfirmed (with much higher accuracy) by WMAP spacecraft and many ground based experiments. He predicted as well thermal and kinematic effects making microwave observations of clusters of galaxies a powerful tool of observational cosmology. He introduced the beautiful Zeldovich approximation, demonstrating that the formation of the structure of the Universe should produce 'pancakes,' sheets and filaments due to the intersection of trajectories of particles. And we observe them today, not only in the real Universe but also in Cold Dark Matter numerical simulations, which today have become a great industry of modern cosmology. He was interested in the physics of the Λ-term in the Einstein equations and its importance for the evolution of the Universe. And I often think how excited he would be to hear about the recent discovery of dark energy being the main constituent of the surrounding Universe. He liked the idea of dark matter. It was an interesting revolution for a person who told me for 10 years that the mass of neutrino should be equal to zero because this is beautiful and natural, what else! But once he said that anything is possible which is not forbidden. How interested he would be today when we know that known types of neutrino might contribute to the average density of matter in the Universe more than all visible stars.

Unbelievable changes have occurred not only in his science but also in the country where he grew up and spent his whole life. The Soviet Union no longer exists and his three Gold Stars of a Hero of Socialistic Labor are now gold stars of a nonexistent country, even though they show how important and well recognized his contribution was to the 'Atomic project' in the Soviet Union. He spent more than 20 years working on the problems of weaponry. He participated in the creation of the simple Russian rocket KATYUSHA which became a very effective weapon during World War II. He worked on atomic and hydrogen bombs and believed that the bomb guaranteed the longest peace episode in the history of mankind. He was not able for many years to travel to the West and naively thought that if he were able to travel he would learn a lot. Now there are no longer great restrictions on travel from Russia but the science which he served is in disastrous condition there. The children of many of his friends and many of his pupils have emigrated from Russia. There are unbelievable changes in Russia, there are unbelieveable changes in the whole world and there are great changes in his beloved science — cosmology, and it is a great pity that he was unable to see all these exciting or sad developments.

It is very difficult to imagine his behavior in present day Russia, where there have been tremendous changes (mainly positive), but I don't know how it would be possible for him to be happy with the great decline in the funding and support of fundamental science. He served science his whole life and obviously it would be very difficult for him to watch the increasing lack of interest in fundamental science in Russia. However, I remember him saying that Alexandr Friedmann found his solution for the expansion of the Universe in the hungry and frozen Petersburg of 1921 when living conditions were incomparably worse than in present day Russia.

Next year we will commemorate the 90th birthday of Yakov Zeldovich, and it would be nice if reading this book people could see how complicated, and at the same time exciting, the life of this tremendously talented man, great scientist, fascinating person, and great Mentor was.

Rashid Sunyaev
Moscow–Garching–Pasadena
2000–2003

Preface to the Russian edition

This book contains reminiscences about Academician Yakov Borisovich Zeldovich by his students, friends, and colleagues. Similar collections about a number of eminent scientists have been published in recent years. What is their value? In our view, it is not simply to pay homage to people who have forged new paths in science. Being written by a set of completely different people, they allow us to create an objective view of a scientist and his characteristic personal traits — to peek into his scientific laboratory and become acquainted with his methods and approaches to problems in his scientific work and his life. This is extremely useful for those who are interested in science, and especially for those who are just beginning a career in science or are preparing to begin such a career. Since they inevitably touch upon circumstances and stages in the solution of important scientific and technical problems, scientific memoirs are of undoubted interest for the history of science. Together with this, they are concerned with the work of a scientist, the form of his thoughts, and the general mood of the surrounding society, so that they recreate the epoch in which he worked. In this sense, they are of interest (as are any memoirs) for future generations.

This aspect of the collected reminiscences of Yakov Borisovich takes on special meaning if we recall the problems on which he was working in the 1940s and 1950s. It is for precisely this reason that we have not tried to 'clean up' individual contributions, and have restricted our editing to shortening some of the unavoidable repetition in cases when the same episodes are described in chapters written by different authors. The book also includes some documents related to Yakov Borisovich: official reviews, letters, and so forth. The article *In Memory of a Friend* by Zeldovich himself is likewise included, since it seems to us that it reflects well his style and temperament.

The layout of the material is primarily chronological, and reflects the major areas of science with which the name of Ya.B. Zeldovich is associated first and foremost.

Corresponding Member, RAS, S.S. Gershtein
Academician R.A. Sunyaev
1992

Acknowledgments

This book is based on the Russian edition which appeared in 1993. We edited the Russian version with Prof. Semyon Gershtein and many people who had known Zeldovich for many many years wrote their memories immediately after his death. Natalya Morozova was a compiler of the Russian version of the book and made the Russian text much more pleasant to read.

Some of the photos for the Russian book were taken from the unique archive of the Institute of Chemical Physics organized by Igor Bavykin. Igor Bavykin prepared all the photos in the Russian edition for printing.

For this English edition of the book we were requested to supply originals of the photos. The possibility was open to double the number of photos. Zeldovich's daughters Marina and Olga chose additional photos and provided all the requested original photos from the family archive.

Denise Gabuzda translated the majority of the papers and Rachel Harrison did a good job of editing the manuscript. Victor Selivanov translated part of the last contributions and has managed the progress of the project during the final stages of preparation for press. Janie Wardle managed the project for six years.

Today some of the authors of the Russian version are no longer with us. Some emigrated and are working in the United States and in different areas in Europe. It was not simple to find all of them and ask them to read the proofs of the translation. I'm grateful to all the authors for replying and sending their corrections to the text. Several people added new parts to their old texts. Ten new papers were written especially for this edition. I'm especially grateful to Jim Peebles for the long afterword to his report written initially in 1987. From this report we see how the contribution of Zeldovich to cosmology was seen in 1987 and then in 2002. Ed Salpeter and Bill Press sent photographs for this book. Marina and Olga helped proofread the manuscript.

I am grateful to everyone mentioned above for their contributions and to the copyright holders of the Russian edition of the book: Division of Physics of the Russian Academy of Sciences, Natalya Morozova and Igor Bavykin for supporting the idea of the English edition.

Rashid Sunyaev

*A Short Biography**

Born 8 March 1914 in Minsk. Died 2 December 1987 in Moscow.

Father, Boris Naumovich Zeldovich, lawyer, member of the collegium of barristers; mother, Anna Petrovna Zeldovich (Kiveliovich), translator, member of the Writers' Union.

From mid-1914 till August 1941 lived in Petrograd (later Leningrad); till the summer of 1943, in Kazan'; from 1943, in Moscow.

In 1924, entered the third grade of secondary school, which finished in 1930. From the autumn of 1930 till May 1931, studied at courses and worked as a laboratory assistant at the Institute of Mechanical Processing of Mineral Resources. In May 1931, was enrolled as a laboratory assistant at the Institute of Chemical Physics, USSR Academy of Sciences; was associated with this institute till his last day.

Having started working at the Institute of Chemical Physics without higher education, was self-educated with the help and under the leadership of Institute's theorists. From 1932 till 1934, studied at the correspondence department of the physico-mathematical faculty of the Leningrad University, from which he did not graduate; later attended lectures at the Physics-Mechanics Department of the Polytechnical Institute.

In 1934, was admitted to the postgraduate studies of the Institute of Chemical Physics; in 1936 defended his PhD thesis, and in 1939 the DSc thesis.

From 1938, headed a laboratory at the Institute of Chemical Physics. In late August 1941, was evacuated to Kazan' together with the Institute. In 1943, was transferred to Moscow together with his laboratory. From 1946 till 1948, headed the Theoretical Department of the Institute of Chemical Physics. Simultaneously, till 1948, was professor at the Moscow Engineering Physics Institute.

From February 1948 till October 1965, dealt with defence studies on the atomic problem, in connection with which was awarded a Lenin Prize and, three times, the title of the Hero of Socialist Labor; was head of the department and deputy head of the facility that later became known as the All-Russian Research Institute of Experimental Physics, Sarov or Arzamas-16.

From 1965 up to January 1983, headed a department of the Institute of Applied Mathematics, USSR Academy of Sciences. From 1965, was professor

*Compiled on the basis of Ya.B. Zeldovich's autobiography from his dossier stored at the Archive of the Russian Academy of Sciences.

of the Physical Faculty of the Moscow State University and headed the Department of Relativistic Astrophysics of the Sternberg Astronomical Institute.

From 1983, was head of a department of the Institute of Physical Problems, USSR Academy of Sciences, consultant of the Board of Directors of the Institute of Space Research, USSR Academy of Sciences. From 1977, headed the Scientific Council on Combustion, USSR Academy of Sciences.

In 1946, was elected corresponding member of the USSR Academy of Sciences; in 1958, was elected academician.

Major Directions of Scientific Work:

• Heterogeneous catalysis and adsorption (experimental and theoretical works, 1932–1936); problems of adsorption also served as the subject of PhD thesis

• Oxidation of nitrogen in combustion and explosions — experimental works at laboratory setups and larger-scale installations, and theoretical works, 1935–1940; oxidation of nitrogen was the subject of DSc thesis

• Theory of combustion, inflammation, and propagation of flame, 1937–1941, then after the war, 1945–1948

• Shock and detonation waves, gasdynamics of explosion — from 1938

• Theory of uranium fission — theoretical works published in 1939–1940 together with Yu.B. Khariton; elucidation of conditions for stationary fission in energy installations and explosive fission

• Internal ballistics of new weapons and the theory of gunpowder combustion, 1941–1948; theoretical and experimental work on gunpowders was carried out, besides the Institute of Chemical Physics, at the Department of the Moscow Mechanical Institute, 1945–1948

• Participation in the research and development of atomic and then hydrogen weapons, 1943–1963

• Research in the field of nuclear physics and of the theory of elementary particles: works on μ-catalysis, prediction of new isotopes, in particular, ^{8}He, new types of particle decay ($\pi^+ \rightarrow \pi^0 + e^+ + \nu$), properties of vector current, a pioneering work on the theory of heavy mesons — from 1952

• Works in the field of relativistic astrophysics and cosmology: studies on the theory of formation of 'black holes' and neutron stars in the evolution of normal stars, energy release and X-ray radiation due to accretion of matter onto black holes; development of the theory of the hot model of the Universe, properties of relic (Cosmic Microwave Background) radiation, the theory of formation of galaxies and of the large-scale structure of the Universe, inflation theory of the early Universe, 1965–1987

In 1943 was awarded a Stalin Prize for works on combustion and detonation. In 1949, 1951, and 1953, was awarded Stalin Prizes for weapons-related works. In 1957, was awarded Lenin Prize for weapons-related works.

In 1945, was awarded an Order of Red Banner; in 1949, was given the title of a Hero of Socialist Labour and presented with an Order of Lenin and a Gold Hammer and Sickle Medal.

In 1953 and 1957, was awarded the second and third Gold Hammer and Sickle Medals and was presented with the title of Triple Hero of Socialist Labour. Simultaneously with each Gold Medal, the Order of Lenin was given.

In subsequent years, was awarded Orders of Labour Red Banner (1964), two Orders of Lenin (1962, 1974), and an Order of October Revolution (1984).

Was elected a foreign member of the Royal Society (London), National Academy of Sciences (USA), German Academy of Natural Scientists Leopoldina, American Academy of Arts and Sciences, Hungarian Academy of Sciences; was elected an honorary member of several physical societies and universities.

Was awarded honorary medals: a Manson Medal in 1972 and a Lewis Medal in 1984 for works on the gasdynamics of explosions and shock waves; a Kurchatov Medal for discoveries in nuclear physics in 1977; a Catherine Wolfe Bruce Gold Medal of Astronomical Society of the Pacific in 1983; a Gold Medal of Royal Astronomical Society in 1984; and a Dirac Medal of International Centre of Theoretical Physics in 1985.

List of Contributors

Altshuler, Lev Vladimirovich (b. 1913), Professor, Doctor of Science in Physics and Mathematics (high-pressure physics); Institute of Thermal Physics of Extreme States, Joint Institute of High Temperatures, Russian Academy of Sciences, Moscow

Arnol'd, Vladimir Igorevich (b. 1937), Full Member, Russian Academy of Sciences (mathematics); V.A. Steklov Mathematical Institute, Russian Academy of Sciences, Moscow

Barenblatt, Grigory Isaakovich (b. 1927), Foreign Associate of the National Academy of Sciences of the USA, Professor, Doctor of Science in Physics and Mathematics (hydrodynamics and mechanics of continua); P.P. Shirshov Institute of Oceanology, Russian Academy of Sciences, Moscow; University of California, Berkeley

Bavykin, Igor Borisovich (b. 1937), collector of a unique photographic archive on scientists; head of the photographic laboratory at the N.N. Semenov Institute of Chemical Physics, Russian Academy of Sciences, Moscow

Bisnovatyi-Kogan, Gennady Semenovich (b. 1941), Doctor of Science in Physics and Mathematics (theoretical astrophysics); Institute of Space Research, Russian Academy of Sciences, Moscow

Bonnet, Roger M. (b. 1938), Director of Science, European Space Agency (solar-terrestrial physics and astrophysics), Paris

Cherepashchuk, Anatoly Mikhailovich (b. 1940), Corresponding Member, Russian Academy of Sciences (astrophysics); Director, P.K. Sternberg Astronomical Institute, Moscow

Chernin, Artur Davidovich (b. 1939), Professor, Doctor of Science in Physics and Mathematics (theoretical astrophysics and cosmology), P.K. Sternberg Astronomical Institute, Moscow

Dolgov, Aleksandr Dmitrievich (b. 1941), Doctor of Science in Physics and Mathematics (physics of elementary particles and cosmology); Institute of Theoretical and Experimental Physics, Moscow; University of Ferrara

Doroshkevich, Andrei Georgievich (b. 1936), Doctor of Science in Physics and Mathematics (theoretical astrophysics and cosmology); M.V. Keldysh Institute of Applied Mathematics, Russian Academy of Sciences, Moscow; Theoretical Astrophysics Center, Copenhagen

Dubovitskii, Fedor Ivanovich (1907–1999), Corresponding Member, Russian Academy of Sciences (chemical physics, combustion processes); N.N. Semenov Institute of Chemical Physics, Russian Academy of Sciences, Chernogolovka

Feoktistov, Lev Petrovich (1928–2002), Corresponding Member, Russian Academy of Sciences (nuclear physics); P.N. Lebedev Physical Institute, Russian Academy of Sciences, Moscow

Frenkel, Viktor Yakovlevich (1929–1997), Professor, Doctor of Science in Physics and Mathematics (theoretical physics, history of science); A.F. Ioffe Physical Technical Institute, Russian Academy of Sciences, St.-Petersburg

Fridman, Alexei Maximovich (b. 1940), Full Member, Russian Academy of Sciences (astrophysics, plasma physics); Institute of Astronomy, Russian Academy of Sciences, Moscow

Gelfand, Boris Efimovich (b. 1941), Professor, Doctor of Science in Physics and Mathematics (physical chemistry, combustion, and explosion processes); N.N. Semenov Institute of Chemical Physics, Russian Academy of Sciences, Moscow

Gershtein, Semyon Solomonovich (b. 1929), Full Member, Russian Academy of Sciences (nuclear physics, μ-catalysis); Institute of High-Energy Physics, Protvino

Ginzburg, Vitalii Lazarevich (b. 1916), Full Member, Russian Academy of Sciences (physics); Nobel Prize for Physics in 2003; P.N. Lebedev Physical Institute, Russian Academy of Sciences, Moscow

Gol'danskii, Vitalii Iosifovich (1923–2001), Full Member, Russian Academy of Sciences (chemical physics, nuclear physics); N.N. Semenov Institute of Chemical Physics, Russian Academy of Sciences, Moscow

Golitsyn, Georgii Sergeevich (b. 1935), Full Member, Russian Academy of Sciences (physics of atmosphere); Director of the Institute of the Physics of Atmosphere, Russian Academy of Sciences, Moscow

Grishchuk, Leonid Petrovich (b. 1941), Professor, Doctor of Science in Physics and Mathematics (theoretical astrophysics and general relativity); P.K. Sternberg Astronomical Institute, Moscow; Professor of the University of Cardiff, Wales

Gurevich, Isai Izrailevich (1912–1992), Corresponding Member, Russian Academy of Sciences (experimental nuclear physics); Russian Research Centre 'Kurchatov Institute,' Moscow

Ioffe, Boris Lazarevich (b.1926), Corresponding Member, Russian Academy of Sciences (physics of elementary particles); Institute of Theoretical and Experimental Physics, Moscow

Khariton, Yulii Borisovich (1904–1996), Full Member, Russian Academy of Sciences (chemical physics, nuclear physics, and technology); All-Russia Research Institute of Experimental Physics, Sarov

Khlopov, Maksim Yurievich (b. 1951), Doctor of Science in Physics and Mathematics (elementary particles and cosmology); M.V. Keldysh Institute of Applied Mathematics, Russian Academy of Sciences, Moscow

Kirzhnits, David Abramovich (1926–1998), Corresponding Member, Russian Academy of Sciences (theoretical nuclear physics); P.N. Lebedev Physical Institute, Russian Academy of Sciences, Moscow

Konstantinova, Nataliya Aleksandrovna, English language teacher, Chair of Foreign Languages, Russian Academy of Sciences; daughter of the known radiophysicist A.P. Konstantinov

Leipunskii, Ovsei Il'ich (1910–1989), Professor, Honoured Scientist (combustion and explosion processes, crystal physics); N.N. Semenov Institute of Chemical Physics, Russian Academy of Sciences, Moscow

Longair, Malcolm Sim (b. 1941), Jacksonian Professor of Natural Philosophy, Cambridge University and Director of Cavendish Laboratory, Cambridge

Manelis, Georgii Borisovich (b. 1930), Corresponding Member, Russian Academy of Sciences (kinetics, combustion, and explosion processes of condensed systems); N.N. Semenov Institute of Chemical Physics, Russian Academy of Sciences, Chernogolovka

Makhviladze, Georgii Mikhailovich (b. 1945), Professor, Doctor of Science in Physics and Mathematics (combustion and explosion processes); Institute of Applied Mechanics, Russian Academy of Sciences, Moscow; University of Central Lancashire, Preston

Melott, Adrian L. (b. 1947), Professor (computational astronomy, hydrodynamics), Department of Physics and Astronomy, University of Kansas, Lawrence

Merzhanov, Aleksandr Grigorievich (b. 1931), Full Member, Russian Academy of Sciences (general structural macrokinetics, combustion theory); Institute of Structural Macrokinetics, Russian Academy of Sciences, Chernogolovka

Mestel, Leon (b. 1927), Fellow of the Royal Society, Emeritus Professor of Astronomy, University of Sussex, Brighton

Moffat, Keith (b. 1935), Fellow of the Royal Society (theoretical astrophysics and hydrodynamics); Emeritus Professor, Department of Applied Mathematics and Theoretical Physics, Centre for Mathematical Sciences, University of Cambridge

Morozova, Nataliya Dmitrievna, member of the Journalists' Union of Moscow, member of the Association of Science Journalists and Writers, head of department of the State Historical Museum, Moscow

Myshkis, Anatoly Dmitrievich (b. 1920), Professor, Doctor of Science in Physics and Mathematics (mathematics); Moscow State Railway Technical University, Moscow

Novikov, Igor Dmitrievich (b. 1935), Corresponding Member, Russian Academy of Sciences (theoretical astrophysics and general relativity); Astronomical Centre, P.N. Lebedev Physical Institute, Russian Academy of Sciences, Moscow; Director of the Theoretical Astrophysics Center, Copenhagen

Okun', Lev Borisovich (b. 1929), Full Member, Russian Academy of Sciences (physics of elementary particles); Institute of Theoretical and Experimental Physics, Moscow

Ovchinnikova, Marina Yakovlevna, Doctor of Science in Physics and Mathematics (atomic and molecular processes); N.N. Semenov Institute of Chemical Physics, Russian Academy of Sciences, Moscow

Peebles, Phillip James Edwin (b. 1935), Member, USA National Academy of Science, Albert Einstein Professor of Science Emeritus (cosmology), Princeton University

Pinaev, Viktor Semenovich (1932–2003), Doctor of Science in Physics and Mathematics; Theoretical Division of the Federal Nuclear Centre, All-Russia Research Institute of Experimental Physics, Sarov

Polnarev, Aleksandr Grigorievich (b. 1949), Doctor of Science in Physics and Mathematics (general relativity); Astronomical Centre, P.N. Lebedev Physical Institute, Russian Academy of Science, Moscow; Queen Mary College, London

Ruzmaikin, Aleksandr Andreevich (b. 1944), Doctor of Science in Physics and Mathematics, Jet Propulsion Laboratory, California Institute of Technology

Sakharov, Andrei Dmitrievich (1921–1989), Full Member, Russian Academy of Sciences (nuclear physics, plasma physics, elementary particles, astrophysics), Nobel Peace Prize in 1975; P.N. Lebedev Physical Institute, Russian Academy of Sciences, Moscow

Sakharova, Tatiana Andreevna, PhD (biology), daughter of A.D. Sakharov, Moscow

Sazhin, Mikhail Vasilievich (b. 1951), Professor, Doctor of Science in Physics and Mathematics (general relativity and gravitational waves); P.K. Sternberg Astronomical Institute, Moscow

Sena, Lev Aronovich (1908–1997), Professor, Doctor of Science in Physics and Mathematics (plasma physics, gas discharge physics); A.F. Ioffe Physical Technical Institute, Russian Academy of Sciences, St.-Petersburg

Shekhter, Anna Borisovna, Professor, Doctor of Science in Physics and Mathematics (kinetics of chemical reactions in a discharge, catalysis); Institute of Chemical Physics, Moscow

Smirnov, Yurii Nikolaevich (b. 1937), PhD (physics and mathematics) (theoretical physics); Russian Research Centre 'Kurchatov Institute,' Moscow

Sunyaev, Rashid Alievich (b. 1943), Full Member, Russian Academy of Sciences (high-energy astrophysics and cosmology); Institute of Space Research, Russian Academy of Sciences, Moscow; Director, Max Planck Institute for Astrophysics, Garching

Thorne, Kip (b. 1940), Member, National Academy of Sciences of the USA; Foreign Member, Russian Academy of Sciences; the Feynmann Professor of Theoretical Physics (general relativity and gravitation), California Institute of Technology, Pasadena

Todes, Oskar Moiseevich (1910–1989), Professor, Doctor of Science in Physics and Mathematics (molecular physics, physical chemistry, combustion processes); A.F. Ioffe Physical Technical Institute, Russian Academy of Sciences, St.-Petersburg

Tsukerman, Veniamin Aronovich (1913–1993), Professor, Doctor of Science in Physics and Mathematics (nuclear physics and high-pressure physics), Honoured Inventor of Russia; All-Russia Research Institute of Experimental Physics, Sarov

Wasserburg, Gerald J. (b. 1927), Member, USA National Academy of Sciences (geochemistry, cosmochemistry); John D. Macarthur Professor Emeritus, California Institute of Technology, Pasadena

Zakharchenya, Boris Petrovich (b. 1928), Full Member, Russian Academy of Sciences (physics); Ioffe Physical Technical Institute, Russian Academy of Sciences, St.-Petersburg

Zakharov, Vladimir Evgenyevich (b. 1939), Full Member, Russian Academy of Sciences (optics); Director of the Landau Institute of Theoretical Physics, Russian Academy of Sciences, Chernogolovka, Moscow Region; Professor of Mathematics, University of Arizona, Tucson

Part I

Early Days at the Institute of Chemical Physics

How Zeldovich was 'discovered'

L.A. Sena

March 1931, Leningrad: the laboratory of Simon Zalmanovich Roginskii at the Institute of Chemical Physics, recently founded as part of the budding Physical Technical Institute. I had been working there for more than six months already, having graduated from the Physics and Mechanics Department of the Polytechnic Institute.

Initially, I worked as a general assistant, but at the end of 1930, Simon Zalmanovich gave me my own research topic, which was perhaps a little odd, and which lay beyond the set of questions that were being worked on in the laboratory. It is known that many organic materials are easily supercooled below the melting point. One such material is nitroglycerine. Supercooled nitroglycerine becomes a viscous mass that can easily be made to crystallize. Upon forced crystallization, two forms can be obtained, with different melting points — unstable and stable. An interesting phenomenon was discovered: when crystallized nitroglycerine was melted and recooled after being kept at a particular temperature, it crystallized quite easily into the same form in which it had crystallized earlier. It was as though nitroglycerine had a 'memory.' Simon Zalmanovich suggested that I work on unravelling the mystery of this memory.

By March 1931, I had already become quite used to working with nitroglycerine, had learned how to crystallize it in any form, and had carried out many experiments on multiple crystallization, and convinced myself of the reality of the 'memory.' I observed the influence of temperature and the duration of heating required for the preservation of this memory, but was no closer to solving this mystery.

Here, I should make a short digression. At that time, excursions to various organizations, including scientific institutes, were very widespread. The identification pass system was very primitive, and was virtually non-existent in institutes. Since a week rarely passed without the laboratory having one or two excursions coming through, we established 'excursion duty.' Each of us in turn was required to fulfil the responsibilities of a tour guide if an excursion came through that day.

It was my turn to be 'tour guide' on that memorable March day. An excursion came from the Institute for the Mechanical Processing of Mineral Resources. The visitors included a young man, almost a boy (as it later became clear, he had turned seventeen not long before). Like every guide, I began with my own research topic. The visitors listened politely, and this young man began to ask me questions that showed he had an understanding of thermodynamics, molecular physics, and chemistry comparable to that of a third-year university student. Seizing a moment, I walked up to Simon Zalmanovich and said, 'Simon! I like this kid a lot. It would be nice to have him working here.' Simon Zalmanovich answered, 'I agree. I've been half-listening to your talk; I'll take over the excursion from here, and you have a word with him to see if he's interested in transferring here. If he agrees, he can work with you.' So I took the young man aside and asked him, 'Do you like the work we're doing here?' — 'Very much!' — 'Would you like to work here?' — 'That was partly why I came on the excursion.' Soon, young Yasha Zeldovich was transferred to our laboratory and began to work with me, since it was I who 'discovered' him.

To this day I don't know who let out a rumour that the lab had received Zeldovich in exchange for a prevacuum pump, though I suspect it was our laboratory assistant Senya, who was a master of invention. Zeldovich himself liked this legend very much. Many years later, remembering our scientific youth together, he said 'Well, that means that even back then I was worth something!'

In this way, Yasha began to work with me. Although he sent me a congratulatory telegram when I turned 70 saying 'To my first teacher,' we worked together as equals from our very first days together. In spite of the difference in our ages (I was six years older), and in spite of the fact that I had graduated from an institute and he only from high school, we were on an equal level at work. We divided journal articles that needed to be read evenly between us. Most of these were in German; when we now and then had an article to read in English, Yasha took it, since he could manage easily in both languages (my English was rather poor at that time). I believe that he also knew some French.

Our equality was not only in scientific matters. Within the first few days of our work, our friendship began to grow; we went together to meetings at the Physical Technical Institute, went to movies on Yashumov Lane (later Kurchatov Street). We established certain working principles. One of these was that it was permitted to make a mistake, but not to repeat it. An offender was awarded the 'Order of the Large Grouse' — a bird that was supposed to resemble a grouse was drawn on a sheet of paper and hung in view of the guilty party.

When to begin work each day was agreed upon at the end of the previous day, and the end of each day depended on how our work was going, our health and moods, whether there were any interesting articles that had to be read before being discussed, or a new movie that had been seen, etc. The duration of our working days ranged from eight or ten hours to only four, when, in our opinion, 'a devil was loose' in the laboratory. In this case, we went out to the library at the Physical Technical Institute or to the park of the Polytechnical Institute, and continued our work there. At that time, the very concept of the beginning and end of the working day was vague. Most institutes (including the Physical Technical, Chemical Physical, and Electrophysical Institutes) worked 24 hours a day. Passing by them late in the evening, it was always possible to find lit windows.

Our research proceeded successfully; soon, we were able to disprove the hypothesis of a volume 'memory' and demonstrate convincingly that this memory 'sits' at the walls of the material. The results that we obtained in our series of experiments were regularly reported to Simon Zalmanovich, and discussed with him in detail. When we had carried out all the experiments proving our point of view many times, we decided to write them up as an article. The task of writing a first rough draft for discussion was assigned to me. I sat down and wrote an entire 'dissertation' (by the way, there was no such thing at that time). Simon Zalmanovich and Yasha acted as editors, mercilessly criticizing, while I clung to every sentence I'd written. I remember that Yasha jokingly said, 'I'm surprised you didn't include a report about the weather during the experiments.' After some long arguments, we were able to come to an agreement about the contents of the final version. Simon Zalmanovich's wife, Anna Borisovna Shekhter, translated the article into German, since we decided to send it to a new Soviet journal published in German, 'Physikalische Zeitschrift der Sovjetunion'. Our article 'Beitrag zum Mechanismus der Erscheinung des Gedachtnisses der wiederholten Kristallisation' appeared in volume 1, issue 5. This was the first scientific publication for both Yasha and myself.

I remember that we celebrated the completion of our work by destroying the entire store of nitroglycerine that had accumulated during the experiments. We made a fire on some stones in a pit behind the institute, placed an iron sheet on it, and added the nitroglycerine. The auto-da-fe was highly colourful and festive!

After the completion of our project, it became clear that Zeldovich could work completely independently, and our scientific paths diverged. After some time, I began to work with Dmitrii Apollinar'evich Rozhanskii, a development which determined my further scientific career.

Yasha, who soon became 'Yakov Borisovich,' began to swiftly climb the scientific ladder. His subsequent biography is well known: to obtain the right to defend a Candidate of Science dissertation, he took some exams at university, but he was then given permission to defend his dissertation without a formal university or institute diploma. There were among us several such self-taught academicians who had no formal university-level diploma (one example is Boris Pavlovich Konstantinov, who later became the Vice President of the Academy of Sciences). When I talked about Yakov Borisovich in my lectures, I always pointed out that he never finished university, and highlighted other examples of 'unfinished academicians.' However, I warned that the absence of higher education was not a necessary condition for becoming a member of the Academy of Sciences.

My friendship with Yakov Borisovich continued after our scientific paths diverged, right up to the Second World War and his move to Moscow. Then, our meetings became much rarer. Yakov Borisovich came to Leningrad only occasionally. Although I visited Moscow much more often, many of these trips were flying visits, and I was able to stop by to see Zeldovich for a few minutes only a handful of times. Our last meeting was at the end of December in 1982. Although this day was my 75th birthday, I had to be in Moscow to see one of my students from Lvov defend his dissertation. The defence took place in the Kurchatov Institute of Atomic Energy under the chairmanship of B.B. Kadomtsev. I then went to pay a visit on Yakov Borisovich. We spent a very warm and cheerful evening, at which A.V. Gaponov-Grekhov was also present.

At the end of 1987, I turned 80, and they decided to celebrate this event in the Institute of Mining, where I worked. I tried to prevent such a fuss, but in the end I had to submit, having made a bargain that the celebration would be based around a scientific seminar. We made arrangements with the speakers (B.M. Smirnov, V.E. Golant, and G.N. Fursei) and arranged the festivities for the beginning of 1988. I telephoned Yakov Borisovich, who said that he would certainly try to come this time. But then there was an unexpected telephone call on December 2. I listened to the voice of Golant saying, 'Yakov Borisovich died suddenly yesterday.'

Considering my age, I had already had to part with a number of close friends; some, like Yakov Borisovich, were younger than I. Each such loss is difficult. However, the loss of Yakov Borisovich, with whom I shared a friendship that endured without a single cloud for nearly 50 years since we met as young scientists, was one of the most difficult to bear.

A passport to science

A.B. Shekhter

One day in 1931, an excursion from the Institute for the Mechanical Processing of Mineral Resources (*Mekhanobr*) visited the catalysis laboratory of the Leningrad Institute of Chemical Physics (ICP) where my husband Simon Zalmanovich Roginskii was a supervisor. The vast majority of participants were adults or even elderly people, with interests far from those of the lab, who clearly didn't understand the explanations about theoretical investigations of catalysis. Among them, a 17-year-old boy stood out. This young man asked many questions, zeroing in on the key points of the work.

Fortunately, the boy caught the attention of the excursion leader, L.A. Sena, who quickly brought him to the attention of S.Z. Roginskii. It

Yakov Borisovich Zeldovich with his father Boris Naumovich and his mother Anna Petrovna.

turned out that the young man had recently finished high school and had been assigned to *Mekhanobr*, where he worked as a laboratory assistant. His name was Yasha Zeldovich.

After a short conversation with Yasha, Simon Zalmanovich, immediately recognizing his exceptional capabilities, got in touch that very same day with A.F. Ioffe, the director of the group of institutes in which the ICP was included, to request his cooperation in transferring Zeldovich to work at our institute. Ioffe, like no one else, loved looking for and 'fishing out' new talent. In this case, he cut through the usual red tape and rapidly established 17-year-old Zeldovich as an employee in the catalysis laboratory. Yasha adapted to the research topics under investigation in the laboratory very quickly, and essentially became a co-author on a whole series of projects on the theory of heterogeneous catalysis.

There were many arguments about whether Yasha should enroll in a university. Many (S.Z. Roginskii, Ya.I. Frenkel and others) were against the idea, on the grounds that they felt Yasha could acquire the knowledge that he required by himself, and that a university would only dry up his bright talent. The future proved that they were right. The start that was given to Yasha in the catalysis laboratory was clearly all he needed.

I don't remember precisely in what year Yasha defended his Candidate of Science dissertation, but it happened quite soon. It was not easy to obtain permission for him to defend, due to his lack of a higher education diploma.

In 1939, Zeldovich defended his Doctor of Science dissertation. At that time, he was 25 years old. On this occasion, there was a small banquet in the apartment he shared with his mother and sister, where there were only a few people (N.N. Semenov, the two opponents — Ya.K. Syrkin and A.N. Frumkin — and Simon Zalmanovich and myself), but the wine flowed freely, and a light-hearted, cheerful mood reigned. Frumkin pronounced a jesting toast and, turning to Semenov, said, 'With you, Kolka, everything is always unusual. Now tell me, who in your institute defended their Doctor's dissertations this year?' And then he answered himself, pointing at Yasha and at me, 'One child and one woman!' (I also defended my doctorate in 1939.)

One more amusing episode comes to mind. About three or four months after Yasha began to work at the catalysis laboratory, his mother called Simon Zalmanovich at his home and asked how Yasha was doing. S.Z. praised him. And then she asked a question about which she was clearly most anxious: 'Tell me, my Yashenka is never rude to you, I hope?' In fact, he was never anything of the sort, and throughout his life, Yasha treated his first teacher with great respect, both when he worked in the catalysis laboratory and when he later went to Semenov's laboratory, having become interested in the theory of combustion.

I should also add that Yasha met his future wife, Varvara Pavlovna Konstantinova, in the catalysis laboratory, where she was also a researcher. In this sense, it would be fair to say that Zeldovich obtained a passport, not only to scientific work, but also to family life during his time at the ICP catalysis laboratory.

'Talent welled up in him'

O.M. Todes

On November 10, 1987, I arrived at Moscow for one day from Leningrad to play the role of an opponent at a defence at the Institute of Problems of Mechanics of the USSR Academy of Sciences. That morning, I stopped by to see Yakov Borisovich, with whom I'd enjoyed many years of friendship, at his home. Yasha was in good spirits, cheerful, as always, and energetic. He shared some new thoughts about fundamental problems of interest to him in cosmology and some family news. We also discussed the dissertation for which I had come to Moscow to act as opponent. It was dedicated to problems in adsorption — a field in which Yakov Borisovich had carried out fundamental work in 1939, and which, though not formally published, was well-known to all workers in that area. Alas, it was less than a month later when the telephone call from Moscow came, informing us of the death of Yakov Borisovich from a heart attack.

There were three years between us in age. I wasted these three years on my university education — I graduated from the Physical Mechanical Department of the Leningrad Polytechnical Institute. Yasha 'saved' these three years by gaining all the knowledge he needed in his work (initially with our help), beginning as a laboratory assistant and ending as a member of the Academy of Science. Talent, multiplied by energy, welled up in him, and N.N. Semenov, the director of the Institute of Chemical Physics, successfully enrolled Yakov Borisovich as a graduate student without any formal undergraduate diploma from a university or institute. Having sidestepped this bureaucratic barrier, Yakov Borisovich rapidly passed through all subsequent degrees. In addition to the work on the theory of adsorption dynamics mentioned above, he became a pioneer in finding solutions for many new and important scientific problems.

This development was facilitated by the atmosphere in the Institute of Chemical Physics at that time, when the mean age of the scientific workers was about 25. There was no trace of division according to rank: the laboratory assistant Zeldovich and the senior scientist Todes discussed vital scientific problems together, and also relaxed and found diversion together. Working in directions begun by the founder and director of the institute (chain reactions, combustion, and explosions), the companionable young researchers each did their bit without any hierarchical friction or arguments, as if they were running a relay race together.

As a theoretician in the gas combustion laboratory of A.V. Zagulin, I was working on taking into account chemical kinetics during the production of a thermal explosion (a concept that had been introduced earlier by Semenov), and developing a theory for non-stationary thermal explosions. In connection with this, Yasha analyzed the behaviour of variations in the induction period with approach to the self-ignition limit. In parallel, Yasha worked with David Frank-Kamenetskii on the development of a theory for the rate of propagation of a flame in gaseous mixtures, taking into account kinetics. This was later used as a basis by Shura Belyaev, who reduced the propagation of flames in solid bodies to the vaporisation of the solid material due to the heat transferred to the surface of the flame front. In Yasha's hands, this mechanism for the combustion of powders had important applications during the Second World War.

The phenomenon of detonation was studied in Yu.B. Khariton's laboratory. The first schemes for the joint propagation of a flame front and shock wave were created much earlier in the work of G. Riemann, B. Jouguet, and P. Hugoniot, but the structure of the front itself remained unknown. The first work on developing these schemes at the ICP was done by Serezha Izmailov on the reflection of a shock wave from an obstacle and the limiting increase in the intensity of the reflected wave. In the theoretical hydrodynamical studies listed above, the alignment of the shock wave and the flame front was determined graphically by drawing a line from the point corresponding to the gaseous state ahead of the front, tangent to the Hugoniot shock adiabat, corresponding to the complete release of chemical energy. The speed of the detonation wave was determined from the slope of this tangent line, and to this day, this is how the speed of a detonation wave as a function of its amplitude is calculated. The only serious physical basis for the choice of the tangent point was that it corresponds to the maximum entropy behind the front. Given the fact that the process of chemical transformation and energy release was continuous, I proposed calculating the front structure described by intermediate points of this tangent line where the tangent intersected the Hugoniot adiabat; these corresponded to a continuous release of chemical energy in the reaction, and to regimes with hydrodynamic propagation with the speed of the shock front (along a contour of constant velocity — an 'isovel').

Having discussed and analyzed this idea, Yasha showed that these intermediate points correspond to temperatures that are too low for the chemical reactions to keep up with the shock wave. He extended this 'isovel' to its intersection with the Hugoniot adiabat corresponding to the chemical reaction. This allowed him to establish that the pure shock wave front is located ahead, and is supported by the reacting gas back along the 'isovel,' which he called the 'Todes curve' in his first publications on this topic.

In the summer of 1932 or 1933, we obtained vacation passes through the Physical Technical Institute, and set off in a group of four people — Serezha Izmailov, Varya Konstantinova (who was also a former fellow student of mine), Yasha Zeldovich, and I — to a sanatorium (run by '*KUBUCH*' — the Commission for the Improvement of Lives of Scientists) in Sudak. Keeping apart from the young students at the resort, we arranged open-air scientific seminars in the evenings. Having carried out a small theoretical calculation, Serezha and I wrote an article and sent it off for publication, indicating the date and place where the work was carried out (Sudak, Sanatorium for the Neurotic); the editors omitted this note. And the four of us carried out one piece of experimental work together.

I'm referring to our verification of the theory of the 'ninth wave,' which, according to tradition, is supposed to always be the most intense. We placed markers made of one, two, three, and more stones perpendicular to the sea shore, and made notes of which waves had made it to which marker over several hours in the evening. From these observations, we made samples of every ninth wave and determined the mean distance for each set. If the 'ninth wave' theory was correct, the distance for one of these samples would have been significantly larger than for the others. In our experiment, however, all the mean distances for all the wave samples were virtually identical, and we considered the 'ninth wave' theory to be completely refuted. We never found a scientific journal, however, in which we could publish the result of such an experiment.

Not knowing yet what would come to pass later, Yasha and I began to play a risky game in Sudak — to marry our two older companions, Varya and Serezha, to each other. In our walks together, Yasha would start talking some sort of nonsense, which Serezha, when he could bear it no longer, would refute, at which point I would nudge Varya, saying 'See what a sharp chap Serezha is!' At the edge of the sanatorium there was a small kiosk in which a young Tatar woman named Fadme sold cream soda and other non-alcoholic beverages. One evening, at the initiative of Yasha, we took this kiosk by storm, closed it early, and took Fadme to the movies with us. During the film, I sat in the middle, with Yasha and Fadme to my right and Serezha and Varya to my left. Yasha demonstratively flirted with Fadme, and I poked Serezha and said, 'Take a look at what's going on to my right and take the same action on the left front.' In spite of all of our contrivances, nothing came of this venture, but a couple of years later, Yasha himself married Varya!

The sanatorium, of course, was a 'dry zone,' but on the beach, at the boundary with the neighbouring sanatorium for the Moscow Military District, there was a small pharmaceutical kiosk. When a known group arrived, the outer doors closed and bottles of wine were fetched from under the counter. However, we had to walk to Sudak to get good Massandra wine. During one such visit, we came across a very good wine that was sold by volume without packaging. Since it was almost time for us to leave, we wanted to get some to take back with us. Right near the store, we found a pharmacy, in the window of which stood large, black litre bottles of 'Essentuki Mineral Water No. 13.' We bought a bottle of this stuff and began to drink it, swallowing with disgust the excessively salty and rather nasty mineral water, until Yasha realized that there was an irrigation ditch running right nearby, where we could pour it out, which we did with great pleasure. Having filled the empty bottle with the tasty port, we returned to the sanatorium, proud of our own inventiveness. When we met the sanatorium doctor, we showed her the black bottle with the label 'Essentuki.' Since this doctor had been a classmate of mine in school, after our departure, I confessed to her about the contents of the black bottle, and she was very sorry that she had not thought to check it.

One day on the beach, Yasha wanted to demonstrate a feat of the sort seen in GTO[1] exercises — swimming with a grenade in his hand. Not having a grenade, he seized the first object that fell to hand — my shoe — and, after swimming with it in his raised hand, threw it onto the beach. However the problem was that when we undressed, I had placed my watch in that shoe, and in the climax of the demonstration, though the shoe made it to the beach, my watch didn't. We searched for it for a long time, and eventually did rescue it from the sea bed, still working. I felt the consequences of my watch's swim about a year later, when it stopped. After examining it, the watch-repair man informed me with amazement that the entire mechanism had somehow rusted.

The events of those summer days at the sanatorium were reflected in our collective composition of a poem based on Svetlov's 'Granada,' which began with the lines:

> We went by train, rushing through the mountains,
> We held his impudent tongue in our hands ...

---

[1] This stands for *'Gotovnost' k Trudu i Oborone'* meaning 'Readiness for Labour and Defence.' GTO exercises often combined some type of sports activity with a display of military-type skills, and constituted a high-profile Communist campaign aimed at the entire Soviet population. [Translator]

and in place of the usual refrain, we sang:

Cream soda,
Crimea, Sokol,
My Genova.

To the left of the resort towered Sokol Mountain, and to the right, on a precipice, stood the ancient Genova fortress.

Many years later, we met in a sanatorium in Livadia. I can confirm that Yasha and Varya's style of active relaxation, with great physical and intellectual demands, hadn't lost its vitality.

Of our city escapades, I remember one time in particular when Yasha and I went to visit two lady friends of ours. We were getting ready to go to the theatre, and each of us invited one of the ladies. After the show had ended, we, proclaiming the 'principle of interchangeability of ladies,' each set off to escort home the other's partner. The young ladies didn't object to this, evidently having established an analogous principle of interchangeability for us.

After the transfer of the Institute of Chemical Physics from the jurisdiction of the People's Commissariat for Heavy Industry (*Narkomtyazhprom*) to the Academy of Sciences, Nikolai Nikolaevich Semenov gave me the responsibilities of scientific secretary of the institute. These included overseeing the organization of the graduate student programme, the education of the graduate students and their exams, and the preparation of scientific yearly plans for approval. In some specialized disciplines, I supervised the teaching of students and the setting of exams together with Lev Gurevich. In place of the standard procedure for setting exams (of the sort used for undergraduate students), we preferred to select some new and interesting article from the literature and assign the student taking the exam to read it, analyse and explain it in detail, and defend it from our questions. This experimental examination was the one sat by Yasha Zeldovich. Once he was a bit lazy, and received the comment 'calculates like a five-year-old,' an occurrence he remembered and recounted even in his last years.

As he fulfilled the requirements for his Candidate of Science degree, his scientific potential rose. When he came home, his mother Anna Petrovna often asked him the same question — 'Well, how is it going — are you a Todes, or not yet?' — since the only reference for scientific standing that was used in their home was me. She got a clear answer to this question in April 1944, when the three opponents for the defence of my Doctor of Science thesis were B.V. Deryaguin, Ya.I. Frenkel (my former teacher), and Ya.B. Zeldovich (my former student).

During my period in Leningrad I very much enjoyed spending time with the family of young Yakov Borisovich, talking with his father, a talented lawyer, and his mother, a well-known translator and member of the Writers' Union. Yakov Borisovich's younger sister, Ira, later became a doctor. This same intellectual and working style was present when Yakov Borisovich started his own family, and he and Varya raised two girls and a younger son.

Anna Petrovna Zeldovich
(1890–1975)

Boris Naumovich Zeldovich
(1889–1943)

His son Boris inherited his father's talents and was made a Corresponding Member of the USSR Academy of Sciences — an event that Yasha sadly didn't learn about before his death.

During the war, when we had been evacuated to Kazan, my catalysis laboratory, which was headed by S.Z. Roginskii, moved from the Institute of Chemical Physics to the Institute of Physical Chemistry (then called the Colloidal–Electrochemistry Institute, directed by A.N. Frumkin). At the end of the war, I returned to Leningrad to head a physics department at a technical university, and left the hierarchy of the Academy of Sciences. At that time, I was interested in the macroscopic physics of dispersion systems, whose hydraulic and thermal processes play an important role in practical chemical technology. It was only in the 1970s, via this work on dispersion systems, that I returned to the physics of combustion, and developed a radiational model for the propagation of flames in aerosuspensions of fuel grains.

After the Second World War, together with the Institute of Chemical Physics, Yasha returned to Moscow rather than to Leningrad. Beginning with the famous papers of 1939–1940 written with Yu.B. Khariton on the theory of chain-reaction fission of uranium, the main focus of his scientific interests moved toward nuclear physics, and later, after having worked on its main practical applications, to cosmology. The 700 km distance between us hindered regular contact, and made our meetings rare. However, in 1981, Yakov Borisovich partially returned to the main theme of the research of our youth, and organized and headed the Scientific Council on Combustion of the USSR Academy of Sciences, in which he invited me to participate.

Leningrad, 1938

Varvara Pavlovna Konstantinova
(1907–1976)

He put the same inexhaustible energy that characterized all his activities into the last scientific work he organized. He generalized his scientific achievements in the fields of combustion and cosmology in a number of monographs and books. For those who were only starting to become acquainted with science, he wrote special manuals of the 'Mathematics for Beginners' type. The omission of rigorous derivations in these books (such as, for example, the theorem of existence) made it easier for a wider audience to understand very abstract concepts, however, they called forth furious criticism from pure mathemeticians. Yakov Borisovich educated the higher contingent of his immediate colleagues on the joint solution of complex problems, especially in the field of cosmology.

Scientist, pedagog, and popularizer — he approached each of these roles with equal interest, energy, and drive. On the basis of principles he was able to convincingly defend points of view that were under attack. I will present just one example of his adherence to principles. During the evacuation of the ICP to Kazan, Nikolai Nikolaievich Semenov, having arranged for the demobilization of one of his talented students from the army, decided to help him skip over one step in his career, and suggested that he be immediately granted the degree of Doctor of Science for the dissertation he had presented for his Candidate of Science degree. Nikolai Nikolaievich had shared a new and original idea with the student, who had then taken an active role in its development; they had published this idea together, and it was then included in the student's Candidate of Science dissertation. In spite of his gratitude and great respect for the supervisor, Yakov Borisovich actively argued against Semenov's suggestion when it was discussed at the Scientific Council, and in the end convinced the Council to award only a

Candidate of Science degree for the work. This result proved to be favourable for the student, whose own independent ideas and initiative later allowed him to become both a Doctor of Science and an Academician!

This once youthful generation of students is now in its 80s; however, many have left life significantly earlier. These include the youngest of our generation — Yasha Zeldovich. Having been passed the baton by our teachers, we, and especially Yakov Borisovich, will leave substantial results and achievements, as well as the talented young people we have educated. The Institute of Chemical Physics, which was small and young in 1933, has now grown into a collective of several thousand workers and has founded a number of large branches in Yerevan and Chernogolovka. We can honestly say that those who have already passed on did not live in vain. This is especially true of Yakov Borisovich.

I can see him in front of me, as if still alive

V.Ya. Frenkel

Memoirs are usually preceded by a standard apology to the effect that the authors cannot avoid telling at least a little bit about themselves and their relationship with the person about whom they are writing. In my case, the appropriate apologies should probably be presented a bit differently. Since my acquaintance with physicists about whom I've been called upon to write was always connected in one way or another with my father, Yakov Il'ich Frenkel, who was their friend, colleague, or teacher; I must also say a few words about him. This will be the case here, as I write about Yakov Borisovich Zeldovich.

I knew him by sight even in pre-war Leningrad, and met him in the Lesnoi, then on the outskirts of the city, where there was a cluster of institutes: Physical Technical, Chemical Physics, Boiler–Turbine, Electrophysics, Musical Acoustics, Agrophysics, not to mention the educational Polytechnical Institute. But I remember most clearly an encounter with Zeldovich in Kazan in about 1943, in the corridor of the university. At that time, practically all of the evacuees from Moscow and Leningrad academic institutions were concentrated in a relatively small three-storey building, which occupied a section of picturesque Chernyshevskii Street (later renamed Lenin Street). You would meet all sorts of people there. Here, in a suede zip-up jacket, with his ubiquitous pipe between his teeth, is P.L. Kapitsa; there, climbing the stairs to the offices allocated to the Radium Institute is its director V.G. Khlopin, tall and erect. L.A. Orbeli, with his unusually gentle and kind face contrasting with his official general's uniform; the legendary O.Yu. Schmidt, the director of the Institute of Theoretical Geophysics; S.I. Vavilov, director of the Physics Institute of the Academy of Sciences, with the flag of the Supreme Soviet on the lapel of his well-fitted suit. The physicists

L.D. Landau, G.S. Landsberg, D.V. Skobeltsyn, I.E. Tamm; the chemists N.N. Semenov and A.N. Frumkin; the mathematician S.L. Sobolev; the humanists N.S. Derzhavin and I.I. Tolstoi. Future academicians — the young employee of the Institute of Russian Literature (Pushkin House) D.S. Likhachev, the very young V.I. Goldanskii, the fourth-year student L.D. Faddeev. The list goes on!

And one day my brother S.Ya. Frenkel and I met a thin, rather short person — Yasha Zeldovich, as he was called in our home. He was in a worn light overcoat and a soft hat. I remember his glasses, with their thin metal frame. The thick, deep voice was incongruous with his then fragile figure. He stopped my brother, who worked at that time in the Colloid–Electrochemistry Institute, and said, 'Serezha, you should try to influence Yakov Il'ich. Why is he messing about with fluid theory? It's time to be working on gas discharges and vacuum electronics.' I probably remembered this encounter well because my brother, in accordance with Yakov Borisovich's suggestion, relayed the content of this conversation to our father that very evening at dinner. I remember that my father burst out laughing and dismissed the idea away with a wave of his hand. In my youth, I was a little annoyed at Zeldovich, supposing that my father could decide himself what he should be working on.

Soon Yakov Borisovich left for Moscow, and if I'm not mistaken, his mother remained in Kazan. I remember her as being very nice in many ways. First, she resembled my own grandmother on my father's side; second, I was touched by the pride with which she told my parents about the successes of her son when they met on the streets of Kazan. Finally, I already loved Dickens, and had also read Chesterton with focused attention. When I learned that Anna Petrovna had translated a biography of Dickens written by Chesterton, my respect for her grew. Much later, when I reminded Yakov Borisovich about this, he proudly stated that his mother had graduated from the Sorbonne.

Among the spectrum of attractive qualities in the character of Yakov Borisovich, I would like to note his devotion to his teachers and older colleagues — those whom he surpassed (the majority) and those with whom he was on a par. As a theoretical physicist, he began work in the theoretical department of the Institute of Chemical Physics (which was then headed by my father). At the anniversary celebrations for other workers in this department — L.E. Gurevich, S.V. Izmailov, O.M. Todes, and L.A. Sena — Yakov Borisovich's written and oral greetings made his gratitude towards them clear. Among the colleagues that he valued highly, I would single out Boris Pavlovich Konstantinov. It seems to me that the dominant role in their scientific and personal contact was played by feelings of friendship, rather than of kinship.

In Boris Pavlovich's last period of activity, which coincided with his responsibilities as director of the Leningrad Physico-Technical Institute, he became very interested in astrophysics, and in particular, in the possible existence of antimatter in the solar system and in the universe as a whole. A

During the evacuation period in Kazan, 1942.

large programme of investigations to test this possibility was developed at LPTI. Boris Pavlovich's hypothesis, which was subsequently not confirmed, was initially met with scepticism by a number of scientists. In spite of his characteristic sceptical or, to put it better, sharply critical mind, Yakov Borisovich tolerated this idea. I remember his opinion that it is sometimes necessary to take risks, even if the probability of success in some ventures was very small; if the improbable proves to be true, the effect can be enormous.[2] Furthermore, as Boris Pavlovich himself loved to say, every question or effect has the right to exist if it does not contradict the second law of thermodynamics! This half-joking argument was the reasoning behind another more fundamental point: Boris Pavlovich and Yakov Borisovich were sure that the astrophysical work being developed at LPTI would ultimately lead to success, though, perhaps, not in the initial field of investigation. In fact, that is exactly what happened, and the Ioffe Physico-Technical Institute became one of the most important centres for astrophysical investigations in the country.

Unfortunately, in Yakov Borisovich's personal relations his respect for older colleagues, devotion in friendship, and tolerance sometimes conflicted with his unwillingness to compromise. There were some cases where he broke off relations with his colleagues, and I'm sure that both the work and the two parties involved suffered. In such situations, it appears that emotion held sway over reason.

[2] Zeldovich himself later wrote about this in an article about Konstantinov, 'In memory of a friend' (see p. 335 of this book).

However, my own relationship with Yakov Borisovich stayed on an even keel over the 20 or so years in which we had contact (in this sense, I was lucky). I can recall only one instance when he expressed unrestrained annoyance. I will begin with this case, though chronologically the episode corresponds to a relatively late period in our professional contact. I believe it was in 1982 that the editors of the *Large Soviet Encyclopedia*, having prepared for the issue of a volume on the Great Patriotic War (World War II), asked me to write mini biographical sketches of several Soviet physicists whose work had contributed to the defence effort. Naturally, Yakov Borisovich was included among them. In a meeting, we agreed on the contents of the sketch: which work from those years I could write about. However, it later became clear that each entry had to include a few lines outlining the scientific interests of each of the physicists. Of course, I knew what Zeldovich had worked on in the past and was working on now, but I couldn't imagine how I could make it fit one or two lines — in fact, I knew that there was no way it would fit! (Even a bibliography of his papers runs to eleven sections.) Consequently, it would be necessary to leave something out, and it seemed natural to me to call him and ask for his help in deciding what. To my surprise, Yakov Borisovich immediately refused rather sharply, commenting with annoyance that I could get the necessary information from any relevant handbook or encyclopaedia. My attempts to find out what the problem was (hindered by the fact that it was a long-distance call) were not successful. However, I came to the conclusion that I may have simply called at an inconvenient time, and also that the question itself may not have seemed entirely ethical to Yakov Borisovich. In short, I saw that the rumours about the sharpness of his character had some foundation.

Chronologically, my first major encounter with Yakov Borisovich was in Leningrad, at the Physico-Technical Institute. He presented a talk about the physics of stars here in the early 1960s, at a seminar in the theoretical department, which already included a small group of astrophysicists. The seminar room was full and not only with theoreticians.

Among the issues discussed by Yakov Borisovich was the theory of white dwarfs. He reproduced the important derivation of the stellar mass that was necessary for a star to be destined to become a 'dwarf.' At the end, he said that it was especially pleasant for him to present such a talk at the LPTI, where L.D. Landau had worked on the theory of white dwarfs in the 1930s. At that time, still fresh in my mind were preparations for the publication of a volume of selected works by my father, including the article 'Application of the electron-gas theory of Pauli and Fermi to cohesive forces.' The last section of this work was entitled 'Superdense Stars' and contained the same reasoning and estimates (the role of the electron gas) that Yakov Borisovich had presented. Therefore, when the talk was over, I went up to the board where he stood, surrounded by a crowd of theoreticians. When my turn came, I cautiously told Yakov Borisovich that, in my opinion, the work of Ya.I. Frenkel in this field from 1928 had escaped his attention. We agreed that I would send him a copy of this paper. Having read my father's work, he agreed that, indeed, it presented some of the same material about which he had given his talk at the LPTI. However, to ensure correctness, he would also like to see Fowler's article on this same topic. I sent him a copy of this article, and after some time, he announced that now things had become clear to him, that his article would soon be appearing in *Uspekhi Fizicheskikh Nauk*,[3] and that he would make appropriate reference there to the work of my father, Yakov Il'ich.

Some time later, I happened to be in Moscow, and stopped by the *Uspekhi* editorial office on business. I had to wait a little while for one of the office staff to arrive, and the editorial secretary, L.I. Kopeikina, gave me the proofs for the current journal issue to look at, to help pass the time. It turned out that this issue (*UFN*, 1965, volume 86, issue 3) included Yakov Borisovich's article, 'Relativistic Astrophysics,' which was written jointly with I.D. Novikov. I quickly found the section on white dwarfs, and, to my surprise and distress, could not find any reference to my father's work. That same day, I called Yakov Borisovich and told him about the omission. His reaction was forceful: 'That can't be!' I answered in kind: 'But it is!' As a result, he assured me that the corresponding correction to the proofs would be made, as indeed it was, to my great joy. Later, in 1975, an article entitled 'The stars tell their tales' written by Zeldovich, Novikov, and Sunyaev appeared in the March 9 issue of *Pravda*. Zeldovich (no offence to his co-authors, but clearly it was Zeldovich) notes that 'the Soviet scientists Ya. Frenkel and L. Landau have made large contributions to the theory of white dwarfs.'

Soon after this episode, Zeldovich again came to the LPTI and, in the institute-wide seminar, discussed successes in astrophysics and cosmology, and the influence on progress in these fields of the fundamental works of A.A. Friedmann. Using the work of W. MacCrae and E. Milne, he obtained the main results of Friedmann in 'Newtonian' language. I remember that, having finished his calculation, he stepped away from the board, looked at the chain of formulae, and, turning to the audience, said something like, 'Such possibilities are included in Newtonian mechanics! You know, I've worked in many fields in my life, but I think I love classical mechanics most of all!'

At the end of the 1970s, I started to become interested in the scientific biography of G.A. Gamow. A colleague from the United States sent me a page from Gamow's autobiographical book *My World Line*, on which there were two caricatures of I.F. Ioffe and Ya.I. Frenkel, my father. This, of course, boosted my interest in the book, however, it was not to be found in the libraries in Leningrad. Someone told me that Zeldovich had a copy. This turned out to be true, and he kindly agreed to show it to me.

'Call tomorrow morning,' he said on the telephone, 'and we'll agree on a time to meet.' 'Morning — that would be at about ten?' I asked. There followed a short pause, which I would now call ironic. Then, Yakov Borisovich explained that he gets up around six o'clock, but I should call him

[3] Known as *Physics Uspekhi* in its English translation. [Translator]

at about half past seven. This I did, and we agreed on an evening meeting that same day. Yakov Borisovich's flat was in number 2b Vorobiev Road, a building I knew, where many workers from the Institute of Chemical Physics lived. That was the first time I crossed the threshold of flat 47 and, having taken off my coat in the hallway, I entered Yakov Borisovich's large study. It could be that my memory is at fault, but it seems to me that there was no desk in it. The dominant place in the centre of the room was occupied by a large table (if it were a dining room, I'd call it a dining table), on which books, journals and manuscripts were scattered in disorder. On one of the walls hung a slate blackboard, and in one corner, I saw dumb-bells. Yakov Borisovich got Gamow's book and gave it to me to look at. I found the page with the caricatures and found some words about A.F. Ioffe, but nothing, unfortunately, about my father. Nevertheless, even a glance through the book indicated that it was extremely interesting, and I asked if I could take it with me, promising to return it by an agreed date. I noted that Yakov Borisovich hesitated. He then handed me a sheet of paper, which I immediately understood had been taken from the book. It said that, in accordance with his order, the book was being sent to Academician Ya.B. Zeldovich, but could he please not lend it to anyone. Finally, he said, 'Well, take it, but please don't show it to anyone else. You can return it the next time you're in Moscow.'

I then read the book with great interest. Today, a number of segments of it have been published, and preparations are underway for its full publication. In those days, it was hard even to imagine this!

At that time, I got the impression that there was a certain 'law-abiding quality' in Yakov Borisovich. Although he disobeyed instructions, he did this not without some difficulty, and clearly only after having taken control of himself. I have been told that this sort of cautiousness in small affairs was to some extent characteristic of him. For example, another instruction forbade using a third person to send letters from the 'object'[4] to the 'outside world.' He followed this rule strictly, and refused innocent requests to take letters back to Moscow and toss them in a post box, so that they would arrive more quickly at their destinations. In fact, it could be that such behaviour was displaying not cautiousness, but rather discipline. He was used to working in a disciplined way, and could not change this. I witnessed another example at the beginning of the 1980s. An editor of one of his articles insisted that something be removed on censorship grounds. In my opinion, it was not appropriate to agree to this, and Yakov Borisovich was more than sufficiently protected by the establishment should he wish to refuse. However, he decided not to refuse: 'It's not worth getting involved in!' His decision can be explained in this way: the question was not fundamental, and removal of those few lines did not in the least affect the meaning of the article.

---

[4] This was a sort of code name used to refer to the highly classified establishment where work was being conducted for the Soviet atomic weapons programme. It is also known as Arzamas-16, an official classified name, which gave rise to the nickname 'Los Arzamas,' and obvious reference to the American equivalent. The real name of the site is Sarov. [Translator]

In about 1974, I began to collect reminiscences about my father. Naturally, Yakov Borisovich was among those I contacted during my research. Of course, I had not forgotten the Kazan episode of 1943, and also clearly remembered other events. For example, in 1947, in response to my father's receiving a State Prize for his work on the kinetic theory of fluids, Yakov Borisovich and David Albertovich Frank-Kamenetskii sent the following congratulatory telegram:[5]

> We send our best wishes by telegraph switch.
> Greetings! Two hundred thousand to you, Yakov Il'ich!
> To Madame Frenkel glory, honour to Frenkelettes
> Members of the family of the great laureate.
> Your portrait is in *Pravda*. A nice portrait du jour
> Forever crowned with your fine chevelure.
> And it seems you're not only an eminent scientist,
> But also de facto a fiddler and artist.
> Going without sleep, devoted to your cause,
> You've become well acquainted with all fluid laws.
> He who understands fluids far from the least
> Will invite a big crowd to a jolly great feast.
> We would like to be among the guests in your hall
> But we're afraid there might not be enough fluid for all!

Frank-Zeldovetskii

A few words of explanation: in addition to his certificate and medal, the winner of a State Prize received a cheque for 200,000 roubles. Announcements of the prizewinners were published in all the main newspapers, together with a photograph of the laureate. Evidently, out of respect for Yakov Il'ich, they retouched his bald spot in the photograph published in *Pravda*. This photograph was then circulated among the other publications.

Soon, there arrived from Moscow an envelope with a photograph taken by Yakov Borisovich, I believe in Vorobiev 2b in the flat of A.S. Kompaneets. In this photograph, Yakov Il'ich was shown together with Kompaneets and Frank-Kamenetskii, and on the reverse was written 'Dear Yakov Il'ich, from a Photo-correspondent Member.' This was, of course, the signature of Zeldovich, who was chosen as a Corresponding Member of the USSR Academy of Sciences in 1946.

As I expected, Yakov Borisovich and many others readily met my request for reminiscences about my father. At that time, these people were all no longer young and very busy. For all their full and heartfelt willingness to participate in the project, as a rule, they didn't manage to submit material in accordance with the agreed deadlines. I often had to remind them — in a word, to pester them — a rather unpleasant task. However, with Yakov Borisovich, it was quite a different story. Right on time, he sent his article, written

[5] This telegram is archived in the Ya.I. Frenkel foundation of the Leningrad division of the Russian Academy of Sciences archive.

cordially, temperamentally, and with a literary flourish (to say nothing of its historical and scientific content). Yakov Borisovich was certainly a marvellous writer; this is clear from other memoirs that he has written (there are quite a few), and also from his books and reviews, including those of a popular scientific nature.

Our contacts and meetings became more frequent. In October of 1980, the USSR Academy of Sciences decided to conduct a session of the Presidium of the Academy of Sciences in Leningrad, dedicated to the 100th anniversary of the birth of A.F. Ioffe, the founder of the Physical Technical Institute. Among the various events, there were two in which I actively participated. The Presidium resolved to conduct annual lectures in memory of A.F. Ioffe, and to publish the talks given during these lecture series. The organization of the lectures and their publication was assigned to me. In the first set of lectures, there were four speakers — Zh.I. Alferov, B.P. Zakharchenya, Ya.B. Zeldovich, and G.N. Flerov. Yakov Borisovich's talk was entitled 'Astrophysics of Our Day.' I have already said that he was a brilliant speaker. Soon, based on a transcript of the talk, A.D. Chernin prepared a typewritten version, and it proved easy to find agreement on the details.

At the same time, work was underway on collecting materials for another publication — *Problems in Modern Physics*, which was also timed to coincide with the 100th anniversary of Ioffe's birth. This was a thick volume divided into nine sections. The section on nuclear physics presented an article by Yakov Borisovich and Yulii Borisovich Khariton entitled, 'The Role of A.F. Ioffe in the Development of Soviet Nuclear Physics and Technology.' Much of this was written jointly; after all, the nuclear activities of Ioffe unravelled before their eyes. However, the last pages allowed each of the co-authors to give their own impressions. Yakov Borisovich wrote about the role played in his life by A.F. Ioffe, who wrote the letter to the Institute for the Mechanical Processing of Mineral Resources, requesting that the 17-year-old Zeldovich, who was working there as a laboratory assistant, be transferred to the Physical Technical Institute.

In 1984, there were two anniversaries to celebrate together: Khariton and Zeldovich had birthdays at nearly the same time, with a difference of ten years (Yu.B. Khariton was born on February 27, 1904 and Ya.B. Zeldovich on March 8, 1914). We prepared for the events in good time. On the initiative of A.P. Aleksandrov, Ya.B. Zeldovich and some of Khariton's colleagues from his defence work, it was decided to issue a special collection of articles to celebrate Khariton's 80th birthday. A board of editors was created, in which I was included as an editor and compiler. All questions about the structure of the collection and potential authors were decided by Yakov Borisovich. This was determined not only by the fact that he had worked in the same fields of physics as Khariton (molecular physics, physics of explosions, chemical kinetics, nuclear physics and technology), but also because, according to my own understanding and impressions formed during our conversations, Zeldovich considered Yulii Borisovich as the first teacher or older physicist to have influenced him. He also mentioned L.D. Landau in this capacity. Again, Zeldovich was one of the first to present his contribution for the book — an unusually vivid article called 'Yulii Borisovich Khariton and the Science of Explosions.'

It was decided to include some of Yulii Borisovich's own papers in the book — old works, from the 1920s and 1930s, and also newer ones. I was assigned the job of providing commentary for each cycle of Khariton's works (chemical kinetics, explosions, nuclear physics). My meetings with Yakov Borisovich took place comparatively early in the morning, in his flat. At that time, after the untimely death of his first wife Varvara Pavlovna, Yakov Borisovich was married to Anzhelika Yakovlevna Vasil'eva. Always rather lively, she was very hospitable to guests. Our conversation began at the table: they invited me to share their breakfast, and Yakov Borisovich never ceased to surprise me in his ascetic moderation with regard to food, and his ability to resist the temptation of trying tasty dishes. From the cozy little kitchen cum dining room, we moved to his study, and worked there on the manuscripts. Yakov Borisovich examined the presented material with such a good-natured, attentive, yet demanding approach, recommending that various things be clarified or expanded, and adding his own commentary.

In one case, this commentary was especially important. The work in question was the pioneering research of Khariton on the possibility of using a centrifuge to separate isotopes. This work was carried out in 1937, long before the task of separating isotopes of nitrogen and oxygen became so obviously relevant. The question has an interesting history, which culminated in the 1940s. At that time, the idea of using a centrifuge for the industrial separation of isotopes of uranium (rejected, though seriously considered by both the Soviet and Anglo–American uranium projects) had recently acquired a new life. I more or less knew about all this. However, it was Yakov Borisovich who told me of the importance of the conclusions drawn by Khariton in his work, and he wrote the following lines in his commentary: 'In essence, they developed methods that later formed a separate area of the thermodynamics of weakly non-equilibrium processes. In his small note, he established a strict non-equilibrium at the largest possible output of the apparatus. In its generality, the result of Khariton ... can be compared with the second law of thermodynamics.'[6]

Such a strong assertion would ring true only coming from the lips of Zeldovich, and not from mine. Therefore, I asked Yakov Borisovich for permission to place his name after the commentary. He categorically refused. After repeated attempts to persuade him, he said, smiling slyly, 'Don't worry, people will recognize my hand in these lines even without my signature at the bottom — *Ex ungue leonem* (the lion is recognizable by his claws).' Nonetheless, we arrived at a compromise: in place of the ellipsis in the citation given above is written 'as noted by Ya.B. Zeldovich'.

Yakov Borisovich was equally attentive to the commentary for the cycle

---

[6] *Voprosy sovremennoi eksperimentalnoi i teoreticheskoi fiziki* (*Topics in modern experimental and theoretical physics*), Leningrad, 1984, p. 82.

of his collaborative work with Khariton on the chain decay of uranium. After we had finished discussing this work, I discovered several minor misprints in the original (journal) text of one of their articles. I corrected them, but wanted to verify my corrections. At that time, I had to make a trip to Moscow; Yakov Borisovich was at that time on an extended trip elsewhere, so I contacted Yulii Borisovich, and we got things sorted out. I mention it here because, upon his return, having learned about my meeting with Khariton, Yakov Borisovich lamented, 'You should be ashamed to have disturbed Yulii Borisovich!' This testifies to just how protectively he acted toward his teacher and elder colleague.

In connection with the publication of the Khariton book, there was one other noteworthy episode. Following Yakov Borisovich's advice, I sent an official invitation to Academician E.I. Zababakhin, requesting his contribution to the book. On time, he sent an article entitled 'Some Cases of Motion of a Viscous Fluid.' The crystal clarity and physicality of the article made a strong impression on me — perhaps especially because I was working closely at that time with several articles by Einstein on similar topics. I shared my impressions with Yakov Borisovich. I remember his reaction well. On his mobile features, even before he began to talk, there appeared an expression of extreme satisfaction. He then said, 'Zabakhin! What's surprising here? He is a physicist of the highest class, a physicist with a capital P!'

Since I worked on the book dedicated to the 80th birthday of Yu.B. Khariton, I wasn't surprised when I was asked if I would prepare an anniversary article about him for *Uspekhi Fizicheskikh Nauk.* Of course, I discussed this article most of all with Zeldovich. When, at almost the same time, his colleagues asked me to compose the non-astrophysical part of his own scientific biography for a similar article in *UFN*, I was both flattered and surprised (the astrophysical part of the article was prepared by R.A. Sunyaev).

When work on the article was finished, the question arose of whose signatures it would be nice to see at the end of the article. I think that I won't be disclosing a frightful secret if I say that Sunyaev and I asked Yakov Borisovich's advice about this; I still have his list, composed on a page from a desk calendar. This list coincided almost exactly with the one we had compiled ourselves earlier, with one exception, about which I'd like to say just a few words. Yakov Borisovich suggested that the list should include L.D. Faddeev. I think that this was not only because he valued Faddeev's work highly. The point was that some mathematicians had sharply criticized Zeldovich's books for schoolchildren and young students, with a heated passion that was quite inappropriate to the issue. I accepted this simple mission with enthusiasm, all the more so given that I myself had been delighted with both *Higher Mathematics for Beginners* and, especially, *Elements of Applied Mathematics*, which had been written together with A.D. Myshkis. In this latter book, I was especially impressed by a masterful explanation of the basis of Friedmann's theory of the universe, undoubtedly written by Zeldovich. As expected, L.D. Faddeev willingly signed the article.

The serious part of the anniversary took place at the end of February, in the form of a session of the Scientific Council of the Institute of Chemical Physics. This was a double celebration — for Khariton and Zeldovich, teacher and student, two co-authors of classic work on the theory of uranium fission, two triple Heroes of Socialist Labour, who deeply respected and loved each other. The first to speak, due to his seniority, was Yulii Borisovich, after which there were presentations related to his speech. Then came the second part of the programme, dedicated to Yakov Borisovich. He prepared for his speech carefully, and I could sense that he had worried over it. The session was lightly, cordially, and wittily led by N.M. Emanuel.

The affair ended with a small buffet reception, which took place in a room adjacent to the assembly hall. Here, we could drink to the health of the two heroes of the day, exchange a few words with them, and see the cream of Soviet physicists, primarily from the fields of nuclear, chemical, and astrophysics.

The festivities continued about two weeks later, at the Institute for Physical Problems, where Yakov Borisovich headed the theoretical department. First, there was a small show in the hall in which the scientific council met. The Nikitins, of whom Yakov Borisovich was an admirer, were invited, and they performed several songs from their repertoire, supplemented by a potpourri of popular tunes from Zeldovich's youth in the 1930s, and followed by dancing. Here, Yakov Borisovich, sitting with his wife Anzhelika Yakovlevna in the front row, eventually couldn't hold out, and the two of them slipped out during a tango circle, along a narrow path between the stage and the front row. Later, the spectators and performers went to the foyer, where Zeldovich and his guests could make suitably witty presentations with toasts around a festive table.

The last intensive stage of my interactions with Yakov Borisovich was connected with preparations for the publication of a book of reminiscences about Boris Pavlovich Konstantinov, his brother-in-law. How much labour, energy, and love did Yakov Borisovich put into this venture! He wrote a vivid essay, 'In Memory of a Friend.' He not only suggested many contributors, but also actively urged us on in our work, and then acted as an unofficial editor.

I would like to note the fact that the talented Konstantinov family was a delight to Zeldovich. On his suggestion, one of the sections of the reminiscences is entitled 'The Konstantinov family.' This section includes an article about the elder brother of Boris Pavlovich — Aleksander Pavlovich who was a prominent physicist and radio engineer, and who headed a laboratory at the LPTI. However, perhaps the main object of care for Yakov Borisovich was an article by Varvara Pavlovna Konstantinova ('My Brother Boris'). Zeldovich devoted much time and energy to the preparation of his first wife's manuscript, and worked on it with love, careful attention, and, especially, protectiveness.

When the book had come out, Yakov Borisovich and I discussed the circle of people to whom copies should be sent. He then was seized by the idea that a review of the book should appear in one of the major literary magazines. He proposed a very successful choice of author for this review — the writer D.A. Granin. At Zeldovich's suggestion, I sent Granin a letter asking if he would be willing to do this, and his interesting review soon appeared in the pages of *Neva* — to the great satisfaction of Yakov Borisovich.

****

I have written primarily about my professional contacts with Yakov Borisovich. It was also sometimes the case that we saw each other under different circumstances. For example, I remember an unusually cheery evening in Moscow, when I was the guest of Yu.N. Semenov and his wife T.Yu. Khariton. This was around the spring of 1957. Those present were mostly physicists, including Yakov Borisovich and his friends A.S. Kompaneets, V.I. Gol'danskii and his wife, Lyudmila Nikolaevna, and N.M. Emanuel. These were all very witty people. The son of the splendid Soviet film and theatre actor B.V. Shchukin was also there (unfortunately, I can't recall his name). He brilliantly directed a merry game of charades, in which the most lively participant was Yakov Borisovich.

One more memory of our later years of contact: in 1980, when the 100th anniversary of the birth of A.F. Ioffe was being celebrated at the LPTI, the most honoured guests were met at the Moskovskii railway station when they arrived in Leningrad. I was assigned to meet Yu.B. Khariton. We had not even finished getting him settled in his room at the Evropeiskaya Hotel when there was a knock at the door, and in came Yakov Borisovich. 'Well, what do you think, Yulii Borisovich, are we going to wear our stars tonight?' They decided that they would. And so, soon, the guests of the fashionable hotel could observe an impressive spectacle: two ageing, rather short men, each displaying three gold stars on his chest, descending the staircase into the cafe. Even the hotel manager, a woman who had seen a lot in her time, was truly stunned!

At the end of 1980, I presented a doctoral dissertation for defence, sent Yakov Borisovich the abstract, and asked if he could respond with an evaluation, if possible. He kindly agreed, and so it happened that I was waiting for him one gloomy November morning at the entrance to one of the auditoria of the Moscow State University Physics Department, where he gave lectures on astrophysics to the students. The bell rang, and the audience began to come out of the hall. I saw Yakov Borisovich talking to a group of people — not only students, but also a number of people who were old enough to be Candidates or even Doctors of Science. Once he was free, he came over to me and confessed that he had not yet managed to write the evaluation. 'But that's okay,' he said, 'we'll go over to the Astronomy Institute, and I'll do it there.' We put on our coats and went out of the building. The snow wasn't just falling — it was coming down in buckets! The windscreen of Zeldovich's 'Volga' was thickly covered. Yakov Borisovich opened the car door, got a brush, waved it toward the windscreen, and said, 'Okay, now, earn your thesis evaluation!' He warmed up the motor, and a few minutes later we parked at the entrance to the Sternberg Institute and went inside to his office.

Zeldovich settled at his desk and, without a single correction, straight off, wrote a two-page evaluation of my work in about five minutes. 'Come on, we'll have to type it up.' We went through various corridors to a room, where Yakov Borisovich asked a young woman who had just finished bashing out something on a typewriter to type the evaluation. I myself sat down to do the job, using a second typewriter, and simultaneously read his unusually good-natured evaluation. Then, Yakov Borisovich signed it, and we once more set out through the corridors, this time to the institute's administrative office, where Zeldovich's signature was certified. When we returned to his office, there was a Ukrainian television crew (from Kiev, I believe). Yakov Borisovich had promised them a ten-minute interview. I asked his permission to watch this distinctive live transmission, or, more precisely, reception. The director, a woman with a sweet appearance, asked, 'Yakov Borisovich, perhaps we should practise your interview a bit?' — 'Why? Are your questions ready?' — 'They're ready.' — 'Well, then, ask them!' And without any preparation, calmly, Yakov Borisovich answered all their questions concerning the modern state of astrophysics and cosmology. I remember that, at the end of the recording, the director commented that she had never before seen anyone who could hold himself so naturally and in such a relaxed manner in front of a camera.

In 1985, a two-volume set of collected works by Ya.B. Zeldovich was published under the editorship of Yu.B. Khariton. Shortly after this, having found myself in Moscow at the Institute of Physical Problems, I dropped into the theoretical department and, passing Yakov Borisovich's office, whose door was open, saw him at work. He sat at his desk in a deep, high-backed armchair. He glanced by chance in my direction and, seeing me, invited me in. 'Have you received my book?' he asked. 'No, Yakov Borisovich.' — 'That's odd, I clearly remember that I sent one to you. Oh, well.' And he got up from the armchair and went to a corner of the office, where a stack of handsomely bound books lay next to a blackboard: *Ya.B. Zeldovich. Selected Works. Particles, Nuclei, and the Universe.* He took the top copy and inscribed with his bold handwriting, 'Dear Vitya Frenkel, with love, Ya. Z.' I was deeply touched. When I returned to Leningrad, a package from Yakov Borisovich was waiting for me. This time, the inscription in the book was more official, but unusually pleasing to my self-esteem.

During one of our subsequent meetings, having already become acquainted with both volumes (and knowing Yakov Borisovich's popular science books, which were not included, rather well), I told him that I was convinced that he should collect his writings addressed to a wider circle of readers and his memorial articles and publish them in the academic series *Science, World View, and Life*. He answered, 'No, I'll have to wait.'

It later became clear that the publication of this two-volume collection of his works was not without certain difficulties, which were solved due to efforts of Zeldovich's colleagues at the Academy. The resistance came from those same circles who had made unfounded criticisms of his popular books on mathematics.

In an article published in *Uspekhi Fizicheskikh Nauk* for Zeldovich's 70th birthday, I wrote: 'The years do not have power over Ya.B. Zeldovich. He continues to work with youthful enthusiasm, provoking surprise and delight at his capacity for work. How does he find the time to be up to date on literature, the theatre, and, more broadly, cultural life? His workload, years of being widely known, and his popularity have not changed his democratic character. The sharp mind of a theoretician combines in him with inexhaustible human wit, making him an extremely interesting and cheery conversationalist.'

Indeed, Yakov Borisovich did not change at all with the years, and I never heard him complain about being tired or not feeling well. Only after he had passed away did I learn that he had diabetes. And it really seemed as if he could never wear out. Lively, lean, and muscular, he looked much younger than he actually was, even in 1987, by which time he was getting on in years.

I remember well December 3, 1987. In the morning, I went to work as usual at the institute. At that time, I was working with A.D. Chernin and E.A. Trop on a book about A.A. Friedmann: the 100th anniversary of his birth would be in 1988. He was very highly regarded by Yakov Borisovich, who proposed that he participate in the anniversary sessions and kept track of our work. On that day, I wanted to fix a meeting with him during my next trip to Moscow, and was climbing the front stairs of the main wing toward a telephone booth with a direct connection to Moscow to dial his office. On the stairs, I saw D.A. Varshalovich (in 1966, Yakov Borisovich had insisted that Dmitrii Aleksandrovich be immediately awarded a Doctor of Science degree for the dissertation he presented for his Candidate of Science degree). Something in Varshalovich's face looked unusual.

'Have you heard the horrible news? Yakov Borisovich has died!'

It was difficult to believe!

I can see him now, as if he were alive, standing in front of me with his thoughtful look. He remains in our memory, 'Alive, alive, and only alive — to the end.'

'Don't forget to write about me'

N.A. Konstantinova

For me, it is a great honour to write a few words about Yakov Borisovich Zeldovich, whose fate was bound with that of my uncles and aunts for many years. One of these aunts was Varvara Pavlovna Konstantinova — Yakov Borisovich's wife and the mother of Olga, Marina and Boris Zeldovich.

Yakov Borisovich was a multifaceted person. Everyone who writes about him sees him a bit differently. My view will reflect impressions from my contacts with Yakov Borisovich, his home, and his family, which were not frequent, but which extended over many years. I will try to reflect the warm feelings and love that all my Leningrad relatives felt towards him, and which has always been shared by me.

Varvara Pavlovna became acquainted with Yakov Borisovich (then an 18-year-old) in 1932. He soon began to visit the Konstantinovs' large flat on Malyi Podyacheskii Street. The family at that time was a sort of commune of the five Konstantinov brothers and two sisters (only one of the brothers was married and had children). The family was surprisingly friendly and close. Of course, the core was formed by the two sisters — Varvara and Maria. Having lost their father and then their mother at an early age, they accumulated a great deal of kindness, which they radiated, bringing happiness for many years to their children, nephews and nieces, and all who met with them in their daily life.

The Konstantinovs always spent summer at the family dacha in Tyarlevo (between Pavlovsk and Detskoe Selo); Yakov Borisovich's parents also often stayed there in the summer. He began to come more and more often to the volleyball court near the dacha, where there were always hotly contested battles between the teams from the 'upper' part of the area (where the Konstantinovs lived) and the 'lower' part. There were also bicycle trips from Leningrad, past Pulkovo to Tyarlevo on old bicycles which constantly broke down. He was a first-class cyclist — that was the first impression my Aunt Varya's fiance made on my childhood memory.

Many years later, Yakov Borisovich loved to recall that he very much liked the family of Varvara Pavlovna, their youth, cheerfulness, rather disorganized way of life and ascetism. All this was a pleasant contrast to the well-organized home of his parents, to which he was still bound. He was taken with the idea of becoming a member of this large, young 'commune.' Yakov Borisovich often talked about his pre-war life in Leningrad, when he already had two small daughters, and about his relations with the Konstantinov brothers (one of his stories was about how they made a trip to Ladozhskoe Lake to get firewood). These tales contained much mischievous cheerfulness.

His relations with Boris Pavlovich Konstantinov were especially close and fruitful. For many years, they were united by their professional interests, as well as by the bonds of kinship; by their common love of science, passion for research, and their deep and lifelong friendship. Boris Pavlovich was a frequent and very welcome guest at the home of Yakov Borisovich. I think that Yakov Borisovich has described this in his article 'In memory of a friend'[7] much better than I can.

I was always both agitated and delighted that Yakov Borisovich had united his fate with that of Varvara Pavlovna and had entered her family at

[7] Included later in this book (see page 335).

a most frightening time for them — the autumn of 1937. Not long before this, the Stalinist butchers had taken the life of Varvara Pavlovna's older brother — my father A.P. Konstantinov, who was a scientist and radiophysicist — and nobody could predict the fate of his relatives. In that distant time, when it was not unheard of for people to denounce their own close relations, and repudiation of parents, brothers, sisters and friends was taken as the norm, Yakov Borisovich's choice can be considered an act of courage. He acted very nobly toward my sister and me when we were left without our parents.

I recently asked my sister E.A. Konstantinova, 'What do you remember most about Yakov Borisovich?,' and the first thing that came into her head was the following. In 1940, it became necessary to pay for school, and this caused my sister considerable distress. The next day, Yasha arrived and brought the necessary sum of money. Such deeds were typical of Yakov Borisovich in the future as well. He displayed similar acts of kindness to practically all of our relatives. Mariya Pavlovna recalled: 'Only the insistence of Yakov Borisovich convinced me of the need to evacuate, and this allowed us to avoid the fate of many Leningraders, who perished in the blocade.'

Thinking of Yakov Borisovich, I can't help remembering the atmosphere of his parents' home on Marata Street in Leningrad, then of the writers' home on the Petrogradskii side where his mother lived with her daughter in the years after the war.

I've heard many good things about Yakov Borisovich's father, Boris Naumovich, with his peaceful, moderate character, but I didn't know him personally. However, I shared a long friendship with his mother, Anna Petrovna. She had a broad education, which she obtained at the Sorbonne, and became a translator and member of the Writers' Union. I remember how she actively and ardently worked on translations of the works of Balzac and Zola. Small, lively, energetic, and emotional, she was a very kind and compassionate person. Due to poor health, Anna Petrovna spent her last fifteen years 'in her dressing gown,' that is, she spent most of her time at home. But it was something to see the frenetic activity she undertook there, trying (and usually succeeding) to help relatives, close friends and mere acquaintances to solve their problems.

When Yakov Borisovich came to Leningrad, Anna Petrovna always arranged a modest reception for a small circle of Leningrad relatives. I had the honour of being among them.

Anna Petrovna loved her children very much: they were different, somewhat diametrical opposites, and this was due not only to pedagogical influences, but also to their natural characters. Anna Petrovna was very proud of her son and his successes in science. Yakov Borisovich had much in common with her. His interest in people, his eagerness to help them — this clearly comes from her. I won't hide it — Anna Petrovna loved to praise her son, his wife, and, of course, her grandchildren. However, one has to confess that there were good grounds for this. Her grandchildren reflected in a balanced way the best qualities of their parents: seriousness, love of work, responsibility and conscientiousness, not to mention their capabilities.

I have always loved Moscow very much. However, many of my friends believed that I love not so much Moscow itself as my dear relatives on Vorobiev Road. It is possible that they were largely right, and that it was not only the home of my mother's ancestors in Zamoskvorechye that drew me to Moscow. It was always pleasant for me in the Zeldovich home, with its changing number and frequently changing street name. But what didn't change was the creative spirit and atmosphere in which the parents lived and in which the children and grandchildren live now. The home of Varvara Pavlovna and Yakov Borisovich was the most hospitable of those that I visited. There was everything: books, the joy of socializing, and arguments that passed into vivid discussions. I remember 1957 and 1958, when the first satellites were launched and the mass enthusiasm in connection with this. It was a festive time for science. We all remember the much-talked-of article by Yakov Borisovich in *Izvestiya*, in which he discussed the need to raise the level of physics and mathematics education in our country and to found specialist schools. I remember a heated discussion in the kitchen of the flat on Vorobiev Road, in which I formed an absolute minority, having been brave enough to champion the possibilities and means of humanitarian education (instead of physical and mathematical education) to achieve a very high level in the development of the human intellect.

In our student years, my sister and I were often in Moscow visiting friends or in transit to other places. I remember especially well the summer of 1949, which Yakov Borisovich's family spent in the village of Dunino on the banks of the Moscow River. There were few cars at that time, and it was not easy to get to what was then a relatively remote village. One had to go by train to Zvenigorod, then walk about six kilometres with backpacks, as Varvara Pavlovna often did. Occasionally, it was possible to get a ride to the village on one of two small vehicles whose lucky owners spent their summers in Dunino.

Our relatives rented a small room with log walls, divided by a partition. A dacha-type veranda had been attached to the side of the cabin. Hospitable Yakov Borisovich often slept there on a camp bed. The village of Dunino was located on a high bank of the Moscow River, surrounded by fields and woods, in which 30 to 35 years ago there grew an abundance of choice wild mushrooms. Not long before this, the writer M.M. Prishvin settled in Dunino. When walking through the woods, we often met him and his beautiful dog. This was the last summer before the move to the new dacha in Zhukovka.

In Zhukovka, the houses were built right in the woods; however, this academic settlement is not best known for its landscape, but for the people who lived there. Life at the Zhukovka dacha was associated with many significant events and encounters. Here, more than one generation of Zeldoviches were raised. I remember receiving a postcard each spring, often followed up by telephone calls, inviting me to visit the dacha. I wasn't always able to accept these invitations, but how pleasant and cherished they were! I missed them for a long time afterwards.

Many years have passed since the death of Varvara Pavlovna. Her grandchildren have grown up. In the spring of 1987, my cousin V.B. Konstantinov met with Yakov Borisovich during one of his trips to Moscow. During their conversation, they talked about a project to write a book about A.P. and B.P. Konstantinov. Yakov Borisovich ardently supported this idea, and said something which, at that time, seemed somewhat strange. He said, 'Don't forget to write about me.' These words surprised me, since not long before, in March of 1984, the 80th birthday of Yu.B. Khariton and the 70th birthday of Ya.B. Zeldovich were widely celebrated. There were many articles in the central papers and in various journals. And all the same, something prompted Yakov Borisovich to pronounce these words in the year that was to be his last .... .

Thinking of Yakov Borisovich, I feel constantly grateful for all his kindness toward those close to me and to me myself. He was a member of our family in the fullest sense.

Part II

The Move to Moscow: Chemical Physics

Forty-five years

V.I. Gol'danskii

I have taken it upon myself to write about Yakov Borisovich on March 8, 1989, his 75th birthday. I have just seen and heard him for the first time since his death — the film *Scientific Roles of Academician Zeldovich* was shown on television, and once again, his death seemed completely wild and implausible to me.

Forty-five years passed between the time I first saw YaB in Kazan to the evenings of November 29 and 30, 1987 — the last nights of his life, when he, with his usual youthful lightness, ran up the steep staircase to our flat to look over some new American journals with articles about his works from the 1930s and 1980s, and to gossip about the approaching elections to the Academy. Consciously thinking over this long yet short time, which covers the whole of my adult life, sorting through messages from YaB, his joking verses, various rough notes, I feel especially keenly what an emptiness was left by his unexpected death.

I first heard of YaB from one of his closest friends and colleagues, Ovsei Ilich Leipunskii, soon after I arrived in Kazan from blockaded Leningrad, and began working as an assistant in the laboratory of S.Z. Roginskii in April of 1942. I could tell many tales about life at the Institute of Chemical Physics (ICP) during the evacuation ('The 40s, deadly, leaden, filled with gunpowder, war is taking a romp through Russia, and we are so young ...'). But that will have to wait for another time.

Leipunskii talked about Zeldovich with unbridled delight, and it was difficult to reconcile the image conjured up by these tales with the quick-moving young fellow in glasses whom I met on the stairs, in the library, in the queue at our bread stall, and whom everybody addressed simply by his first name. In November, I first heard YaB speak at a session of the Scientific Council of the Institute of Chemical Physics — this was the first time I saw a thesis defence.

Thirty-four-year-old A.B. Nalbandyan was defending his doctoral thesis and twenty-seven-year-old N.M. Emanuel his Candidate's thesis. However, all the opponents (there were three, as is usual for a doctoral thesis defence) were in favour of granting Emanuel a Doctor of Science degree. His work, which showed the determining role of the intermediate product — mono-oxide series — in the kinetics of the oxidation of hydrogen sulphide, was undoubtedly of high quality, and N.N. Semenov very insistently pressed it forward as the work of a Doctor of Science. I think that it was precisely this insistence that called forth opposition in the Council members, and YaB became one of two leaders of the opposition. He spoke several times, and argued against the idea that the student should skip the degree of Candidate of Science, citing as part of his argument the names of leading chemical physicists who had not yet become Doctors of Science. My own sympathies were on the side of Emanuel. The Council ended by proclaiming him a Candidate, not a Doctor, of Science, and Semenov went into the empty institute corridor to telephone P.L. Kapitsa and bitterly complain about the results of the voting. I was on duty in the institute that night, and accidentally overheard this; I must say that I also sympathized with Semenov.

To finish this episode — and for the edification of current young scientists — I'll add that subsequently, my two friends and neighbours Zeldovich and Emanuel were bound for many years by a close and warm friendship, which ended only with the death of Emanuel in December 1984.

The year 1943 began. Zeldovich, like Semenov, Khariton, and other leaders of the ICP, spent more and more time in Moscow. The so-called 'atomic problem' began, and the question arose of moving the entire institute to the capital. A mock-angry telegram from YaB sent to Kazan, urging haste in the presentation of urgent materials, provoked much laughter at the ICP: 'The lack of arrival of your grades is worrying your mother. Transpose. Zeldovich.'[8] YaB possessed an excellent sense of humour, as I myself have seen on many occasions.

In the spring of 1943, YaB was awarded a Stalin Prize for his work on the theory of combustion and detonation. Such a high honour for such a young fellow (he was 29 years old!), working not as part of a collective, but independently, was at that time highly exceptional.

In the summer of that same year, in accordance with the wishes of my superior, S.Z. Roginskii, I also ended up in Moscow, where my formal acquaintance with YaB soon began. A large proportion of workers in institutes of the Academy of Sciences were still evacuated, and there were few of us in Moscow. We all got dinner using our coupons in one of two academic cafeterias, and we therefore all knew each other, at least by sight.

In the autumn, the entire Semenov family appeared in Moscow, and lived in room 1211 in the Moskva hotel for more than a year. When I stopped by to see my friend Yura, the son of Nikolai Nikolaievich, I met YaB from time to time.

In February 1944, we were able to get tickets to a concert in the Scientists' House concert hall, given by Aleksander Vertinskii, who was a sensation at

[8] A play on not exactly decent expressions beginning with 'Your mother ...'. [Translator]

Group of researchers of the Institute of Chemical Physics. From left to right: N.N. Semenov, O.I. Leipunskii, P.P. Shater, N.N. Simonov, B.V. Aivazov, E.N. Leont'eva, Z.I. Kogarko, M.A. Rivin, Ya.B. Zeldovich. Moscow, 1946.

that time. I learned to imitate his singing tolerably well, and YaB frequently requested performances of songs that he liked.

At one point in the summer of that same year, having met by chance at Semenov's place at Uzkoe, we returned to the institute together — a journey that took an hour and a half by tram and on foot. Our long conversation about Leningrad, the war, books, and mutual friends first gave an element of warmth to our acquaintance. At the end of 1944, I moved to the newly constructed institute building as a postgraduate student of Semenov. I slept first in the accounting office, then in the laboratory, where I put up a camp bed next to the laboratory desk each night. My days began by switching on the electric hotplate (for morning tea) and the vacuum pump. YaB stopped in to see me quite often, and our topics of conversation at that time were roughly equally divided between science and the fairer sex. As a first test in science, he gave me a task — to find and correct all the mistakes or misprints in his joint article with Yu.A. Zysin about heat transfer during exothermic reactions in flows. With regard to the other theme of our conversations, V.L. Ginzburg many years later teased me by telling me about the time he first heard my name. He was walking with YaB along a corridor of the only wing of the ICP that was then active, and a young female colleague asked YaB where Gol'danskii's office was. 'It's this one here,' he said, indicating my door, 'but you should be aware that any woman who goes through that door is as good as married.' (I can't remember the exact lines from one of my favorite poems of Yunna Moritz: 'When we were young, and talked wonderful nonsense ... .')

The unforgettable days of May 1945 passed, and in June, the USSR Academy of Sciences celebrated the 220th anniversary of its founding. The Joliot–Curies, Langmuir, Hinshelwood, Svedberg, Auger, and many other famous guests came to our institute. YaB (who, by the way, obtained at that time his first award — an Order of the Red Banner of Labour) was one of the main centres of attraction for these guests. Entering his laboratory, two of the guests tried to allow the other to go through first, so that the whole group of six had to come to a halt. From childhood, YaB could speak German decently, but at that time, speaking German was not something that was accepted, besides which, we didn't feel like it. In broken English, he laughingly told the guests of the famous dialogue between Chichikov and Manilov in a similar situation. Both guests, without listening to the end of the story, threw themselves forward, and, of course, got wedged in the doorway. YaB festively ended his commentary with the words, 'With Gogol it ended in just the same way!'

In 1944, following the birth of two daughters, the Zeldovich family finally produced a long-awaited heir, Boris (YaB joked about this: 'one, two, three — and it's done'). They moved to the first floor of a residential building in the yard of the ICP, into apartment 11. YaB frequently invited me to stop in for a bite to eat, and would explain to his little daughters Olya and Marina, who observed these visits with curiosity, 'Do you think he's really eating? Sillies, it's just make-believe.' This expression became a catchphrase with us for a long time (for example, 'Let's drink some make-believe tea').

At that time, I was sweating over the writing of my first article — 'Indications of Chain Reactions of Nitrogen Oxide,' which appeared in January of 1946 in *Uspekhi Khimii* — and couldn't think of how to begin it. When he learned about this, YaB said: 'There's nothing simpler. You have to begin with the words "The vigorous development." What's the article about? Okay, so you should start it like this: "The vigorous development of methods for the indication of chain reactions ..."'

At the end of 1945 or the beginning of 1946, YaB came to me with an unexpected suggestion — to write together a comical play to present at Semenov's 50th birthday celebration, which was coming up in April 1946. We set about this task with passion. We worked on some scenes together; others we wrote separately and brought to each other for judgement. YaB urged me to introduce as many Freudian overtones as possible into the play. The play depicted (in a humorous way) the events of 1924 (the wedding of Semenov) and 1926 (the discovery of the lower limit for ignition of phosphorus by Khariton and Valta) up to 1986 (the final scene entitled 'Forty Years After'). One of several interludes between the two main acts of our play was dedicated to the first appearance of the young Yasha Zeldovich at the ICP. I present this interlude below, with a brief explanation: the unknown character who figures in this skit appears with a request to 'put the glass blowers in order' in all the acts, which take place over several decades.

Interlude 1. 1931.

(Semenov sits in his office. Enter a boy with spectacles.)
Boy: *Uncle, are you the director here?*
Semenov: *Yes. What do you need?*
Boy:. *I want to become a professor.*
Semenov: *Is that so? How much is two times two?*
Boy: *Are you kidding, Uncle? Give me an integral to solve!*
Look — your subscripts aren't written correctly!
Semenov: *Indeed, they aren't. Interesting. What's your name, boy?*
Boy: *Yasha Zeldovich.*
Semenov: *Well, let's see, come on with me to the Scientific Council!*
He goes to the door. The Unknown comes to meet him.
Unknown: *Nikolai Nikolaiovich! You need to put the glass blowers in order!*
Semenov: *Later, later! We're going to the Scientific Council now.*
Unknown: *There won't be any Council. Somehow, things won't work out with my equilibrium.*
Semenov: *Outrageous! You haven't yet learned how to do your equilibrium calculations!*
Boy: *Such a big boy, and he can't calculate an equilibrium.*
(Unknown gives the boy a clip on the back of the head. Exit Unknown. The boy again approaches Semenov.)
Boy: *So, can I be a professor?*
Semenov: *Well, we'll talk about it later.*
Boy: (crying) *No, take me on, I want to be a professor.*
Semenov: *Hmmm, we do need to help youth advance. Okay, we'll take you on at the institute.*
Boy: *I'll come tomorrow. And I'll bring Dody Frank [D.A. Frank-Kamenetskii], he's also good.*
He and I will play leapfrog.

(Runs off. Screams loudly: 'I'm going to be a professor! I'll go tell the guys!')

Having dealt with the playwright's responsibilities, YaB and I now became directors, selected a cast, and conducted several rehearsals. The role of Semenov was successfully played by the future academician V.V. Voevodskii, Natalya Nikolaevna (Semenov's wife) was played by their daughter Mila, and the role of Yasha Zeldovich was played by the son of one of the oldest researchers at the institute, B.S. Kogarko (now a Doctor of Science). We also did not shun our duties as actors: for example, YaB appeared in the scene entitled 'Morning of a Postgraduate Student' in a dialogue between the postgraduate student (who was clearly meant to be the author of these reminiscences) and Semenov. This scene was written almost entirely by YaB, in full agreement with his instructions about Freudian overtones. Our play was successfully presented on April 21, 1946 in the hall of the Institute of Physical Problems, with lighting and sound effects provided by A.I. Shalnikov.

Yakov Borisovich in the year of his election as a Corresponding Member of the USSR Academy of Sciences. Moscow, 1947.

In the last lines of the final scene of the play, I finished with a report on the life and work of Semenov, read 40 years later, at his 90th birthday in the presence of YaB. Alas, in that very year, Semenov died, and Zeldovich left us in the following year. Now, the cheerful couplets bring forth only grief about past times and people who will never return.

In the autumn of 1946, I submitted my Candidate of Science dissertation to be typed up, and Semenov at the same time decided to transfer me from catalysis to nuclear physics. He temporarily transferred me on a half-time basis to the theoretical department under the supervision of YaB. I remember the day of the elections in the Academy at the end of November, when YaB was a candidate to become a Corresponding Member of the USSR Academy of Sciences. YaB, who was nervous, kept leaving the room, disappearing for an hour and a half or so, then returning. Each time he came back, I asked, 'Well, how are things?' until he finally answered in a fit of temper, 'Go take a crap.' The elections finished with victory for Zeldovich, and a few days later, a banquet was given on the third floor of the institute, in the space into which the theoretical department was soon to move. Let me present several verses of the comical poem that I read at that banquet:

Oh, what a festive moment!
Zeldovich is a member correspondent,
And Khariton a corresponding mem —
Let us sing praises to the two of them.

Doing our job in two bursts,
We'll tackle Yasha first.
He is witty and ... fantastic.
Plump, pleasantly stout,
To women terribly attractive,
Due less to his looks than his clout.
Leaving his soother, from the age of two
He solved integrals, easy as two plus two.
At twenty-two he became a Candidate
(Even the Devil doesn't know his final fate.)
Corresponding member at the age of thirty-three
But still, Yasha, you take care to see
That you don't put on any airs
Or Christ won't want you among his heirs.
He was your age, just abouts
When carried up into the clouds.
(To complete the rhythm of the strophe,
Let's not forget about Golgoth!)
No, Christ won't drive you from us
Though we have nails, we have no cross.
And there's no reason to heaven to ascend,
On Earth, there are plenty of aims to tend.
In our glasses are institute spir'ts
Leave aside for now other flirts!
Our toast for our pride and joy
Our toast for Yasha — our wonder boy!
Our Yakov has all that it takes:
In the fields of science, volleyball,
Drama, tennis and all,
An anti-aircraft fighter in Kazan,
In short — Yasha is a great man!

I spent several months in 1947 in the theoretical department of the Institute of Chemical Physics. There, under the leadership of YaB, worked such individually bright people as D.A. Frank-Kamenetskii and A.S. Kompaneets; the still very young shining talents of Serezha Dyakov (who tragically died three years later) and Kolya Dmitriev; and also F.A. Blyumberg, who helped with the computational work.

In the department, we had a so-called 'door newspaper' (a nearly word-for-word translation of the Chinese 'Daotzibao;'[9] each of us could compose something, in verse or prose, and tack it to the doors for general survey and

[9] This refers to one form of rather unsubtle Chinese propaganda used in the period of the Cultural Revolution in the 1960s and 1970s, in which huge banners with slogans were hung in conspicuous places. [Translator]

discussion. I was able to gain acclaim as a local Mikhalkov,[10] having composed the following anthem for the theoretical department:

We're the navel of the institute,
The longing of all others.
Our great corresponding member
Tosses a hundred ideas each minute.
We carry the light of learning to the plebs
Who sing the praises of experiment,
In front of Semenov, we are not shy,
There are none braver than we.
Landau himself carries our banner
We are powerful, healthy, strong,
Only our asses suffer, as we sit
And bake top-secret reports like pancakes
Glory to physicists, to the poets
Who surround Yakov's throne,
We are content in the calm of our office,
We're not intimidated by neutrons.

Along the way, there were various practical jokes, which were not always completely harmless. At one point, a reprint from some foreign journal, from some unknown author with a Russian name, came to the institute addressed to YaB. Kompaneets and I wrote on it, 'To my dear friend Yakov Borisovich, from Staff Captain[11] So-and-so' and put it on YaB's desk. You have to understand how dangerous it was at that time to have ties outside the Soviet Union (especially for a person with a high security clearance) to imagine how anxious YaB was about this, and how much we later regretted this joke.

Our entire lives were concentrated around the five-storey institute building: here, in the yard, were our apartments, here we played volleyball, went on morning jogs with YaB, Frank-Kamenetskii, and Voevodskii, threw each other medicine balls. As in science, YaB engaged in sports with passion, at times driving himself to the brink of exhaustion.

He fell in love — partly due to my influence — with dancing, and nearly every Saturday we went to take a turn at the ICP or the neighbouring Institute of Physical Problems. YaB clearly derived pleasure from dancing, though, in spite of my encouragement, he never tried to master the art. He simply let himself be carried by the music, deliberately ignoring any

[10] S.V. Mikhalkov, who is well known in Russia as the author of the lyrics for two different versions of the Soviet national anthem, reflecting the ever-changing ideological priorities under the Communist regime. [Translator]

[11] This military rank existed in the Russian Tsarist army, but not in the Red Army of the USSR. [Translator]

'regulation' steps. For him, dances were a sort of game of chase-and-catch, with pleasant partners — what could be better?

In October 1947, YaB hurried to my assistance at a difficult time in my life. Mila Semenova and I had to announce our marriage to her parents after the event (toward which they (especially my mother-in-law) had been far from favourably inclined). YaB took this task upon himself. He invited Semenov to his office, talked with him for a long time in private, and finally came out to me, satisfied and smiling, so that I immediately felt a relief in my heart. And indeed, the very serious conversation I subsequently had with Nikolai Nikolaevich proved to be very warm.

More and more often, YaB left Moscow for the 'object' in Ensk,[12] and it gradually became more accurate to say that he rarely came to Moscow. I can't say if January 1948 was still the end of his Moscow period, or whether his time in Ensk had already begun, but in any case, in this month the death of Mikhoels became imprinted in my memory.[13] YaB came into my room without his usual smile, serious and preoccupied, and proposed that we go to the Jewish theatre, where there would be a farewell for the deceased — or rather, murdered, as we soon learned — performer. Leaving the tram near the Nikitskii Gates, we saw a huge crowd blocking the street in front of the theatre, and were taken aback for a moment. YaB mournfully joked, 'We can't hope to get through without waiting in a huge queue — they're all Jews'[14] and dug into his pocket for his pass certifying him as a laureate of a Stalin Prize. This pass cleared the way for us — and not only into the theatre, but into the guard of honour, and we could not help noticing the huge bruise near the temple of the deceased. We returned home silent, immersed in sad thoughts. Jumping ahead, let me say that the steadily growing problem of anti-Semitism, clearly demonstrated in this incident, worried YaB very much, and caused him much grief.

The anti-Semitic mood was a clearly hidden motive in the attacks on YaB's book *Higher Mathematics for Beginners*. Two days before his death, he told me about a vile anonymous note in which, in addition to the usual anti-Semitic abuse, they called him a Satan. Sensing that the upper echelons of power in the years of stagnation were home to conscientious gardeners who

---

[12] The 'object' was a name used to refer to the highly secret Soviet atomic weapons development project. The location of this project was classified, and was said to be in 'Ensk,' which is the Russian equivalent of 'City N' or 'N-ville.' This city is now known to have been the old Russian town of Sarov. One nearby city was Arzamas, and one of the joking names for 'Ensk' among physicists working on the project was 'Los Arzamas,' an obvious reference to their American counterpart. [Translator]

[13] An outstanding Russian Jewish actor and theatre director whose assassination in 1948 is widely believed to have been ordered by the KGB. This assassination marked the beginning of a well-orchestrated anti-Semitic, and more generally, anti-intellectual, campaign during Stalin's regime. [Translator]

[14] Allusion to a widespread assertion in Russian jokes that Jews are so smart that they can bypass any queue. [Translator]

were interested in cultivating future neo-Nazis, YaB and A.N. Frumkin wrote about this account to Andropov, met with Trapeznikov, and requested an audience with Suslov and Zimyanin — but all in vain.

Let me return to my chronological outline. Beginning in the autumn of 1947, the small group under my supervision was assigned the task of investigating the passage of neutrons through a shell of exploded material, which was supposed to surround the nuclear charge of the future 'article.' We imitated the composition of the exploded material (TNT with RDX) and conducted experiments for over a year — first with a neutron (Ra–Be) source, then in the reactor and cyclotron of Laboratory No. 2 (the future Kurchatov Institute of Atomic Energy). At one point, YaB led a group of generals to see the experiments, and completely seriously, without a hint of a smile, explained to them, 'And here we place sheets made of indium, which are, therefore, called indicators,' and gave us a provocative wink.

Memory usually separates out the lighter, joyful, cheerful remembrances. In fact, times became more and more difficult and base. The resolutions of the Central Committee of the All-Union Communist Party on literature and art came one after another; in science, they held a court of honour for Klyueva and Roskin, and sent Parin, Balandin, and Parnas to prison. They decided to turn the courts of honour into omnipresent spectacles (comrades' lynch trials, so to speak). They chose YaB and Leipunskii as the most suitable defendants at the ICP. They were required to apologise for trying to push through some of their work on powders as an open publication, which had been hurriedly signature-stamped. M.A. Sadovskii (the less guilty culprit, so to speak) had to apologise for the fact that, as assistant director, he had signed the authorization for this work to be published. However, YaB firmly stood his ground, and refused to acknowledge any guilt in what he had done. This shameful 'court of honour' accelerated YaB's official departure from the ICP and his transfer to Khariton's group in Ensk.

A year or so later, after the first Soviet atomic explosion (August 29, 1949), a 'golden rain' of medals was poured on nuclear scientists. When YaB was decorated with the first of the three stars of a Hero of Socialist Labour, he invited to a party at his home not only all his friends, but also his most ardent persecutors. I won't name them here — not according to the principle *aut bene, aut nihil* (of those who have since departed life), but for the sake of the living and innocent children and grandchildren of the faithful servants of the Stalinists in our institute. When they began to say that they were prepared to ask forgiveness of YaB and call him back to the ICP, he presented to me his favourite couplet from the 'Allaverdi' mischievous rhyming alphabet: 'The handsome captain takes a modest bow, You can't shove cowshit back in the cow.'

In the period of YaB's transfer to Ensk, there were also attempts by the so-called 'secretaries' (three KGB men who took turns to relentlessly watch over leading nuclear scientists round-the-clock) to stick to him. Among these, there were, by the way, some who became genuinely attached to their wards, and gradually became their indispensable helpers. However, YaB (like Landau) categorically repudiated the trouble taken by the KGB. From 1949, we all lived in a single area, next door to each other, and on several nights, some strange people spent the night in our stairway standing on the stairs or sitting up on the windowsill. At first, my wife and I didn't understand who they were, and YaB wouldn't allow them to come into his apartment. In the end, the 'secretaries' weren't able to withstand working under such conditions, and YaB claimed victory.

In the 1950s, fate separated us — YaB went to Ensk; in 1950–51, I was working in Dubna on experiments on the absorption and multiplication of high-energy neutrons (at that time, 120 and 380 MeV were considered high energies). At the beginning of 1952, as part of a battle with nepotism, I was transferred from the ICP to the Lebedev Physical Institute in Moscow, where I worked for nine years. Here, I must express my gratitude to the Lebedev Institute, thanks to which I expanded the circle of my knowledge and interests, and could therefore later enjoy new scientific contacts with YaB. However, that is several years further in the future. At that time, our meetings became rare, though we were neighbours. Working at the 'object,' YaB clearly felt the growing severity of the regime during the last few years of Stalin's rule. He suffered very much over the troubles of the Korean War, though he didn't like to talk about it at that time. It was only many years later that he told me about his anguished sufferings — if our first atomic bomb had not been built in 1949 with the active participation of YaB, the Korean War would probably not have begun in 1950. Sharply feeling the danger of being shadowed, he reprimanded A.S. Kompaneets and myself severely for our love of idle talk. During one of his visits (on the eve of which there was a small fire in our courtyard), I responded to these reprimands with some verses, which are literally overflowing with our silly carefreeness and inability to comprehend the full seriousness of the dangers felt by YaB:

Not for nothing, back a few days
Moscow met Zeldovich with a blaze,
Honour to Yasha the regal.
Of golden stars he's got a pile,
And they'll be more in a while!
They're singing songs in Russian style
About our eagle.
He's not yet constipated,
and already a hero celebrated,
Reproach to skeptics far and near.
For the younger generations,
You're a genius at elaboration,
We remember your orations,
Our Corresponding Member dear.
No more verses will we rhyme,
nor will we joke to pass the time,

We'll put up solid gates.
At you, banisher of prattle
All of Rus' and Europe do marvel
Your eyes fresh and full of battle ...
How feverishly he creates!

For five or six years, beginning in 1952, I believe, we had a tradition — to see in the New Year the pleasant company of friends and colleagues at the Institutes of Chemical Physics and Physical Problems, and also the Lebedev Physical Institute. Then, on the following evening, January 1, we would gather again for a 'leftover holiday.' In this way, over the course of several years, we went round to the flats of the Zeldoviches, Ryabinins, Landaus, Ginzburgs, Emanuels, Semenovs, as well as to our own and many others. We prepared for the festivity in a big way, and with inventiveness, planning gifts, raffles, and charades (in which YaB was especially active). Once (at Landau's place), Kompaneets and I even wrote and performed the play *Day of a Man of Science.* For YaB, these New Year festivities, like the birthday of Kompaneets and Leipunskii — January 4 — and the birthday of Stanyukovich — March 3 — were times when he traditionally came to Moscow; for us — apart from the general merriment — they were days of joyful meetings with YaB, who was much missed by all of us.

After 1956 — already a three-star hero — YaB moved back to Moscow, although he didn't return to the ICP, and instead he went to work for M.V. Keldysh at the Institute of Applied Mathematics of the Academy of Sciences. He became my neighbour again, and we began to exchange joking rhymed messages, which we tossed in each other's post boxes. Being worried by my considerable weight gain, I now and then stopped by the Zeldoviches' in the morning in my shorts to weigh myself on their scales. One morning I discovered in my post box the following document:

Office of the Moscow Bureau for the
Weighing of Live Goods (OMBWLG)

To our honourable clients Lyudmila Nikolaevna and Vitalii Iosifovich,

Determination of the longitude, latitude, lunar phase and acceleration of gravity on the night of March 5–6 revealed a systematic error in weighing of −0.85+0.07 kg. Indications of live weight stamped on the stomach before the 6th of March are invalid, and must be replaced.

Chief astronomer of OMBWLG,
Ya. Zeldovich

And here's an even earlier message from YaB, after he bought me a book of Simonov's poetry at the academic book store and, to my chagrin, thoroughly stained this long-desired tome with butter:

Account

for his Excellency Monsieur Gol'-donskoi[15]

0.	*Delivery by private car*	*20.00*
1.	*Simonov*	*3*
	Butter on the cover (full cream)	*1.40*
2.	*Payment on account at the grocery*	*56*
3.	*The guard 'didn't see' the vodka*	*100*
4.	*... or confiscate the herring*	*200*
5.	*The Princess of Dollars*	
	tickets	*46*
	vodka 2 × 100g	*15*
	sandwiches	*10*
	moral damages	*100*
Please pay in total		*551.40*

We put into each other's post boxes a whole pack of verses and prosaic messages. Their genre was rather varied — from quite unrestrained (most often clearly marked: 'Mila! Don't read this, I can't vouch for the decency of the contents') to rather serious: 'Vitya! In connection with the work of K, it would be interesting to consider the formation of NO in the tracks of fission products. The kinetics in this case are understood. There is an interesting technical task — to obtain NO thermally in a reactor. In particular, P. is working on this, but not doing a good job. What do you think — do you want to get into this?' And after one of the anniversary evenings at the ICP, a minor worry about his presentation: 'What a joke I let slip! Engels: labour created man from an ape ... I became a man working at ICP.'

I keep each scrap of paper with YaB's sweeping lines as a dear memory — but, of course, there's no point in sharing all of our serious and light-hearted notes and sketches. I'll limit myself to telling about one last episode, which is especially amusing.

Just before the 1950s, during the reconstruction of the Vorobiev buildings, they had to urgently tear down a group of wooden single-storey huts next to our building (later, the residence of A.N. Kosygin was built on that site). The destruction of these huts was a genuine disaster, for the abundance of bedbugs that lived there and migrated without delay to our building (it's a pity that we weren't able to see with our own eyes the procession of their close ranks).

One far from fine night, we couldn't get to sleep because of furious itching, and when we turned on the light, we were horrified by the horde of bedbugs on the wall. We had to throw out the bases of our divans the next morning, but we couldn't get rid of the mattresses — we couldn't sleep on

[15] Play on the name of the author of this chapter: *gol* means *beggar* and *Donskoi* is the name of an old and highly esteemed noble Russian family. [Translator]

the floor, after all. Having thickly smeared these mattresses with kerosene, we set them to air out on the hall stairway. After two or three days, there appeared on the mattresses placards made by YaB. Arrows, on which he had depicted nasty-looking insects, were directed toward our doors, at their other end there was a 'No Entry' symbol, meaning 'entrance to YaB's flat forbidden', and at the top of the picture was the caption, 'Bedbugs, go home [in English].'

At about this time, YaB and I began to have scientific meetings nearly every day — we were interested in the properties of nuclei that were far from the region of beta stability. It began with YaB's search for an underlying behaviour in the shell-filling energy in light nuclei, and led to the prediction of the existence of superheavy helium — ^{8}He. I refined the expected properties of this nucleus, based on the non-existence of superheavy hydrogen ^{5}H. YaB's prediction, which at first provoked a rather sceptical reaction, was soon brilliantly confirmed.

Comparing various ways of estimating the properties of light nuclei, I came across the work of A.I. Baz', who had opened broad possibilities for various predictions based on the isotopic invariance of the nuclear force. YaB and I joined with Baz' and continued the work together, each of us in his own direction. YaB predicted the existence and properties of dozens of new neutron-abundant isotopes and analyzed the possible existence of neutron fluid, Baz' studied in detail the problem of the dineutron; and to me fell the job of predicting the properties of many new neutron-deficit isotopes and of two-proton radioactivity. Pooling our results in these three areas, in October 1960 the three of us published a review article entitled 'Undiscovered Isotopes of Light Nuclei' in *Uspekhi Fizicheskikh Nauk*. Twelve years later, we also wrote a monograph in collaboration with B.Z. Goldberg.

Thirty years has passed, but I vividly recall the joy and instructiveness of my everyday scientific and personal contact with YaB in those years. It was an enviable existence. Periods of getting so wrapped up in work, of such impatience in the evenings, wishing it would soon be a new morning, bringing new thoughts and discussions, are rather rare — you can count them on the fingers of one hand — and I'm glad that one of these periods in my life was closely connected with YaB. I still remember some of his favourite sayings: 'I verified harmony by algebra' and (addressing those who tried to pry into his private life) 'I'm a poet, which makes me interesting'[16] (the rest is not relevant here).

Science was science, but sports — especially skiing — and various jokes or pranks invariably occupied a prominent place in YaB's life. I remember when we lived in neighbouring rooms at the Uzkoe Resort, and he and Varya (his wife, Varvara Pavlovna Konstantinova — a woman of rare spirit) went racing ahead on their skis, tearing down the hills, while I crawled along somewhere behind them. On one ski trip in the winter of 1957, when my wife

[16] A quote from Mayakovskii. [Translator]

Mila (then expecting our second son) joined us, having watched my cautiousness for a while, YaB said, 'When we look at the two of you, it seems like it is Vitya who's pregnant!' In the evening, we had various medicinal drinks placed for us on our nightstands. Once, after draining in a gulp the measuring glass marked with its usual label ('valerian tea — Goldanskii'), I coughed with surprise — YaB had replaced my mixture with a drink of nearly the same colour — cognac. A large number of the participants of the Kiev conference on high energy physics in 1959 have reminded me of the swimming competitions on the Dnieper that YaB organized for me and his older daughter Olya. I can swim quite decently, and I couldn't understand why I couldn't break out ahead of Olya for at least a few strokes, and why all the spectators were laughing. After I was all knocked out from my furious swimming, I finally learned that YaB had simply given Olya a pair of flippers.

In 1963, we moved to another flat in the same building, and my encounters with YaB ceased to be daily. At the same time, however, an important new topic of conversation arose — the affairs of the Academy of Sciences, since I became a Corresponding Member in 1962. From time to time, we organized family trips to the cinema, and several times, we saw each other at New Year's evenings at the central Writers' House, though at different tables. One special source of pride for me was YaB's reaction to my talk 'Studies in Gamma-Resonance Spectroscopy,' which I gave in February 1966 at the annual General Assembly of the USSR Academy of Sciences. This talk went quite well, YaB liked it very much, and that same day proposed that we address each other using the informal word for 'you' — *ty* — rather than the more formal *vy*. I won't try to hide the fact that the possibility of addressing a person whom I revered as a living legend as *ty* — especially in the presence of a third party — rather flattered me for a long time, until I became accustomed to it.

In December 1974, YaB made a sharp presentation at the podium of the General Assembly of the Academy of Sciences, in connection with the fact that the Academy administration had not made a vacancy for me to become a full member, although I had for the second time received the qualifying number of votes in the Division (the first time was in 1968). This provided at least some balm for my wound.

As time went by news of difficult losses became ever more frequent. In the summer of 1969, YaB's close friend and brother-in-law Boris Pavlovich Konstantinov (about whom he left written memoirs) died suddenly. In 1970, D.A. Frank-Kamenetskii passed away after a long illness. In 1974, we were literally dumbfounded by the sudden death of Kompaneets, and in August of 1976, while on holiday in the Crimea, YaB's wife Varvara Pavlovna died of a heart attack. I remember how we met the plane carrying her coffin at Vnukovo Airport, and YaB's parting words at the morgue. Looking only at Varya, as if he had moved off somewhere and left us behind, he told, as if to himself, the story of their acquaintance and of their love. The 'twilight of the gods' had begun.

During the shooting of a film on the life of N.N. Semenov. From left to right: Ya.B. Zeldovich, V.I. Goldanskii, and A.E. Shilov. Moscow, Institute of Chemical Physics, 1984.

However, ahead lay international appreciation of his work, international visits — first only to socialist countries, then to Greece, Italy, and finally, the USA in April 1987. In April 1980, YaB's talk on cosmology at the 'Leopoldian' General Assembly in the German Democratic Republic was very well received. He spoke in an odd mixture of German and English, and, to explain the term 'big bang,' expanded his cheeks and loudly depicted an explosion with a cry of 'Poof-f-f!' After I arrived in Washington and entered the building of the National Academy of Sciences, almost the first thing I saw was YaB sitting at a table with drinks and sandwiches, during a coffee break. This encounter was unexpected for both of us, and therefore all the more joyful.

Of our meetings in the last year of YaB's life, I remember well a video evening at our home — in particular, a joint viewing of *Bridge over the River Kwai*. About two weeks before his death, YaB, his wife, Inna Yurevna Chernyakhovskaya, and Mila and I made a cultural trip to the Oktyabr Cinema to see *Tomorrow was the War*.

Of course, I'm omitting a lot here — for example, the celebration of YaB's 70th birthday at the Institute of Physical Problems in March, 1984, which is well worthy of description. However, I think that many of the people contributing to this collection of reminiscences will want to talk about this anniversary, and in general about YaB's last years of scientific creativity, associated with the Space Research Institute and the Institute of Physical Problems. I'm naturally drawn to subjects that were further in the past, and from the last period in the life of YaB, I'll turn now only to one episode that occurred shortly before his death.

It happened that YaB and I saw each other and talked together just two evenings before he died — November 29 and 30, 1987. Among the latest foreign scientific journals I'd received were several containing material about YaB and his work: *Science News* of May 30, 1987, which described a lighthearted dialogue between Zeldovich and Gell-Mann at the working group on quantum cosmology in Batavia during YaB's first (and only) trip to the USA in April and May of 1987; *Nature* of October 29, 1987, which was a special issue about Soviet science containing the article 'Great Men and Barons' about YaB; and finally, *Chemical and Engineering News* of August 31, 1987, which had an excellent review on chemical combustion by Fisk and Miller. Having learned that I had copies of these journals, YaB immediately came running to my flat to look them over, then, on the following day, came to borrow them to read more carefully. He was especially interested in the review on chemical combustion, which gave him great satisfaction as it demonstrated that even after 40 years, YaB's classical work on the formation of nitrogen during combustion, the mechanisms for the combustion process itself and the propagation of flames, chemical reactions in flames, and detonations remained evergreen.

We didn't see each other on December 1. Only later, I learned from my son that YaB had asked to be driven to the Institute of Physical Problems — and that clearly meant that he hadn't felt well. At about midnight on December 2, we heard a siren in the yard and some noise in the neighbouring stairway. Mila ran to find out what was going on, and then told me that YaB had been rushed into hospital. I then took it upon myself to call I.K. Shkhvatsabaya (now deceased) with a request that he take care of YaB's affairs. We went to sleep rather alarmed, but still far from being fully aware of the impending tragedy. But just 14 hours later ...

Several days after YaB's death, I learned that Rudolf Mössbauer wanted to nominate him for a Nobel Prize and hesitated only over the choice of science — chemistry or physics. Even before this, I heard from a number of first-class scientists in Western Europe and the USA that YaB was considered a worthy candidate for a Nobel Prize. I think that it was precisely the surprising diversity of his work that made the decision difficult — physics or chemistry? In both the novelty and brilliance of his ideas and in the importance of his results, YaB was unquestionably a representative of scientists of Nobel quality in both these fields.

The further we move from the day when we were parted from him, the more we feel the enormity of the loss to science. As far as we, his friends, are concerned — he will remain with us until the end of our own days. We simply cannot come to terms with the idea that he is no longer.

Fragments of wall newspaper articles dedicated to the memory of Ya.B. Zeldovich

O.I. Leipunskii

Ya.B. Zeldovich worked at the ICP for sixteen years — from 1931 until 1947 — and performed research on adsorption, catalysis, phase transitions, hydrodynamics, shock waves, combustion and detonation theory, and combustion of powders. On the basis of this work, research in these areas has continued in the Combustion Department of the ICP over many decades.

In the obituary in the newspaper, it said, 'We cannot overestimate the contribution of Ya.B. Zeldovich to the defence capability of our country.' In particular, this contribution includes the development of the theory of powder combustion and, on the basis of this, the theory of the internal ballistics of solid-fuel rockets. Such rockets ('Katyushas') developed in the 1930s, were used during the Second World War, and the charge and engine weighed 10 kg. It proved impossible to develop the power of the engine further, since the weight of the charge had to be chosen via testing, and it was not possible to calculate the effect of new engines. During variation of the engine parameters, unexpected phenomena appeared: the sudden decay of the powder charge or an extreme growth in the internal pressure, capable of destroying the engine. Thus, while it was possible to select a stable fuel charge with a weight of up to 10 kg, extending this to weights of 100 kg was empirically nearly impossible.

Yakov Borisovich began to develop his theory for powder combustion at the end of 1941, working it out over a period of several months, so that his publication had already appeared in the *Zhurnal Eksperimental'noi i Teoreticheskoi Fiziki* in 1942. He discovered a new form of powder combustion — with a non-stationary combustion rate. Together with other new phenomena discovered in Zeldovich's laboratory, this formed the launch point for the physical basis for the modern internal ballistics of solid-fuel rockets, enabling the prediction of the required combustion charge for any weight, and making it possible to construct modern rockets with charge weights, not of 10 kg, but of many tons. As a witness of Zeldovich's creation of his theory of powder combustion, I obtained aesthetic pleasure from the uncommon simplicity and the speed with which he worked. I believe it to be one of the most beautiful among his many hundreds of research accomplishments.

After Yakov Borisovich left the ICP, he worked for 16 years on the foundation and creation of various new techniques, the theory of elementary particles and nuclear physics. He then worked for 22 years in astrophysics, cosmology, and the origins and structure of the universe. His activity in all these fields did not just touch on various interesting physical phenomena, but rather was concerned with fundamental, ground-breaking work, similar to his combustion research.

With O.I. Leipunskii at the Institute of Chemical Physics. Moscow, 1980.

In all the world of science, there is no other physicist with such a wide grasp of investigations in physical phenomena, leading to deep and fundamental results. Zeldovich was a unique phenomenon in science. He provides an instructive and interesting example for studying the properties and capabilities of the human intellect.

When filling in forms, under 'social status,' Yakov Borisovich wrote 'civil servant.' In print, the term 'intelligentsia' is often used. Yakov Borisovich was an intellectual — a member of the intelligentsia — in the deepest meaning of this word. In Russian, there are two meanings for 'intellectual.' In one, the dictionary meaning, the defining characteristic of the intelligentsia is their acquisition of knowledge and use of it professionally. The other meaning flows from Russian classical literature, which draws attention to and describes many intellectuals whose defining characteristic is their conscience and decision to act in accordance with this conscience. Naturally, there are cases where these two meanings are merged in a single person. Yakov Borisovich is one example of such a merging of these traits. His dominating force was not only the acquisition and use of knowledge, but also the possession of an active conscience: scientific (protesting against incorrect scientific views, irreproachable scrupulousness in questions of authorship, etc.), civic (doing applied research when required), and personal (actively helping those unjustly harmed by circumstance, etc.).

Yakov Borisovich loved his work very much, and worked from early in the morning until late in the evening; his work was part of his feeling of being alive. He was an affable, kind, good, responsible friend, undarkened by any shadow of self-importance. He had all of the qualities that are included in the concept of a simply good person.

Tyutchev painted a fundamental and gloomy picture of the relations between man and nature:

> However fierce the fight, long the battle,
> Take heart, struggle, o friends brave,
> Above, the silent stellar circles,
> Below, only mute, deaf graves.

Yakov Borisovich, in essence, was a man engaged in a fierce battle with nature. However, he substantially improved on the gloomy mood of Tyutchev: the battle above 'mute, deaf graves' can become the joy of life and be filled with happiness. For Yakov Borisovich, the 'stellar circles' were not 'silent' — he understood them and responded to them. For him, the battle was not 'fierce,' for he was one for whom 'there is rapture in the battle and at the edge of the wild abyss.' In my view, the words of Tyutchev speak precisely of him:

> Let the Olympians look with envious eye
> At the battle, as unbent hearts stand.
> He who, conquered only by fate, fires,
> Tears the crown from their deathless hands.

F.I. Dubovitskii

The 1930s was a period of vigorous development in the new Institute of Chemical Physics of the USSR Academy of Sciences, under the leadership of its founder, N.N. Semenov. At that time, Semenov took active measures to select and educate young scientists. He sent letters to provincial institutions of higher education, urging them to send graduates inclined toward scientific work to the institute. In this way, in 1931, N.I. Chirkov and I came to the Institute of Chemical Physics from Voronezh State University, A.B. Nalbandyan from Erevan, A.Ya. Apin and others from Kazan, and O.I. Leipunskii and A.F. Belyaev from the Leningrad Polytechnic Institute.

However, the most momentous event was the appearance of Ya.B. Zeldovich in the institute. Though he had not graduated from university, he immediately became actively involved in scientific research on a par with the leading scientists of the institute. Today, it is difficult to imagine the talent, strength, and breadth of intuition possessed by the 17-year-old laboratory assistant Yasha Zeldovich. He was proficient in the mechanisms for a wide variety of phenomena and in methods for the study of chemical processes — and not only in the narrow field of catalysis, in which he began his work with S.Z. Roginskii; he actively participated in the institute seminars, helping to understand complex scientific questions in other fields as well. He made presentations easily, freely, ardently, and self-confidently. When he was occasionally late to a seminar, while still walking in, he would ask questions of the speaker, who had already made nearly half his presentation. Yasha Zeldovich was very gifted and possessed the highest degree of intelligence and talent. Of course, the difference in age between him and his colleagues was important at times (after all, he was between 7 and 10 years younger than the other workers in the laboratory), but in science, he was mature beyond his years. Scientific researchers in all departments of the institutes in the area, including the strong theoretical department headed by Ya.I. Frenkel, treated him as an equal.

YaB rapidly grew and reached manhood. He understood the necessity of applying the results of his scientific investigations to support the national economy and industry. In 1937, under his supervision, a group studying high-temperature reaction kinetics during explosive processes was founded, and just before the Second World War, the laboratory of combustion physics was set up. His creative activities flowed into exclusively friendly collaboration with the founders of combustion and explosion physics (of course, Zeldovich himself also deserves this title) — N.N. Semenov, Yu.B. Khariton, D.A. Frank-Kamenetskii, I.I. Shchelkin, A.F. Belyaev, and others.

In 1942, I lived together with him in the Moscow Hotel on the 12th floor, in a two-room deluxe suite. I.V. Kurchatov also lived with us there. I was a direct witness of YaB's selfless labour. Yasha worked until late, lying in bed with a notebook in his hand. He tried not to disturb the sleep of the other colleagues living in that room with him (G.A. Varshavskii and I. Makarov). Kurchatov and I slept in the adjacent room in the suite.

I have kept a letter from YaB to Semenov, sent from Kazan on March 1, 1943. In this letter, YaB talked about the laboratory in Moscow and the state of affairs in the part of the Institute evacuated to Kazan. I present this letter in its original form:

1 March, 1943

Dear N.N.

I am waiting impatiently for news from you, or for your arrival, so that I can learn how things are with the laboratory. All of us have a fighting spirit. Ovsei [Leipunskii], having carried out his plan for the first quarter, discovered a very interesting ignition effect; Barskii only recently returned, but has already been able to confirm our main result in two new cases. I really want to immediately develop work along both practical and fundamental theoretical lines, to help overthrow our enemies and establish the truth.

We could do a lot now — experiment and prepare the project for an apparatus for future work in Moscow. Apart from the usual failure to fulfil orders in the workshops, apart from excusable reasons such as the failure of electrical power this time, work is also hindered by the following: (1) they took Grabovetskaya away while Barskii was ill, and now won't let her go without your permission; (2) the lack of staff (laboratory assistants, technicians, young or old) — we could expand the front and increase the speed of our work, we have many pressing questions and preparatory work to do, but no one to do it; and (3) the sad condition of Leipunskii. Fedor [Dubovitskii] has probably told you that he has myocarditis; in a letter to Fedor, I suggested he request to be sent to a sanatorium. The situation is aggravated by the fact that Leipunskii is in a very difficult position financially — 800 rubles per month. Sergei Petrovich recommends the path laid out in the enclosed paper, composed by him. I'm sorry to take up your time, but it will be easiest for you to obtain on paper a resolution from Bruevich, and for the 'object' — Ovsej — for both work and fairness, this is very important.

If we are able to work, after 1–2 years, the question of internal ballistics will be developed experimentally, physically and chemically, and mathematically, like a candy better than any other in this field, better than the topic of my doctoral work — it's a pity we didn't work on this before the War.

Yours, Ya. Zeldovich

P.S. Everyone is asking about our move, when, how. What should I tell them? The institute has awfully few people, now the most active wild heads are away; isn't it time to add some young workers in the proper sense (our generation with ten years of experience can only conditionally be considered young; many are ill and are stooped over); Kuyumdzhi and Shchelkin aren't bad; by the way, a number of things (scientific examination of the plan) have been set aside until your return.

The war years brought me especially close to Yakov Borisovich. He was a real, simple, faithful man without a personal agenda, a hard worker, and patriot. I thank fate that I was thrown together with Ya.B. Zeldovich, and I am thankful to him for all the time we have spent together, from our youth to the last days of his life.

A.G. Merzhanov

I had the pleasure of becoming acquainted with Yakov Borisovich Zeldovich (YaB, as his students loved to call him) at the very beginning of my scientific activities, in 1956. Having carried out my first scientific work on a quasi-stationary theory of thermal explosion, I showed it to Yakov Borisovich, on the recommendation of N.N. Semenov. I'll never forget the impression YaB made on me, a young researcher just starting out. I hadn't even had time to open my mouth and explain the problem being studied, when YaB interrupted me and began himself to talk about my scientific work, just as it had, in fact, been done. At times, I got the (false) impression that he had studied my work earlier, in preparation for our discussion. I listened in bewilderment — after all, before this, I had tried to discuss my results with many colleagues, but they had not always understood. When he had finished, YaB asked what questions I had, and said, 'I advise you to calculate the equations numerically. I could do this in two or three hours, but it will take you several days. But don't be lazy — this should be done.' I was so astounded by the strength of YaB's talent, the speed and accuracy of his thoughts, that immediately after this meeting, I began to study all of his work attentively.

The Ya.B. Zeldovich medal, established in 1990 by the American Institute of Combustion and the Council on Combustion at the USSR Academy of Sciences. The first laureate of this medal was Corresponding Member A.G. Merzhanov.

In this way, Yakov Borisovich burst into my scientific life, and his ideas lay at the foundation of our further studies in the fields of combustion and macroscopic kinetics. The theory of flame propagation concentrated in narrow zones, analyses of combustion limits, the stationary theory of ignition, the establishment of the non-uniqueness of stationary regimes for a proton reactor — without these and much other work by YaB, it would be difficult to imagine our own work on the thermal theory of combustion and explosions in condensed media, and on the macrokinetics of chemical-engineering processes.

The method of narrow reaction zones became a working tool for every theoretician studying self-sustained waves. The elegant theory of flame propagation developed by YaB in collaboration with D.A. Frank-Kamenetskii was immediately recognized as classical work. At the meeting of the Scientific Council of the USSR Academy of Sciences on Combustion dedicated to the 70th birthday of Zeldovich, I gave a talk about the development of this theory. When preparing for this talk, I came to understand very clearly that all six postulates of ZFK theory (as it is called, based on the first letters of the surnames of its two creators) were subsequently investigated intensively by various researchers, so that they could be refined in application to varied conditions. This was all done on the basis of the ideology and scientific apparatus of ZFK theory. It is a great joy for a scientist to be the author of such a work, and Yakov Borisovich felt this.

It is impossible to carry out an investigation in combustion theory that is not connected in one way or another with the name of Zeldovich. For example, our work on gasless and filtration combustion, self-sustained processes with broad reaction zones, elementary models for type II combustion, spinning waves and auto-oscillations, and interaction between individual stages in wave combustion are, at first glance, not directly related to

YaB's research, and sometimes seem even to contradict its basic concepts (broad reaction zones!); in fact, this work is penetrated through and through by the ideas of Zeldovich.

In combustion theory, Yakov Borisovich is unquestionably the leader, but in spite of this, he understood that science develops as a result of the mutual interaction of input and feedback. He was able to learn from his students and colleagues. It is a pleasant thought that some of YaB's more recent well-known work, such as his research on non-stationary ignition theory and the accelerated propagation of flames at high initial temperatures, and also his original model for spin combustion, was done under the influence of our results.

The collective at the Department of Macrokinetics of the Institute of Chemical Physics of the USSR Academy of Sciences (now the Institute of Structural Macrokinetics of the Russian Academy of Sciences) makes thrifty use of YaB's creative legacy, continuing to develop his ideas and concepts. Many of our researchers were personally acquainted with YaB, and that is a great happiness for us. Those who know of Zeldovich only from the literature, who were not subject to the charm of his bright and strong personality, have not truly passed through the school of YaB.

G.M. Makhviladze

This note is about my first memories when I received news of the sudden death of Yakov Borisovich Zeldovich.

1968: The first meeting with YaB at his home. V.B. Librovich, G. Sivashinskii and I arrived together, as a trio, at the third floor of the building on Vorobiev Road, well known to many theoreticians and experimentalists. YaB opened the door and quickly led us into his home study, a modest, businesslike setup. In the centre was a round table, on it a telephone, many journals, reprints, primarily on astrophysics. On another small table, a turntable and phonograph records. In the corner, on the divan, lie some weights. On the wall there is a small blackboard with formulae and several names.

I can tell that Vadim Bronislavovich was anxious. Grisha [Sivashinskii] and I were as if in a mist. We push forward our prepared plots, the first results confirming YaB's idea of the possibility of detonation in a non-uniform preheated reacting gas. YaB was happy, laughed, talked about the 'machines of hell' — neighbouring gas layers that self-ignited independently one after the other. We began to feel more sure of ourselves and began to ask questions. YaB answered rapidly, simply, wittily. Each answer ended with a question: think how that happens, why it is like that. He gave us questions to puzzle over: could temperature non-uniformity give rise to 'knocking' in an internal-combustion engine? He encouraged us to write an article and to think about other technical applications. It is interesting that this idea of YaB's has, in fact, been worked out quite recently — temperature-concentrated inhomogeneities were being examined more and more as an origin of spontaneous explosions.

I was still a student at PhysTech (Moscow Physical Technical Institute), and was struck by YaB's complete democracy, the absence of 'academism' and of any ostentation, his absorbing thirst for new information, his instantaneous reactions.

Working together on our book *The Mathematical Theory of Combustion and Explosion*, YaB many times reread individual bits or entire sections from beginning to end. During each meeting, he gave me two or three fully filled school notebooks, commenting in detail on each section. The number of comments did not decrease, and even seemed to increase, with new comments each time. I received the last set of comments the day before I had to submit the manuscript for publication. I told YaB that it was too late for me to take them all into account. 'If you can incorporate 20% of them, that will be good. The rest can wait for now.'

YaB had a sort of special pictorial-geometrical way of seeing the solutions of equations and their properties. It was as though he took the solutions to be live, palpable forms and moving, breathing trajectories. He interpreted the most complex theorems simply and clearly. Therefore, many mathematicians were eager to discuss their work with him.

There's no question that the theory of combustion was his favourite creation. In his work of 1979, devoted to the hydrodynamical stability of flames, he wrote: 'The tongues of flame characteristic of a campfire, the irregular, random properties of the process of free combustion in the atmosphere, full of inexplicable charm, have long riveted our eyes and thoughts. I remember the seriously ill Shostakovich, walking with difficulty toward a campfire near his last home not far from Moscow. "I love fire," he told me.' These same words could apply to YaB himself.

During one of our meetings, YaB asked 'Have you ever done any experiments? Do some experimental work, it's extremely useful for a theoretician.'

Almaty, 1984: a symposium on microkinetics. After a well-received talk by one of his young colleagues YaB said, 'It's nice to listen to someone who really does understand what he's talking about.'

Tashkent, 1986: a symposium on combustion and explosions. In the hall, there are approximately 800 people — specialists in chemical physics, combustions, and explosions. YaB is giving a presentation entitled 'Combustion Theory — Yesterday, Today, and Tomorrow.' At the end of the talk, these words rang out: 'If anyone wishes to discuss new results with me, call me at the Institute of Physical Problems.' This was not just a beautiful phrase. It was amazing that YaB did find time to meet and listen to anyone who wished to talk to him.

My colleagues and I often took advantage of this. It was sufficient to call and say that you wanted to discuss some new results, and you were greeted by a prompt answer — a meeting time was named for the next day or at most one day later. When YaB is no more, I involuntarily thought, who will there be to call tomorrow? With whom will we be able to discuss our new results?

YaB never did anything superficially. He never simply formally signed off reviews of work or referee's reports. The articles he recommended for publication in *Doklady Akademii Nauk* he read carefully, often asking that something be improved, making suggestions. He didn't immediately agree to head the Council on Combustion of the Academy of Sciences; he said that he might not have enough time to work on this seriously.

Reading books and articles was for him too time-consuming a way of becoming acquainted with work in some field. He preferred to talk with someone specializing in the given area, and he valued live communication. A continuous file of chemical physicists and nuclear physicists, combustion experts and astrophysicists, specialists in mechanics and mathematics came to his study. If someone arrived and YaB had not yet finished discussing everything he wanted with the previous visitor, he presented the newcomer with a heap of journals — science and popular science — and led him into the neighbouring room: 'Sit here for now in my grandchildren's storage room.'

YaB lived science. He rose very early and, until the calls began, he worked, solved problems. He often called at six or seven in the morning: 'Tell me, has anyone ever solved such-and-such a problem ... Can you give me references? Call me back, I'll be waiting,' or 'I read such-and-such an article. And what if ...?'

His enormous capability was staggering. But this was not some sacrifice on his part. He did what he loved, and without which he could not live. Any new result — whether obtained by himself or somebody else — gave him aesthetic pleasure. He wrote his articles immediately after obtaining his results, rapidly and almost without need for revisions. He wrote with the same speed in English.

YaB loved modern literature. Often, one would hear unexpected quotes from him. One scientific text called forth in him the following association: 'We always write that Sartre is "not unknown." As you can see, a double negative can fully distort the original meaning.'

One of his favourite sayings was: 'You'll never ask an assistant at a soda fountain for soda without a certain type of syrup in your water — without cherry or without raspberry. Never begin an article by telling what you haven't done. First write what you have done, then discuss the rest.'

At some point during a break in a meeting at the Institute of Chemical Physics, one of the participants from Leningrad began to tell him that one well-known Leningrad scientist, in spite of his advanced age, preserved his surprising memory and freshness of thought. YaB agreed and, smiling, said: 'I'm struck not only by the freshness of his head, but by the fresh covering on it.' And added, after falling silent: 'What we all fear often occurs very suddenly.'

'Where is that presented in the literature, by whom? If it's not published, all the same, say whose work it is.'

'It's good to work not according to the scheme introduction – calculation – conclusions, but rather introduction – results, with the calculations given as a mathematical appendix.'

YaB thought somehow in his own way. This could be felt when he suddenly became thoughtful during a conversation, then rapidly, rapidly began to explain. At such times, it was difficult to understand him. Understanding that he had lost contact with his listener, he would stop and, after a short pause, begin to speak more slowly, smoothly, without jumps. Someone once joked, 'He realized that he hadn't turned on his translator.'

He didn't like to waste time on just anything connected with science. When our book was published abroad, I received a call from the All-Union Agency for Authors' Rights with a request to organize a meeting with YaB, so that the administration of this respected organization could solemnly present him with the first copy of the book. I called YaB — he, not thinking it over for a second, decidedly refused: 'No solemnity for me! Get this copy from them yourself!'

Here are some of YaB's comments from the margins of drafts and manuscripts:

'The style is anti-me!'

'The aim of this book is to show what strictly exists.'

'This type of formalism won't be in any book with my signature!'

'The experimental material carries an "apologetic" character, and is presented as a confirmation of the theoretical calculations?! Theoretical calculations, like the wife of Caesar ...'

'Too scientific!'

'This should be a separate figure! Down with petty economizing that fogs up the brains of the readers.'

This memoir presents only fragments. Maybe they'll remind someone of a close and dear person; perhaps someone will see something in common in the individual episodes.

To each of those who had the joy of working with YaB, his unique talent was revealed in a different way. Each has his own memories. But we are all united by our blessed memory of an eminent scientist and person.

G.B. Manelis

Yakov Borisovich occupies a special place in the life of our generation. We arrived at the institute when Zeldovich's theories of combustion and detonation had not only been formulated, but had obtained brilliant experimental confirmation, after YaB's formal affiliation with the Institute of Chemical Physics had ceased. From the beginning, YaB was for us a legend from the glorious time when chemical physics was founded in the 1930s and 1940s. We were impressed and fascinated by the unity of his views on the complex world of phenomena, the seeming simplicity and obviousness of his concepts, and his approaches to understanding and describing varied processes.

One of the main features of Zeldovich's theory of combustion was that, as well as explaining a huge number of phenomena and observed behaviours, it opened wide possibilities for further investigations of combustion and associated processes, since it formulated a methodological

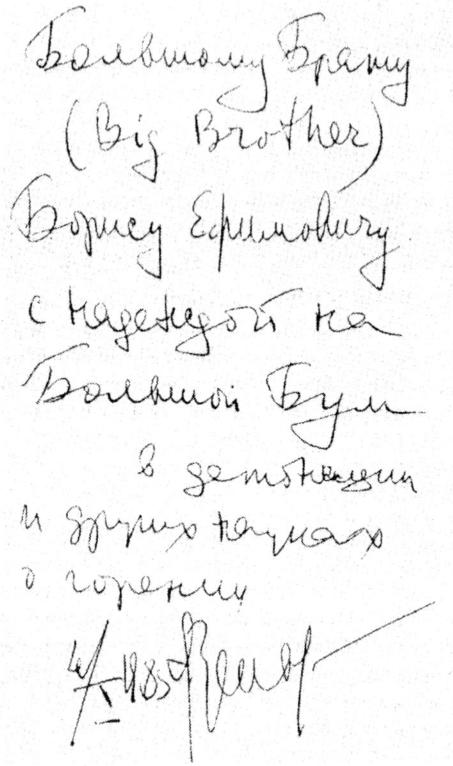
Большому Брату
(Big Brother)
Борису Ефимовичу
с надеждой на
Большой Бум
в детонации
и других науках
о горении

Inscription written by Ya.B. Zeldovich in a book on combustion given to B.E. Gelfand. The inscription reads:

To Big Brother Boris Efimovich with hopes for a big boom in detonation and other sciences of combustion.

approach and described the necessary accompanying apparatus. And in this vein, over the subsequent decades, the main research on the theory and mechanisms of combustion developed, both in the Institute of Chemical Physics in Moscow and in Chernogolovka, after the foundation of the ICP branch there.

The work on the development of thermal physics and stability of combustion and non-stationary processes was extremely fruitful. We concentrated our attention on the chemistry of combustion, first and foremost in condensed systems. Based on numerous experiments in the framework of Zeldovich's thermal combustion theory, it was possible to understand and quantitatively describe the rate and regular behaviour of the combustion of individual substances and systems of substances. Quantitative investigations of the real kinetics of chemical transformations, the role of non-equilibrium processes, and also of melting, dispersion, and sublimation, made it possible to organically incorporate new concepts into the framework of the existing theory. This led to further development of the theory, and also to the creation of methods for controlling the properties of combustion, the discovery of frontal polymerization and the construction of a theoretical basis for understanding it, and the development of new applied areas — the catalysis of combustion in condensed systems, and so forth.

In spite of the fact that YaB no longer worked at the ICP after the end of the 1940s, our interaction with him was often more frequent than with many colleagues at the institute in neighbouring departments or laboratories. We met, argued, discussed, consulted, and simply learned from him at conferences and seminars, in hotel rooms, and at YaB's home, at the famous blackboard in his study. In this way, the theory of homogeneous–heterogeneous reactions with the formation of solid products (the theory of soot formation) was born, based on YaB's well-known work on condensation theory. As a result of these conversations, completely new aspects of the theory of progressive relaxation and of reactions in shock fronts came to light. YaB exerted a strong influence on the development of our work on superadiabatic processes, which interested him most recently. He pointed out a whole series of paths for the possible application of such effects, especially in energetics, which, in my opinion, will find its practical realization in the near future.

YaB's personality, his views, behaviour, and humour exerted a huge influence on many of us, but that is a separate topic of its own.

During my talk on the theory of combustion of condensed matter at the Mendeleev meeting in Baku, Yakov Borisovich, who was the chairman for that session, stopped me at the beginning of my presentation and asked me to throw Zeldovich and his theory out of my subsequent narrative. The only correct answer, as I still feel today, was, 'If we throw out Zeldovich's theory from our combustion research, there won't be much left.'

B.E. Gelfand

In his last years, the founder of the theory of combustion and detonation was plagued by concern over the state of these branches of learning, so valued by him. In connection with this, Yakov Borisovich most often compared himself with the popular character from the film *Frozen*. In this film, a man who has spent a long period in an anabolic, frozen state, begins to zealously court his own granddaughter after his thawing. Yakov Borisovich wanted very much to see new ideas and young forces capable of further developing the ideas and hypotheses he first expressed. Up to the last, he did not agree with the idea that he had already done everything, and therefore there wasn't much left for other researchers to do. He was especially surprised by the sometimes dogmatic interpretations of hypotheses and views he had developed many years before. For me, it was very striking that, before his last holiday, after his return from a colloquium in Poland on the gas dynamics of explosions and reactive systems, he asked me to collect all the latest books on combustion and detonation for him, so that he could study them in detail. Apparently, he was not satisfied with the current situation in this field of science that was so close to him.

In his latest presentations and discussions, Zeldovich on more than one occasion called for research on the conditions for the existence of combustion regimes with velocities between those for detonation (1.5–2 km/s) and deflagration (less than 10 m/s). He never believed that there was an insurmountable division between detonation and deflagration, and that no other regimes were possible. He pointed out a number of systems in which rapid regimes of explosive transformation were possible, clearly driven not by thermal conductivity, nor by shock waves, but rather by other combinations of chemical and gas dynamical processes. Some of his opinions in this connection are of fundamental importance for our understanding of a number of unexplained, at times catastrophic, explosive phenomena.

As a person well versed in gas dynamics, Zeldovich in his last years continuously called for analyses of the action of more than just the thermal energy release in chemical reactions on the motion of a reactive medium. He felt that multifaceted analyses of the gas dynamics of reactive systems taking account of many possible aspects (friction, thermal exchange, mass input, mass output), which became possible with the availability of computers, could substantially augment, and in some cases replace, earlier intuitive concepts. At the same time, Zeldovich firmly supported the action of common sense, and demanded in-depth verification of numerical calculations and comparisons with the results of experiments, to ensure that the work of computers did not become closed fruitless cycles, giving rise to conclusions that were clearly *a priori*, rather than new results.

Yakov Borisovich's ability to denounce concepts that proved to be not entirely precise was very instructive. He insisted that the acknowledgement of the error be public, that is conveyed to the scientific community in some convenient way (printed, oral, etc.).

In conclusion, I would like to describe an episode told to me by I.S. Zaslonko. During a scientific trip to the Federal Republic of Germany, he was supposed to meet with another scientist whom he had not met before. Not knowing how they would find each other, he held in view one of Zeldovich's books (his last academic publication). Without hesitation, his colleague came up to him and said that YaB's book had served effectively as an identifying mark for their meeting. This symbolic case says much.

L.D. Landau's Reference for the elections of Corresponding Members to the USSR Academy of Sciences

June 6, 1946 *To the USSR Academy of Sciences*

Ya.B. Zeldovich is undoubtedly one of the most talented theoretical physicists in the USSR. His large cycle of works in the area of theoretical investigations of combustion processes is especially noteworthy. This research has been the best and most important in this field, not only in the USSR, but in the entire world.

Characteristic of Zeldovich's work is the broad use of hydrodynamics together with the methods of 'usual' theoretical physics. This parallel mastering of these two fields — extremely rare among theoretical physicists — is a characteristic and very valuable trait of Zeldovich, which allows him to attack questions that would be inaccessible to either hydrodynamicists or to more 'traditional' theoretical physicists.

It should be noted that the scientific activity of Zeldovich is still far from reaching its peak. On the contrary, his work displays unceasing scientific development.

L.D. Landau

From I.V. Kurchatov's Reference to the USSR Academy of Sciences

One characteristic feature of the scientific creativity of Ya.B. Zeldovich is the extremely broad range of questions on which he has worked and is now working. He is the author of a substantial number of both experimental and theoretical publications devoted to the phenomena of adsorption, chemical kinetics, the theory of combustion and detonation, gas dynamics, and nuclear physics.

A second characteristic is his ability to find comparatively simple approximation methods for solving complex problems.

A third feature is the high theoretical level for his solutions of varied questions, which earlier had been found intractable via quantitative analyses.

Finally, a fourth and important property of the creativity of Ya.B. Zeldovich is his exclusive ability to direct the strength of his rigorous and precise theoretical analyses toward the solution of problems of important practical significance.

... Ya.B. Zeldovich is a scientist of precisely the type needed in the Academy of Sciences. His election to the status of Full Member of the Academy in the physics–mathematics section will undoubtedly facilitate the further improvement and revival of the work in that division, and help direct the division's research toward themes that are the most urgent and important for our country.

1953 *I.V. Kurchatov*

Enchanted with the world

M.Ya. Ovchinnikova

Deciding to write about my father, Yakov Borisovich Zeldovich, was not easy for me — it was hard to believe that I had a right to do this, and that it might be of interest to somebody. Here, the support of my brothers, sisters,

Zeldovich with his son Boris, 1954.

and the widow of Yakov Borisovich, Inessa Yurevna Chernyakhovskaya, has been very important to me. I am especially grateful to my sister O.Ya. Zeldovich: some parts of this work were, in fact, written by her hand.

I only have vague memories of YaB's father — my grandfather, Boris Naumovich Zeldovich — who was a lawyer by profession. He died young (in 1943). I know only that YaB respected him highly. On the other hand, YaB's mother — Anna Petrovna Kiveliovich, a translator from French, small and energetic, sharp and quick-witted, our restless grandmother — was always present in the life of our family, even though from 1944 until 1975 she lived in Leningrad and we in Moscow. Her sayings still serve as standards for jokes in our family. For example, she would say concerning our home, 'Why go to the theatre when we have a circus at home' or about my father, 'When a fool goes to the market, the bazaar rejoices,' and so forth. Her visits to Moscow, which turned our entire home upside down, had their own special flavour. Her frequent packages to us, with the most improbable and unpredictable contents, demanded large efforts from the very busy colleagues and friends of Yakov Borisovich, who were travelling from Leningrad to Moscow. The main person who was recruited for this purpose was my mother's brother, Boris Pavlovich Konstantinov, who was much loved by all of us.

Only legends concerning the pre-war lives of Papa and Mama in Leningrad have survived: about how Yakov Borisovich got acquainted with the large, friendly family of the Konstantinov brothers and sisters; about bicycle trips, gymnastics, tennis, and volleyball in Tyarlevo, near Leningrad; about how Papa and Mama once by chance both ended up at work with bandaged arms (my father after being burned by liquid nitrogen, and my mother by acid).

Then, the birth of a daughter, my sister Olya, my own birth a year later, the appearance in our family of our nanny, 'Aunt Shura' (A.N. Lavrinova, who has spent her entire life with us from that time to the present). Bustle, worries, the dictatorship of Anna Petrovna, which my father avoided by taking his wife, little daughters, and their nanny to the noisy Konstantinov flat. This was followed by presents from Anna Petrovna and a return to Marat Street. And it was this type of nomadic life that we lived with prams and potties everywhere. At that time, my father was 26 years old, was a Doctor of Science, and had already succeeded in accomplishing much in science, and Mama was also a Candidate of Science. Then came the second World War, the evacuation to Kazan together with the Institute of Chemical Physics, a difficult time for a large family (his father, who was already seriously ill, mother, sister, wife, two small daughters, and Aunt Shura), and extremely stressful work. One can get an impression of our life in Kazan from photographs of my father from that time. In 1944 came our move to Moscow, to the Vorobiev Hills and the 'chem–physics' institute building, in which YaB lived from 1944 until 1987, and with which we all associate many happy years. It was here that my brother Boriska was born in 1944, always an object of special tenderness, caring, and concern for my father and all of us.

In our childhood and through all of our subsequent lives, our Mama, Varvara Pavlovna Konstantinova was the good fairy of both us children and and I'm sure my father as well. She was a person of tremendous tact, gentleness, and respect for everyone with whom she came into contact, including us (both during childhood and as adults), her sons- and daughters-in-law, and her grandchildren. She created around herself an atmosphere of kindness, warmth, and high spirit, free of pettiness and mercenary spirit. This was the case everywhere: at home and at her work in the Institute of Crystallography. At the same time, she was a very strong and masculine person, and her life (in particular, her life with my father) could not be called easy. One of my father's standard jokes was: 'Varya throws a shadow of respectability on me,' or, a more serious phrase: 'Varya is a person without faults.'

It is probably thanks to my mother that there formed a cult for my father in our home, and it was Mama who guarded and preserved our love for him. 'Papa is working' — the magic words that stopped any fussing or noise. However, in our daily life, the main focus was Mama; she defined everything, and worked intensively through it all. Life was very friendly and warm for all of us, and also for our many friends and acquaintances. In our large family, my father often joked, 'See, you're going to grow up and take your parents for rides in a carriage.' Now, it's a bitter feeling, that we never had chance to 'push them in a carriage,' and to express our love and devotion to our parents, having received everything from them and not having enough time to give more than a little in return.

In our preschool and early school years, I remember active and, to us children, reckless games undertaken by my father, always with elements of humour, sports, competition: 'kangaroo goes to dinner' or 'tsop', which required a tennis ball; 'bagel–baranka', which was a carriage; skipping-ropes; swimming races; skiing down a hill through a gate made of ski poles; and many, many others. Later, there were competitions to see who could most quickly and attentively devour large books, such as *Two Captains, Treasure Island, War and Peace, The Forsyth Saga*, or, later, works of Maupassant and Bulgakov. We were asked about tiny details regarding the minor characters. Here, of course, the 'readers' Olya and Borya were victorious, and I always lagged behind. My father himself read a great deal, rapidly and attentively (after all, he was able to examine us in such detail). He read books in the original in German and French, and, later, in English. He, and especially Mama, knew a lot of poetry by heart (Mama knew and loved Pushkin and Zhukovskii, and my father favoured Pasternak), unlike us, with our restricted school education. Any holiday from work for my father was always active — hikes in the mountains or canoe trips, or life by the sea with long-distance swims, when he and Mama went on holiday together.

Later (from 1947 until 1963), my father lived much of the time at the 'object' (in Ensk), working on the atomic problem, while Mama and us children stayed in Moscow. They wouldn't take Mama to work there because of her older brother Aleksander Pavlovich Konstantinov, who was executed by the Stalin regime in 1937, and her sister Katya, who, due to circumstances, ended up in America. In his short trips to Moscow, father tried to teach us or check my sister's and my successes in mathematics, without the least bit of patience or leniency toward us. On the blackboard he would put up a problem, and with a cry of 'Go!,' we were supposed to immediately give the answer. After a delay in answering came the accusation 'Fools!,' and from us instantly came tears, which in our family were called 'pearl tears.' Our education ended not long after it began. I should say that later my father had more patience and pedagogical tact with Boriska. It was initially precisely for Borya that he wrote his *Higher Mathematics for Beginners*.

Many years later, in his conversations with his grandchildren (and these were always full of content, whether talk was of a simple school problem, or a possible experiment, or an explanation of a complex scientific problem), my father showed himself to be a very patient and delicate teacher, at times defending our children from their parents. He always said, 'Children must know that they are loved,' and the grandchildren were unrestrained with Grandpa and Grandma, and flourished in their atmosphere of love.

However in our childhood, we were rather afraid and shy of our father; of how unusual and unpredictable he was, of how relaxed he was with his colleagues, and finally, of his complete orthogonality to what we did at school. In particular, one of our nightmares was the idea that he might go to the parents' meeting at school (fortunately, Mama took all such duties on herself). Sometimes, it seemed that he didn't have time for direct contact with us. But then my father's letters from Ensk, which each of us kept, read, and re-read, conveyed much happiness, warmth, and certainty in his love for us. The last letter from Papa, like the light from a dying star, Olya received a week after his death; in it, was a copy of an article on double beta decay, with which she was working. That was very typical of Papa.

The post-war years, when my father worked on the 'object,' were far from easy. Later, my father told us that many physicists (for example, his beloved friend D.A. Frank-Kamenetskii) were saved from persecution by working on the atomic project. But at the same time, before the first experiment with an atomic bomb, each scientist had a double, intended to replace him in the event of disaster. Many scientists, including my father, had ever-present armed guards, the so-called 'secretaries.' In the light of this, it surprised me very much to read in one of the chapters in this book criticism of my father for refusing requests to convey letters from the 'object' to Moscow against orders. I think one had to be cautious of precisely those people who would be willing to perform similar missions, or a large fraction of the applicants — such was the time. And the special character of that time touched my father rather closely. His youngest daughter, Annushka, was born in Kolyma (1951). My father became acquainted with her mother, O.K. Shiryaeva, in Arzamas-16, where she, being a political prisoner, worked as an architect (many of the buildings in that town were built according to her designs). When she refused the KGB's request to 'collaborate,' they promptly sent her to Kolyma,[17] where she and her daughter miraculously survived.

Of course, only the participants of such events have the right to talk about the problems of that period. Nonetheless, I would like to touch on a question that is often asked by people who didn't live through that time — about the moral responsibility of the scientists for the creation of Soviet atomic weapons. For everyone living in that post-war time — a time of opposition between our country and the USA, which had already used atomic weapons in Hiroshima and Nagasaki — there was only one moral responsibility: to establish a new equilibrium of forces in the world as rapidly as possible. This, together with the entire situation at the 'object,' has been described rather well in the memoirs of A.D. Sakharov (*Znamya*, 1990, No. 10–12; 1991, No. 1–5), the stories of V.A. Tsukerman and Z.M. Azarkh (*Zvezda*, 1990, No. 9–11), and an article by L.B. Altshuler (*Literaturnaya gazeta*, 6.6.1990). Re-evaluation of the situation and reasoning came later. In any case, my father, greatly distressed, said that if Stalin had not had nuclear bombs, he would not have unleashed the war in Korea. This may be why my father was the first of the leading participants in the atomic project to leave the Ministry of Medium Machinery[18] system (six years earlier than A.D. Sakharov was forced to leave this system for his writings), and fully put his energies into science, which he never again put aside.

[17] Kolyma is the name for a river and a region in Eastern Siberia, known for its harsh climate. It was a traditional place for forced labour camps and sites of exile in both Russia and the Soviet Union. [Translator]

[18] Also known as *Minsredmash* — the Soviet ministry that oversaw the atomic industry. [Translator]

Although that time was neither easy nor amusing, our parents protected us — the children — and did not expose us to the anxiety and tension of those years. It may be due to this, and possibly due to the inexhaustible, infectious optimism and good cheer of my father and the golden character of my mother, that only amusing episodes from those years remain in my memory, for example, connected with the 'secretaries' There were many funny stories due to their eagerness to always be with my father, especially during his holidays. It sometimes happened that he swam too far from them in the sea. There was one time when thieves managed to steal the watches of both the guardian and the guarded from the beach. Or another time when one of the guards left his pistol in the hotel, and secretly asked my father for leave to fly back to the hotel to fetch it. Once, my father was in a canoe with me and Boriska and my father's cousin — our beloved Aleksander Grigor'evich Zeldovich — was in a second with his children; we got stuck in a severe storm on Pleshcheev Lake, giving the unsuspecting 'secretary' no choice but to guard our tent camp by the shore. But at one point, the lack of freedom pressured my father so much that he went to a psychiatrist, to demonstrate that he couldn't endure this shadowing. The poor doctor didn't know what to do: it would be awkward to refuse my father's wish, but on the other hand — what about the department? In the end, he diagnosed my father as being very sociable and loquacious. I don't know what helped — this visit to the doctor or the intercession of Yulii Borisovich Khariton (a friend, companion-in-arms, teacher, a man greatly respected by my father) — but the secretaries were taken off duty.

From 1963, my father again began to live permanently in Moscow and to work at the Institute of Applied Mathematics. I remember that, getting up at seven in the morning, together with everyone else, I would find him already sitting with his notebook on the divan in the study: he usually began work at five (during the day, there were people, seminars, talks, lectures ...). And the evidence of his titanic labours — notebooks both standard, school-size and thin, filled with his accurate handwriting — have all been preserved. Each cover is labelled: a number, the year, a problem or question. From these, it is possible to recreate how he studied from books and journal articles (and he studied, entering into fields of science that were new to him, all his life), how he formulated problems and how he performed calculations; how he wrote the final versions of scientific and popular articles, monographs; how he prepared for lectures.

Upon examining these notebooks, one is struck by how highly organized his mind was. While others needed to write multiple drafts, he wrote straight off, and virtually without subsequent corrections. Although YaB was a very good speaker and very often gave lectures, he always prepared for them carefully. The numerous notebooks containing his lecture notes serve as a confirmation of this. Papa's notes of the contents lists of scientific journals have also been preserved (he always subscribed to thousands of journals). Reading each new volume, he would make notes near the articles, he wrote the names of colleagues, students, or even his children, to whom that article might be of interest, and without fail would take the article to show that person.

Nearly all of YaB's six children, and now his numerous grandchildren, have gone into physics — that's how great his influence was. He was always up to date on each of the areas of physics in which we were working. His interest in and understanding of nearly all fields of physics and chemistry was unique. Often, he would ardently try to convince us to set to work on problems that seemed to him to be the most interesting at the time, without realizing that only he who had access to knowledge in all fields of physics, and that each of us must have his or her own work interests. Not recognizing for himself weekends or holidays, Papa grew impatient when it seemed to him that his children were wasting time. Olya's son was only six months old when YaB began to pester her, as she had just graduated from Moscow State University: 'Olga, you'll lose your qualification.' Olya went to work on the night shift at an accelerator with a little child at home. To this day, I don't understand how she had enough strength for it all. About my computer work, looking over my Fortran programs, Papa, half-joking, said, 'Guga, don't be so serious. Stop your schizophrenic scribbling and work on such-and-such.'

I could talk a lot about my father, but others can probably do a better job than I. I do want to defend my father, though I understand that this is a purely subjective feeling, and in fact he has no need of anyone's defence. I admit that it is annoying for me to know that one of my father's colleagues, always a Party member who had permission to travel abroad, now condescendingly says, 'Well, YaB was a bit too cautious.' My father, who was never a Party member and was long denied permission to travel abroad, didn't live to see *perestroika*.

My father, of course, was a realist with regard to the system, and his possibilities within it. However, he never betrayed science. When the subject of discussion was the truth and fundamental scientific matters, YaB was unbending. He left the editorial board of *Uspekhi Fizicheskikh Nauk* when they decided to publish an erroneous article. He was one of only a handful of academicians who never signed a single letter of condemnation of A.D. Sakharov, in spite of the harsh pressure on him from above. He was bound to Andrei Dmitrievich by their long friendship during the years of work in Ensk, and valued very highly the eminent scientific potential of A.D. In 1980, when it was forbidden to make any reference to Sakharov, the translation of S. Weinberg's book *The First Three Minutes* was published, and YaB defended a reference to the work of Sakharov in his supplement to the book. In response to the pressure of the censors, he occupied a clearly defined position: without this reference, he would remove his supplement in full, together with his title as editor of the translation.

Indeed, YaB never joked in public about the system, knowing that he would be acting not only against himself, but also against other people working with him. YaB considered demonstrating against the regime to be less sensible than acting to provide real help and support to those living under the regime, especially by helping capable people join the larger world of science. People very often turned to YaB for help, and he did all that was in his power to do. He helped solve problems with the work placements or university enrollments of unjustly wronged, capable (often Jewish) young people. He made a fuss about medical aid for the families and close friends of many colleagues and acquaintances. Thanks to his efforts, it became possible for the first time to take a child abroad for a heart operation, and the child was saved. He helped colleagues with their attempts to obtain flats. He helped push through scientific trips outside the Soviet Union for many colleagues (he himself was permitted to go to the West only twice, in the last years of his life, though he was showered with hundreds of invitations).

I'm certain that many preserve in their hearts thanks to my father for his assistance in their lives, and an even greater number of people feel gratitude for his support of their scientific work. However, certain people — fortunately, very few of those who had accepted his friendship — found ways to denounce my father. I don't know the reasons for this: whether it was jealousy for his status and awards, or their inability to accept that someone could be more important in science than they. Their snobbishness and unkindness painfully wounded my father. They did not know, and did not wish to know, the price YaB paid for his results. They could not imagine the scale of the load he took on himself, which he was only able to support thanks to his God-given health and energy, together with the strict internal self-discipline he developed. It was precisely this self-discipline that did not allow him to be ill for more than a day, even when his temperature was 39.5 or his blood pressure was 200/100.

I'm not talking now about open enemies of YaB, such as those in the Institute of Chemical Physics in 1947, who tried to force him and his great friend O.I. Leipunskii to repent for their cosmopolitan views, or those who later organized a public pogrom of his *Higher Mathematics for Beginners*. To reiterate, I'm speaking of rare people who were almost like-minded, and for whom the term in use at that time was 'dissidents beneath themselves.' I'll present only one case that cost my father much health — his reading of I.S. Shklovskii's memoirs. I'll touch only on one episode presented there, about a 'dastardly deed' of my father, when YaB was supposed to have deliberately required M.M. Agrest, a practising Jew, to work on Saturday. It is difficult to imagine a more absurd (but far from harmless!) falsehood about my father, who didn't acknowledge for himself any weekend time off, and could not imagine the possibility that others might have restrictions in their work time. It is clear from our copy of the letter from my father to M.M. Agrest (the text is presented below), written in connection with this story, how sorely these, and other accusations made by Shklovskii, wounded my father. He also felt sympathy for many other worthy people to whom Shklovskii was equally unfair.

At the same time, YaB expressed surprising delicacy toward Shklovskii. Not long ago, going through my father's archive, I found a letter dated October 31, 1967, written by YaB and V.L. Ginzburg, about Shklovskii's more than impudent treatment of questions of priority in science. Wishing to avoid the publication of their clearly documented arguments, which would be unpleasant for Shklovskii, they (YaB and VLG) considered it their duty to once again address Shklovskii himself (this was not the first time), this time through their common close friends S.B. Pikelner and N.S. Kardashev. I wanted to present this letter as an example of scientific honesty and personal sensitivity.

There were also cases of rifts between my father and certain colleagues. It's hardly likely that this was good for the work. Of course, I don't know all the reasons for these conflicts. However, one of them was, I believe, the contradiction between the independent solutions of YaB, which were responsible for his convictions, and the expectations of these colleagues, who desired other solutions and who, in some sense, became accustomed to seeing in YaB a 'genie.' It may be that I'm not right about this, and if so, I hope that I'll be forgiven for my opinion.

I never heard about my father's grievances from him, and always learned about them indirectly: from Mama, then after her death, from my father's second wife Anzhelika Yakovlevna Vasil'eva (an outstanding person, with a somewhat tragic fate); and after the death of AYaV, from his last wife, Innochka Chernyakhovskaya, who was respected and loved by all YaB's children. As for my father, in spite of any grievances due to external reasons, periods of dissatisfaction with himself, or remorse, he was the happiest person in the world, throughout his life enchanted with the world, its structure, the beauty of physical theories, seized by the joy of life. It may be that this sounds too loud and banal, but that was how it was, manifest in everything — in his attitude toward science, sports, literature, the theatre, pretty women, and children.

I am not trying to make my father into an icon. He was a rather complex person, but very sincere. And he paid in full for all aspects of his character, including those we were not able to accept in him.

In his last years, he felt himself with satisfaction to be the patriarch of a large family. He loved to visit his children, feast his eyes on his new little grandsons and granddaughters. He and Inna — who always accompanied him on these visits, to the detriment of her own friendships — even had a special term for them: 'walking among the grandchildren.' He was proud of each of his children: Leonid's improbable projects and victories, especially those on the equator; Annushka's large, friendly family and accomplishments in sports; Sasha's beauty, musical talent, and other accomplishments; Borya's discoveries and science, and the thoughtfulness of his children; the kindness, warmth, and coziness around Olga, who most of all inherited my mother's golden character. He was proud of his wife Inna, and her professional work as an immunologist — because of which, however, she faced the responsibility of organizing medical care for all of Papa's numerous descendents.

YaB was also proud of his students. Recently, Olya and I independently and for different reasons re-read the 'Autobiographical Afterword' in the two-volume selected works of Yakov Borisovich Zeldovich.[19] We were once again struck by his kindness toward his colleagues, his painstaking outlines of the merits of his students and his joy at their successes, and his absolutely honest evaluation of his own contribution to science, without any showing off or conceit.

Below are presented several letters which seemed interesting to me. The first three of these — from YaB to M.M. Agrest, from Agrest to the editors of the journal *Khimiya i Zhizn*, and the letter written by YaB and V.L. Ginzburg to S.B. Pikelner and N.S. Kardashev — touch on his relations with I.S. Shklovskii. The last document is a review by Academician A.I. Lurie (a review of a review), which was sent to Yakov Borisovich after the death of Lurie by his son. This last piece sheds light on the methods used in the battle against the book *Higher Mathematics for Beginners* and the attitude toward this of a normal, decent person.

In conclusion, I'd like to express my unbounded thanks to I.Yu. Chernyakhovskaya, my brothers and sisters, all those who agreed to contribute to this collection of remembrances and who wrote so warmly about YaB, and all those who helped to publish this book and found the financial means to do this. We are especially thankful to Yulii Borisovich Khariton, Viktor Nikitovich Mikhailov, Semen Solomonovich Gernshtein, Rashid Alievich Sunyaev, and Nataliya Dmitrievna Morozova. Without them this book would never have seen the light of day. Thanks to all who remember Ya.B. Zeldovich. It is a great joy for us that there are very many such people.

Letter from Ya.B. Zeldovich to M.M. Agrest

14 June, 1981

Dear Mates Mendelevich

We haven't seen each other for a long time. My work in cosmology increasingly leads me to questions of ideology, which it would be interesting to discuss with you. An article by Rees and my commentary will probably appear in the 19/VI, or more likely the 26/VI, issue of Za Rubezhom; *in April or in March, there was an article in* Uspekhi Fizicheskikh Nauk. *Write to me if you have any thoughts about these: Vorobiev Road (ul. Kosygina) d. 2b, kv. 47.*

I won't hide the fact, however, that I took up my pen today for a completely different reason, which is very unpleasant to me.

Shklovskii has written memoirs, which are being passed around in typewritten form. One chapter is devoted to you. He writes about the relation between physical and mathematical knowledge and deep religious beliefs, including the observance of the traditional practices that you follow. Based on your words, he tells how, in those places where we worked together, you came to work on Saturdays, talked to people

[19] See p. 341 of this book.

about scientific questions, but didn't use either a pen or pencil. Your conscience was clear, and in addition, this avoided conflicts with the administration.

Further (also based on your words), it tells how one Saturday, I called you to see me, indicated an error or unclear place in one of your formulas, and asked for an explanation. To answer me, you would have to write something down. But you couldn't do this. The situation is depicted as though I knew about your situation and, in spite of this, didn't let you go. 'This disgusting torture went on for two hours' — Shklovskii writes roughly in this sort of passionate tone, depicting me as a sadist, and a person who would deliberately mock those below him, and mock others' religious convictions.

Here's the main point of my letter. I have completely forgotten details of specific meetings that we had — whether they took place in my office or in yours, on Monday or Saturday, and so forth. I think that you know me well enough: I am an absolute atheist, and all days of the week are completely the same to me. It's quite possible that I could have called you in to see me on a Saturday. At the same time, I respect religious feelings and other views, even if I do not share them.

My question is not only about the factual side of the affair: whether or not we had a professional discussion on a Saturday. The most important thing is the psychological side: if we did have such a conversation, then did you have any grounds to think that I understood your situation and purposefully mocked and tortured you? Is that how you described the incident to Shklovskii? I repeat, the idea that you are not able to work on Saturday, and how you have solved this conflict, both earlier and now, is infinitely far from me. I have never been religious, either in childhood, or in my youth, or later.

And another point: why didn't you tell me about your conflict? We could have shifted our conversation to Monday. Was it that you didn't trust me?

I remember how kindly and warmly you and your family acted toward us in Sukhumi; how we were happy at our meeting on the Crimean road. What am I to think now? That all this time, there was a stone weighing on your bosom, and offence in your heart because of that Saturday? You didn't express this to me — we could have cleared it all up.

I once again repeat my question: why did you talk to Shklovskii about me, when you didn't talk about this with me, or with a common acquaintance, such as Frank-Kamenetskii? Was that honest?

I think that religion should represent not only observance of proper forms, but deep honesty. Therefore, if other arguments are not sufficiently weighty for you, then in the name of your religious convictions, I insist that you give me an explanation for both this episode and for our subsequent meetings, with an evaluation of the literary and slanderous activities of Shklovskii, at least regarding the part of his memoirs that is connected with you and I.

His memoirs defame Landau as well, and many others, both living and dead, Jews and Russians, religious and godless, but that can be dealt with another time. Today, I need to defend my own name.

I have a right to know the truth.

Ya. Zeldovich

Letter from M.M. Agrest to the editors of the journal Khimiya i Zhizn (Chemistry and Life)

13 February, 1992

At the beginning of the 1980s, I reached an agreement with I.S. Shklovskii that his essay 'Our Soviet Rabbi' would not be published without my consent. After the death of Shklovskii in 1985, his widow A.D. Ulyanitskaya and I had the same agreement.

It is extremely unfortunate that the editors of the newspaper Nezavisimaya Gazeta *published this essay without my consent in Issue 13 of January 29, 1991. I have written to the editorial board about this.*

On January 10, 1992, in a discussion with your colleague V.I. Rabinovich, I gave my consent for the publication of the essay 'Our Soviet Rabbi' subject to the single condition that the following footnote should be printed in the appropriate place.

'I told I.S. Shklovskii about this episode in February of 1951. However, at that time, I never believed that Ya.B. Zeldovich knew that I do not write on Saturdays for religious reasons, and that he deliberately tried to force me to disobey this tradition.'

Ya.B. Zeldovich demanded that I lay out mathematical calculations on the blackboard, hoping to rapidly detect the error he suspected had been made. During our long discussion, I kept looking for ways to orally convince him of the correctness of my results.

Having failed to meet his demands, I to this day reproach myself that I did not inform him of my reason for not wanting to use the blackboard, as a result of which I could not endure the ordeal that fell to me.'

M.M. Agrest

Letter to S.B. Pikelner and N.S. Kardashev from V.L. Ginzburg and Ya.B. Zeldovich

30 October, 1967

Dear Solomon Borisovich and Nikolai Semenovich!

We know that you are both close to I.S. Shklovskii. This is why we have decided to write to you now, to demonstrate our sincere desire to avoid aggravating the circumstances unnecessarily, and to attempt to find a way out of the present situation that is most favourable for IS.

Although the behaviour of IS deeply disturbs us, and we consider it objectively completely unacceptable and without any possible justification, we are not like Ivan, forgetting his kindred. Now, as before, we believe that he has many scientific merits, that he has surrounded himself with capable and decent people, that we have obtained a great deal from our interaction with him and his friends. We also suppose that his blunder may largely be associated with his incorrect understanding of questions of priority. For example, whatever a person has done himself, he cannot claim priority if the same result was obtained earlier by others.

We have already on more than one occasion pointed out to IS the incorrectness of his behaviour. For example, one of us (VLG) clearly remembers that IS has had to offer him apologies three times. The other (YaBZ) warned IS about his last article in Astronomicheskii Zhurnal. *But it seems that IS forgets his mistakes, and has been repeating them more and more often. We believe that letting his behaviour go further without paying it the attention it deserves would be an unforgiveable mistake, which would be harmful for science, the relationships between all of us and our colleagues, and finally, for I.S. Shklovskii. It is no secret that both of those who have signed this letter have, more than once, been called upon to firmly defend IS from accusations from various quarters, including reproaches from very decent people from Moscow, Gorkii and other places. That there were grounds for this is clear from the enclosed draft of our proposed letter to the editorial board of* AZh. *We repeat that this is a draft, and rough at that. Don't pay attention to the style and the form, but please look closely at the facts. They speak for themselves, and we believe that publication of such a letter would be a blow to the scientific career of I.S. Shklovskii. That is precisely why, together with the reasons laid out above, we do not wish to publish this letter. In addition, it would give joy to IS's enemies, whom we have no reason to esteem.*

Here, we propose the only possible alternative, in our view — let IS himself write a letter to the editors of AZh *indicating his error ascribing to himself the results of others. We are not implying a repentance in the Chinese style, but do not agree to a merely formal response. In our opinion, the letter can achieve its goal only if it is well-written and clear, and contains acknowledgement of the obvious factors that we have indicated. We believe that nothing could better save the reputation of IS than publication of such a letter, providing it is not delayed for too long. If he understands this — so much the better. We would be only too glad to clear all accounts and maintain good and sincere relations with IS and his colleagues; we intend to push the affair through. However, if IS considers himself to be in the right and will not write such a letter in the nearest future, then we will regretfully be forced to send our own letter to* AZh.

We have nothing to hide, and you may show this letter and the draft of our proposed letter to AZh *to anyone you like. However, in order to avoid complicating the situation, we have decided for now not to acquaint anyone with these letters.*

We hope that you will understand and approve of our actions. In addition, we will always be glad to take into consideration and discuss your own thoughts on this regrettable problem.

With sincere respect,

V.L. Ginzburg
Ya.B. Zeldovich

Letter to the editorial board of the journal Prikladnaya Matematika i Mekhanika (Applied Mathematics and Mechanics), Editor-in-Chief, L.A. Galin

On the anonymous review by Ya.B. Zeldovich's: Higher Mathematics for Beginners.

In works sent for review to the editorial board of a journal, the names of the authors are always indicated. Deviation from this rule is incomprehensible and, I believe, inadmissible. Does this mean that an anonymous review seems to represent the views of the editors? I have been an editor of the journal since prehistoric times (when Pridkadnaya Matematika i Mekhanika *was published by the Leningrad Mechanics Society starting in 1929), and I don't recall any precedents for this; perhaps there weren't any.*

Looking through Ya.B. Zeldovich's book, I was surprised by the variety of the content, the richness of interesting and non-trivial applied problems. In my youth, there were similar books by Lorentz, N.B. Delone, Nernst, and Shenfliss; in these too, the concepts of derivatives and integrals were explained without pretensions to mathematical rigorousness and refined language, and several applications to problems in mechanics, physics, and other natural sciences were presented. Continuing in this tradition, Ya.B. Zeldovich addresses his book to an inquisitive reader who has realized he is unable to understand not only the laws of nature, but also many everyday phenomena, since he does not have at his disposal even a rudimentary understanding of mathematical analysis. The difference is that now, more than half a century after the publication of the earlier works, such books are addressed not to tens of thousands, but to more than a million readers. Correspondingly, the responsibility of the author grows. This could serve as a justification for publishing an unprejudiced and well-grounded review of Zeldovich's book in PMM.

The italic type on page 60 aroused anger: 'The ratio $\Delta z/\Delta t$ approaches a finite value as Δt tends toward zero.' However, nothing is said about the fact that this phrase is preceded by thorough supporting explanations. It is not noted that, on page 255, it says, 'we have, without saying this explicitly, been considering smooth curves,' and on page 257, 'for simplicity in presentation, we have not noted each time that a definite value independent of the way in which a quantity tends to zero (from the left or from the right) exists only for points forming a smooth curve.'

We can make a flurry of accusations about these italic statements (taken from the book's text), but we can also put it differently: 'Ya.B. Zeldovich demonstrated pedagogical tact, gradually preparing the unexperienced reader for refinement of the concepts presented, from less complex to more complex.' It is well known that the concept of the derivative is one of the most difficult in analysis, and a keen mathematician can find insufficiencies in any definition; however, as Kelvin said, 'Leave that to mathematicians — a derivative is a rate.'

The presentation of the basis of mechanics in Chapter VI of Zeldovich's book is unusual. The anonymous review author is correct when he maintains that the use of the concept of a point mass, and the laws of motion would simplify the formulation of the problem. He could have said even more simply: Ya.B. Zeldovich would be better off simply directing his reader to any textbook in mechanics. Apparently, this is precisely what led Ya.B. Zeldovich to decide not to follow the respected canons recommended to him.

I am supposed to provide a review of the review, and not of the book of Ya.B. Zeldovich. In the final lines of the review, it says that the style of the book is 'unduly familiar and not professional.' The reproach for 'undue familiarity,' of course, should be returned to the anonymous reviewer without delay, and we cannot doubt his 'professionalism.'

Reading the review leaves a nasty taste in the mouth.

If the review is to be published, of course, there must be a response from Ya.B. Zeldovich in that same issue of the journal. Naturally, if the review is published, it must not be published anonymously. However, in my opinion, it would be better not to publish it — it is risky to bring dirt into a journal that deserves to be distributed internationally.

A.I. Lurie

Letter from Ya.B. Zeldovich to M.V. Keldysh[20]

25 November, 1975

Dear Mstislav Vsevolodovich!

As you know, the question of republishing my book Higher Mathematics for Beginners *has provoked a stormy reaction from L.I. Sedov and L.S. Pontryagin.*

No one, even a very naive person, believes that they are concerned about the quality of the publication in this case.

This is open persecution. In fact, L.I. Sedov had previously published a review insulting Barenblatt's and my work in an 'abstracts' (!) journal, and did not publish our response. He then tried to smear us in the eyes of American scientists and interfere with the publication of our articles in the USA. The publication of Novikov's and my book on cosmology was delayed for a year; if P.N. Fedoseev had not got involved, the delay would have continued. Slander against me is being spread about the question of priorities in detonation theory.

Do you know that, in connection with a brief reference to the general theory of relativity in the book, my reviewers write about 'religious–cult emotions' and 'terrorist pretensions'?

In another place, they write about the 'author's complete ignorance in questions of the hydrodynamic resistance of a body.' Do you consider such a statement to be acceptable?

I return to the republication of the book Higher Mathematics. *In this question, one decision of the section replaced another. The decision of the plenum of the Editorial Board has been overruled. A commission has been named.*

However, the book was examined in detail back in 1968 by the Scientific Council of the Institute of Applied Mathematics. A positive review was issued with your signature. Under normal circumstances, this would have completely resolved the question of republication. The spiteful reviewer L.S. Pontryagin insults not only me, but also the Council of the Institute of Applied Mathematics.

---

[20] M.V. Keldysh was a well known mathematician; in 1960–1961 he was Vice President of the USSR Academy of Sciences, and thereafter President of the USSR Academy of Sciences until 1975.

If you consider that circumstances have arisen that force you to repudiate your signature, I ask that the question be again presented to the Scientific Council of the Institute of Applied Mathematics. The question of republishing a book that has had a distribution of more than 500,000 and has been translated into many other languages does not worry me (although I, as before, am convinced that there are readers that need it).

However, this does not mean that I am indifferent to the spiteful, slanderous activity of L.I. Sedov. It makes me waste time, strength, and emotional energy, and distracts me from other affairs.

But most of all, I am concerned about your position. L.I. Sedov has been assigned the supervision of the abstracts journal Mekhanika *[Mechanics], he is the assistant editor of* Doklady Akademii Nauk *and chairman of the Section of Physical and Mathematical Literature. In this position, the activity of L.I. Sedov ceases to be his personal affair, and compromises all those who support him in the Academy as a whole.*

I sincerely hope that you, Mstislav Vsevolodovich, will make an appropriate evaluation of what has happened, protect me from persecution, and establish objectivity in the publishing activities of the Academy of Sciences, of which you are the President.

With sincere respect,

Ya.B. Zeldovich

Part III

The Atomic Project

Yakov Borisovich Zeldovich: not one of his many projections

I.I. Gurevich

I knew Yakov Borisovich for nearly a half a century. He was unquestionably one of the most splendid people I have had the pleasure of knowing. It is a bitter feeling to be writing remembrances of him. In my memory, he lives on. His personality is inseparable from the science that he produced. He was unique in the variety of his creativity, from the physics of combustion, through nuclear weapons, to the very depths of astrophysics and cosmology. It wasn't for nothing that one English scientist remarked of him, 'Finally, I've seen Bourbaki in a single person.'[21] Given the depth and degree of Zeldovich's knowledge of each of his scientific passions, these words are one of the best and highest estimations of his scientific creativity. It is not possible for any one person to write a complete memoir about him. Therefore, I think that I will be doing the right thing if I concentrate on one chapter in the scientific accomplishments of Yakov Borisovich, which I observed in detail. This choice is also correct for me because I twice had the pleasure of working in this field with Zeldovich.

I'm referring to nuclear reactor physics. This work of Yakov Borisovich demonstrates well both his creative power and his characteristic scientific style, now so rarely seen. This style incorporates a wonderful ability to reduce any arbitrary scientific problem into a number of subproblems, each of which can be solved via physically transparent methods and without very refined mathematics. This does not mean that Yakov Borisovich didn't like or was not able to use the full mathematical apparatus of modern theoretical physics. He simply considered his method and style for obtaining solutions to be more suited to his deep physical intuition and, I would say, to his very essence. Among twentieth-century physicists, it seems that only Fermi employed similar methods.

I should note that, apart from my delight in Yakov Borisovich's powerful intellect, he was for me a close and very warm person.

---

[21] The 'Bourbaki' was a society of remarkable French mathematicians who published their works under a single pseudonym. Bourbaki himself was a French general at the time of the Franco–Prussian War.

Now, moving on to nuclear reactors. On January 6, 1939, in the German journal *Naturwissenschaften*, an epoch-making article by O. Hahn and F. Strassmann appeared, reporting their discovery of nuclear fission — the formation of radioactive barium during the absorption of neutrons by uranium atoms. A large number of nuclear physics laborabories all over the world (in France, England, Germany, Sweden, the USA) set aside their current investigations and enthusiastically began to study this unforeseen phenomenon. In our country, first and foremost, we should name in this connection the Leningrad Physical Technical Institute (LPTI) — the laboratory of Igor Vasil'evich Kurchatov — and the Radium Institute of the Academy of Sciences (RIAS). Igor Vasil'evich undertook a broad spectrum of investigations into the physics of fission. Some of these were done in collaboration with the physics department of RIAS (L.V. Mysovskii). In RIAS itself, V.G. Khlopin organized and supervised radiochemical studies of fission fragments.

After graduating from Leningrad University, I worked at RIAS, beginning in the spring of 1934. I remember very well the enthusiasm then reigning among physicists and chemists. At that time, a neutron seminar organized by Kurchatov regularly met at the LPTI, to unite physicists at the LPTI and RIAS working under the scientific aegis of Igor Vasil'evich. After the discovery of nuclear fission, this seminar was primarily dedicated to the physics of fission. Reports were made about all known scientific papers in this field, and ongoing work at the two institutes — approaches and results — was discussed.

In April or May of that same year, Yakov Borisovich began to attend the seminar. It was around that time that our acquaintance began. Yu.B. Khariton could also be seen at the seminars, though somewhat more rarely. YaB and YuB's interest in the physics of fission was more than natural. Students of N.N. Semenov, the creator of the theory of chemical chain reactions, could not ignore the grandiose problem of releasing internal nuclear energy. The whole world was talking about nuclear chain reactions following the discovery of instantaneous fission neutrons independently in France, the USA, and the Soviet Union, by F. Joliot-Curie, E. Fermi, W. Zinn, and L. Szilard, G.N. Flerov, and L.I. Rysinov. By 1940, the number of publications on this theme had rapidly decreased, but the enthusiasm grew.

As I was told, YaB's and YuB's interest in this problem began with a conversation between I.Ya. Pomeranchuk and Zeldovich. Pomeranchuk, having briefly talked about the possibility of fission chain reactions, ended with the words, 'Now there's a problem that's right up your alley.' But even if this conversation had never taken place, the result would have been the same. For Zeldovich and Khariton, the discovery of nuclear fission was like the voice of fate, and they were incapable of ignoring that voice.

Their interest in chain reactions was far from purely abstract, and represented active creativity. Their first paper dedicated to the theory of uranium fission by fast neutrons had already appeared in 1939. This work demonstrated that a chain reaction was impossible even in purely metallic uranium with natural concentrations of the isotopes ^{238}U and ^{235}U. The chain was broken by inelastic neutron scattering, which slowed these particles below the threshold for fission of the main isotope. The main conclusion of this work was that to obtain fission chain reactions using fast neutrons, it is essential to use an isotope that does not have a fission threshold (i.e., one that will undergo fission under the action of neutrons of all energies) — nuclear fuel.

The splendid collaboration continued, and two more fundamental papers by these two authors were greeted with acclaim in 1940. The first of these presented the theory of fission chain reactions initiated by thermal neutrons in an infinite, homogeneous mixture of uranium and a moderator. The moderator considered was hydrogen (water). The article followed the history of one generation of neutrons from the creation of a fast neutron in the fission process to its deceleration and absorption in a thermal region of the uranium plus moderator. This was the first time the resonance absorption of neutrons by levels of the main isotope of uranium had been considered, and a full theory for resonance neutron absorption in homogeneous systems was presented, which could naturally be generalized to the case of a moderator with any atomic number A. The history of one generation of neutrons led the two authors to the well-known formula with three factors for the neutron amplification coefficient ($K_\infty = \nu\varphi\theta$). Analyzing the experiments of Joliot-Curie, they showed that a chain reaction is impossible in a mixture of natural uranium and ordinary water, no matter what the concentration of uranium. This important conclusion led them to deduce the following conditions required for fission chain reactions initiated by thermal neutrons. It is essential to use another light moderator that absorbs neutrons less efficiently than hydrogen, or to use uranium rich in the rare isotope ^{235}U. In nuclear reactors, one or both of these fundamental conditions are realized.

For me, this excellent work by YaB and YuB is significant not only from a scientific point of view. During their work, they discussed it with me on several occasions, and I told Zeldovich and Khariton in detail everything that I knew about resonance neutron absorption and the Breit–Wigner formula, which was at the root of their theory of resonance absorption in systems of uranium plus a moderator. In this way, these conversations led to a closer acquaintance with Zeldovich and Khariton.

In connection with the problem of chain reactions by thermal neutrons, I also remember the following. Once, in May or June of 1941, Yakov Borisovich, having met me by chance, lamented that we were living so late. A billion years ago, he said, the relative concentration of the light isotope of uranium was substantially higher, and at that time, a chain reaction could occur in a mixture of natural uranium and ordinary water.[22] Yakov Borisovich didn't say anything at that time about the possibility of a natural reactor, but his reasoning brings us to the natural reactor discovered in Gabon in 1972.

---

[22] See *Uspekhi Fizicheskikh Nauk*, 1983, vol. 139, no. 3, pp. 511–27.

The third pre-war paper by Zeldovich and Khariton constructed the subsequent theory of the kinetics of a nuclear reactor in the presence of deviations from the critical conditions. Its main achievement was the elucidation of the role of lagging neutrons in reactor kinetics under conditions with small deviations from criticality. In this case, thanks to the lagging neutrons, the kinetics becomes very 'soft,' facilitating the regulation of nuclear reactors, and consequently the very existence of controlled nuclear energetics. It is difficult to overestimate the importance of these three fundamental pieces of research.

So far, I have only talked about these well-known works by Zeldovich and Khariton, but without this, the picture of these two scientists' contribution to the problem of nuclear chain reactions would not be complete. I'll now turn to other topics that are less well known. In the middle of the 1940s, YaB and YuB's interest turned to the determination of the necessary quantity of fission material required to bring about a chain reaction when the amplification coefficient exceeds unity. This is the problem of critical size, or critical mass. Here, I was very fortunate. Yakov Borisovich suggested that I join them as a third collaborator in their creative collective in order to work on this problem. Naturally, I was delighted to accept.

All that was then known about this problem was the diffusion equation of Perrin, applicable only for very small amplifications, and the integral equation of Peierls, solved by him for the two limiting cases of very small and very large amplifications for naked fission material (i.e., without a reflector). It is also important to remember that, at that time, all calculations had to be performed using a logarithmic slide rule and mechanical calculators. And we had to solve integral diffusion equations for arbitrary amplifications in the presence of a reflector. It is clear that such a task cannot be tackled using ordinary analytical methods. The decisive role was played by a technique proposed by Yakov Borisovich, which inverted the problem at hand — the neutron-density distribution was given, and the density distribution for the fissioning material was sought. This problem can be solved using a simple quadrature. Various test functions for the neutron density are inserted into the equation, and those for which the density of the fissioning material is nearly constant are identified. As a result of long, but fairly elementary, calculations, we derived the entire curve for the dependence of the critical size of pure ^{235}U on the amplification coefficient for any amplification; first for naked uranium, then for uranium surrounded by a reflector with a scattering cross section equal to that of the uranium. We estimated the known critical mass of ^{235}U, which, in spite of all the uncertainties in the nuclear properties of this fissioning isotope, proved to be a few or a few tens of kilograms.

In the second part of this work, we studied the critical mass of pure ^{235}U in aqueous (usually water) solutions in reactions initiated by thermal neutrons. The lack of an adequate theory for the moderation forced us to adopt the following approach. The experimental curve for the distribution of water-moderated neutrons was interpolated using an exponential relation and the mean square moderation length found experimentally. The exponent was introduced into the integral equation for the density of thermal neutrons as its core, after which it was solved by the technique described above. In this way, we were able to derive the non-monotonic behaviour of the critical mass of ^{235}U, with its minimum at a hydrogen concentration relative to uranium of 200–300; we estimated the critical mass itself to be 1–2 kg.

The fate of this work was different from the first three classical papers by Zeldovich and Khariton, which were published in *ZhETF*[23] in 1939 and 40. Although the main results of our work on critical masses for fast-neutron fission had already been obtained in the autumn of 1940, the work was slow to complete, and was written up only in May of 1941. The war began, and we had no time and energy to worry about publishing it; then, the whole problem became classified, and it has only recently become possible to talk about this work, which closed the pre-war cycle of papers by Zeldovich and Khariton on the theory of nuclear chain reactions. My discussion here shows not only their pioneering contribution to the problem of releasing nuclear energy, but their surprising scientific acumen. It was a pleasure for me to observe further, even if only to a small extent, the fundamental contribution of Yakov Borisovich to this problem of our century.

At the beginning of 1943, Laboratory No. 2 of the USSR Academy of Sciences was created under the supervision of Igor Vasil'evich Kurchatov (this was later to become the Institute of Atomic Energy). The laboratory's task was to solve problems associated with the release of atomic energy and, first and foremost, to establish a Soviet nuclear weapons programme. Igor Vasil'evich attracted the best theoretical and experimental nuclear physicists in the Soviet Union, and assembled a small but highly qualified scientific collective. Naturally, Yakov Borisovich was one of the first scientists to be added to this group. The first task of the laboratory collective was to bring about a fission chain reaction initiated by thermal neutrons in a system of natural uranium plus a moderator. This type of system — called first a uranium cauldron and, later, a nuclear reactor — could be used to obtain nuclear fuel — plutonium. This task was brilliantly carried out by Kurchatov, when he and colleagues brought about a fission chain reaction in the first Soviet uranium–graphite reactor on December 25, 1946.

The path toward this accomplishment was surprisingly short (less than four years), but devilishly difficult. The problem was divided into a number of subproblems of enormous complexity. These included the derivation of extremely pure uranium and the graphite moderator, as well as a series of macroscopic 'exponential' experiments with both pure graphite and subcritical uranium-graphite composites, which were made personally by Igor Vasil'evich. There were measurements of the main nuclear constants for fission isotopes, the construction of the full physical picture of the phenomena

---

[23] *Journal of Experimental and Theoretical Physics*. [Translator]

occurring in the nuclear reactor, the development of an adequate theory for the nuclear reactor, at least good as a first approximation. Here, the full talent of Zeldovich was revealed, uniting the clarity of physical thinking and the scientific intuition of a highly qualified theoretical physicist.

I'll present just two examples. One of the key questions connected with reactor physics was the development of an adequate theory for moderation of the neutrons. Zeldovich started from the natural assumption that the energy of the moderated neutron can be measured in units of the moderation time, or more precisely, the effective moderation time — the age (that is, not in seconds, but in square centimeters); in June 1943, he constructed an age theory of moderation and derived the well-known age equation. When E. Fermi's lectures on nuclear physics came out in 1950, we learned that Fermi had discovered this equation even earlier in the USA, and that it had come to be called the Fermi equation. It would be fair to name this age equation the Fermi–Zeldovich equation, giving credit to both these splendid physicists. I can assert with certainty that, without knowledge of the age equation, our movement toward fission chain reactions would have been substantially hindered. The mean square moderation length in graphite, which determines the critical size of the reactor, could not have been determined by Kurchatov in his 'exponential' experiments without the age equation.

Now for a second example. Here, I had the pleasure, as in 1941, of working together with Yakov Borisovich. The historical background is as follows. The problem of obtaining an amplification coefficient exceeding unity in systems with natural uranium can be solved by heterogeneous (block) placement of the uranium in the moderator. This suppresses undesirable resonance absorption of neutrons by the isotope ^{238}U, due to the self-screening of the inner regions of the uranium block by the outer layers. A theory of resonance neutron absorption in heterogeneous systems, which extended Zeldovich and Khariton's theory for homogeneous systems, had been constructed by Pomeranchuk and myself in September and October of 1943. This theory shows how the resonance absorption decreases with an increase in the block size. However, use of very 'thick' blocks should lead to a decrease in the fraction of thermal neutrons absorbed by the uranium, due to the same self-screening effect. This gives rise to the task of determining the optimal size for the uranium blocks. However, this was not immediately realized, and the formulation of the problem was the job of Yakov Borisovich. Initially, Pomeranchuk and I thought that, due to the very large absorption of resonance neutrons by the isotope ^{238}U and the appreciably lower absorption of thermal neutrons by natural uranium, it would be possible to obtain the necessary increase in amplification coefficient using fairly 'thin' blocks. Such blocks should be 'black' for resonance neutrons (i.e., perfect absorbers of low levels of ^{238}U) — the block effect is present — but nearly transparent to thermal neutrons — the block effect is absent.

After my work with Pomeranchuk, I went to Kazan for a short time. When I returned to Moscow, Yakov Borisovich met with me and said that, to optimize the amplification in uranium grids, it might be necessary to use large blocks, in which we would have to take into account the screening of thermal neutrons. We immediately discussed this question and decided to perform first-approximation calculations of the thermal-neutron utilization factor in a diffusion, single-rate approximation. Before we parted, we had obtained all the formulae for the case of flat blocks (uranium plates). From the outset, this problem was symmetrical, and therefore easy to solve. We then realized that, in reality, there are two block effects. The first effect is associated with the self-screening of the inner layers of uranium in the block, while the second is connected with the drop in the density of thermal neutrons in the moderator near the surface of a strongly absorbing block, compared to the density in regions of the moderator that are far from the block. In heterogeneous systems, resonance absorption decreases with an increase in the block size, but the fraction of thermal neutrons absorbed by the uranium also decreases. There is an optimal block size that yields the maximum amplification coefficient. In the course of several days, we understood how to determine the thermal-neutron utilization factor in heterogeneous systems with any grids and blocks of any shape (cylindrical, spherical), having applied at Zeldovich's suggestion the method of Wigner– Seitz cells, well known in the theory of metals. We found the solutions to this problem in the given approximation. Now, all this seems somewhat naive, but as the groundwork was being laid, that theory qualitatively shed light on the situation.

These two works carried out by Yakov Borisovich at Laboratory No. 2 supplement his pre-war cycle, in which he and Khariton laid the foundations of the theory of nuclear chain reactions. All this research shows the power of his talent.

I won't talk about Zeldovich's other accomplishments, which seem to me to be even more fundamental in the release of nuclear energy. These were primarily associated with nuclear weapons, and I shall leave it to others to write about this.

I am glad that I was able to see the talent of Yakov Borisovich not only from a distance, by studying his work, but also close up, working together with him.

The happiest years of my life

Yu.B. Khariton

I was very lucky in my life: for roughly 25 years, I worked with a fantastically interesting person and absolutely exceptional scientist — Yakov Borisovich Zeldovich. The breadth of his scientific interests is truly amazing: catalysis, the theory of combustion and detonation, adsorption, as well as elementary particles and nuclear physics, astrophysics and cosmology, relativity theory and quantum mechanics. And he was strong, universal in each of these

Ya.B. Zeldovich, Yu.B. Khariton, and N.N. Semenov at a celebration at the Institute of Chemical Physics. Moscow, 1976.

areas. As the 70th birthday of Yakov Borisovich approached, I felt strongly that a collection of his scientific works should be published, because no one had ever before demonstrated such a variety of work as that which he created over the course of his lifetime. The president of the USSR Academy of Sciences, A.P. Aleksandrov, enthusiastically supported my idea, and asked that he be included among the editors of the collection. A relatively large editorial board was organized, due to the need to work with research related to a huge circle of questions. The improbably broad range of topics represented in the two-volume publication, which came out in time for Zeldovich's anniversary, is noted by everyone who reads it. It wasn't for nothing that, when Yakov Borisovich was introduced to the English physicist Stephen Hawking at a conference in Budapest in the 1970s, Hawking remarked that he was now finally convinced that Zeldovich's work had been done by a single person, and not a collective, like the Bourbaki.

The story of the appearance of Zeldovich at the Institute of Chemical Physics in Leningrad in 1931 has already been described in detail by many others, and I will not spend much time on it here. I remember only that he didn't get a university degree — he didn't want to waste the time. His first works were related to adsorption, catalysis, detonation, and combustion. He grew literally right before our eyes. Coming to the institute as a 17-year-old, he soon proved himself capable of independent research. For example, at a meeting of the Scientific Council of the institute, this fellow, still very young, suddenly presented a completely unexpected and very interesting idea. I repeat: he was truly a unique individual, from the very first excelling in the independence of his reasoning. He was able to understand relatively complex things unusually quickly; he felt free and on an equal level with professionals who had worked in the given research area for many years.

The recognition of specialists came to him very soon. As far as I remember, he obtained one of the first Stalin Prizes for his work on combustion and detonation.

We began to work together in 1939. At that time, the first paper reporting observations on the fission of uranium nuclei appeared, by O. Hahn and F. Strassmann. This was followed by the work of L. Meitner and O. Frisch, who explained this phenomenon as fission of uranium nuclei under the action of neutrons. Having read about this, we understood that in this case it was possible that there would be nuclear chain reactions that could be branching, rather than the usual chain reactions, and they could lead to a nuclear explosion with the release of an enormous amount of energy. In our institute, there were many researchers investigating problems connected with chain reactions. The director of the institute, N.N. Semenov, had created a theory for branching chemical chain reactions. Therefore, it was quite easy for us to cross the bridge to nuclear branching chain reactions, and we agreed to work on this in earnest.

Initially, since this work did not figure in our official scientific plans (I was working on explosive materials and organized the corresponding laboratory for this research, and Yakov Borisovich's official research topic was adsorption), we decided to carry out our official work during the day, and turn to investigating the possibility of nuclear chain reactions after the end of the working day. However, we soon understood that we were dealing with a problem that was so important that it was essential that we concentrate on it alone. We shared our results with I.V. Kurchatov, who headed one of the nuclear physics laboratories in the nearby Leningrad Physical Technical Institute. Naturally, N.N. Semenov was also kept abreast of our work; he rapidly appreciated the unquestionable importance of the new problem and the possibility that it might lead to nuclear explosions. Without waiting for our work to be written up in the form of published papers, he wrote a letter to the Science and Technology Department of the Ministry of Oil Industry (which had jurisdiction over the Institute of Chemical Physics at that time), and assigned one of the researchers, F.I. Dubovitskii (now a Corresponding Member of the Russian Academy of Sciences, who continues to work at the ICP), the job of delivering it and trying to 'push it through' as far as possible. In this letter, Semenov laid out the basis for the need to develop this work as widely as possible.

In 1939 and 1940, Yakov Borisovich and I published three papers in *Zhurnal Eksperimental'noi i Teoreticheskoi Fiziki*,[24] as well as a review article in *Uspekhi Fizicheskikh Nauk* [*UFN*],' and submitted a second review article to *UFN*. At that time, we were working in collaboration with I.I. Gurevich. We were able to establish that if 10 kg of uranium-235 could be compressed to high density using ordinary explosives, a branching chain reaction could develop, and if the compression were strong enough, a nuclear explosion

[24] Journal of Experimental and Theoretical Physics.

could occur. Our estimate of the critical mass (10 kg) was included in the second review article in *UFN*, but proved to be rather crude, since we had performed the calculations in a rather primitive fashion — we had no access to electronic computers, not to mention the fact that we didn't have adequate experimental data. As a consequence, our results were out by a factor of approximately five (the correct value is 55 kg). In spite of this error, we were still full of enthusiasm, and undoubtedly would have continued our research on uranium fission if the harsh war with Germany had not started. At the same time, in July, 1940, a letter to the editor by Turner appeared in *Phys. Review*, which suggested that chemical elements with atomic numbers greater than 94 and atomic weights greater than 239 could also produce nuclear chain reactions. One possible way of obtaining such an element was via the action of neutrons on uranium. Subsequently, it was given the name of plutonium. Its critical mass turned out to be a little more than 10 kg. While our article was being edited, the war began in June 1941; publication of *ZhETF* and *UFN* temporarily stopped, and then this work was made classified. In the end, the article, with a brief description of its history, came out in 1983 in an issue of *UFN* dedicated to the 80th birthday of I.V. Kurchatov.

And so, the Second World War began, and we all had to do research that was directly connected to the creation of new types of conventional weapons. (In fact, I had organized the institute's explosives laboratory earlier, observing the rise of fascism in Germany and understanding what that might lead to.) I was assigned to one of the institutes of the People's Commissariat for Munitions (*Narkomat boepripasov*, NII-6), and Yakov Borisovich began to work on powders; we then worked together on some types of weapons. As a result, our research into questions connected with the possibility of nuclear explosions was put on hold for some time. (And in 1942, the well-known letter of G.N. Flerov to Stalin was written.)

During this time, our secret service obtained some very important information. In Germany, before the war, lived a young talented theoretical physicist, Klaus Fuchs, who was a Communist. Fearing persecution, he emigrated to England in 1934, where, after some time, he obtained British citizenship. He continued his theoretical research there. When, after the discovery of uranium fission, work was begun in England on the development of nuclear weapons, one of the project supervisors, the physicist Peierls, invited Fuchs to participate. Disturbed by the fact that all this research was being kept secret from the Soviet Union, an ally of England in the war against Germany, though there was an agreement between the two countries about the exchange of military information, Fuchs went to the Soviet Embassy in London and made a report about the work being undertaken. However, this information apparently didn't get anywhere, and got stuck in the system — from the Embassy, it was passed to the Defence Ministry, where, to all appearances, they were not able to appreciate its importance, since it doesn't seem to have been passed on further. At the same time, the Peierls' group was invited to collaborate on the uranium problem in the US.

There are many books describing work on this project. I'll mention only that the Hungarian emigré L. Szilard also participated in it. It was he who convinced A. Einstein to write a letter to President Roosevelt, which laid out the pressing need of developing research in this direction. The result was the founding of the Los Alamos laboratory in the southwestern state of New Mexico. One of the buildings there was used to construct the first atomic reactor for the production of plutonium.

Soviet intelligence was able to establish contact with Fuchs in the USA, and he began to systematically transfer extremely valuable information about the development of the work. Later, thanks to Soviet intelligence, we were able to obtain information about the principle of implosion, and, at the end of 1945, sketches for the first atomic bomb tested in the USA.

In 1943, our government made the decision to begin work on the creation of nuclear weapons. Upon the recommendation of A.F. Ioffe, the overall scientific supervision of the project was given to I.V. Kurchatov. Knowing our work, Kurchatov proposed that Zeldovich and I supervise investigations into the construction of a nuclear charge. Zeldovich was actively included in the complex of research being undertaken. Like me, he had continued to work on conventional weapons as well for some time.

The information obtained from Fuchs encompassed a wide range of questions; in connection with our own work, it contained a rather detailed description of the composition of the charge of explosive material. The charge created a converging spherical detonation wave that compressed the plutonium located at the centre of the sphere. It was decided to set aside our exploratory investigations and to proceed in accordance with the information obtained. Of course, we could not be sure of the absolute reliability of this data, nor of the absence of misinformation. In order to conduct experiments with certainty, we had to rush through a huge volume of experimental and calculational work, but people were prepared to work day and night to ensure that reliable results were obtained.

Yakov Borisovich very carefully chose the collective of theoreticians. One of the first nominees — Nikolai Aleksandrovich Dmitriev, a student of the brilliant mathematician Academician A.N. Kolmogorov — proved to be an especially fortuitous choice. After two or three years had gone by, and at times after some sharp 'high-level' discussion, Yakov Borisovich would say, 'All the same, I think I'll go get some advice from Kolya [Dmitriev].'

Not long before our experiments, in the beginning of 1949, two groups of physicists were assigned the task of measuring the speed of the products from a detonation of the explosive material that we were using. It was essential to know this in order to calculate the pressure that would develop during the explosion, which would act to compress the sample. According to the data from the first group, supervised by V.A. Tsukerman, everything was fine, but the results of the second group, headed by E.K. Zavoiskii (a very good physicist, who discovered electron paramagnetic resonance, but didn't have as much experience working with explosives), indicated that no nuclear explosion would arise. I remember that, in connection with this,

B.L. Vannikov — a former people's commissar[25] — arrived. By the way, not long before the war, he was arrested, but when the war began, they remembered that he was a prominent specialist, and so returned him to work, having appointed him director of the First Main Directorate of the Council of Ministers, which oversaw all issues connected with the creation of nuclear weapons. So anyway, Vannikov was quite alarmed. It turned out that in Zavoiskii's experiment the plate used to measure the pressure was not light enough, and when the experiment was redone correctly, the results of the two groups coincided. But before this was discovered, we had a bit of anxiety. This story shows once again that it is not sufficient to take and directly apply somebody else's experiments (the American experiments, in this case) — it is essential to understand, calculate, and have a sense for everything oneself.

Since our research involved creating rather powerful explosions, and it was not possible to carry this out in the immediate vicinity of Moscow, a place was sought further away, but still quite close to the city. This was after the war, and many ammunition factories were being closed or converted to peacetime production; we visited a number of such factories. Since it was expected that about two tons of explosives would be detonated, the project required a rather isolated place. Vannikov helped with this. On his recommendation, we set out for the small monastery settlement of Sarov, where there was a small factory producing new types of ammunition for mortars. Next to the factory town was an expansive forest reserve. We were given an area which was large enough for experimental test grounds and a small factory for the production of precision parts made of explosive material. The proposed director of the factory was a specialist who, during the war, had been the chief engineer for one of the ammunition production facilities, a man with whom I had had contact over many years. He accepted the offer with pleasure — after all, this was a completely new application for explosive materials.

Construction began, and since the military factory was small, the project required additional room for production, laboratories, and living space. At that time, no one really appreciated the scale of the project, and when I suggested that we would need to construct a three-storey laboratory building, people did not understand at first, and I was assured that a two-storey building would be quite sufficient. However, our settlement and production began to grow rapidly, and new people had to be found to carry out the work. Some I recruited from those with whom I had associations during the war, among them some young students who had defended their undergraduate theses at military institutions. Yakov Borisovich selected theoreticians, and in parallel Kurchatov formed two groups of Moscow physicists: one under the supervision of L.D. Landau from the Institute of Physical Problems founded by P.L. Kapitsa, and another headed by

[25] *Narkom*, a governmental position equivalent to the rank of minister. [Translator]

I.E. Tamm of the Lebedev Physical Institute, which included A.D. Sakharov. The first atomic bomb was made without their participation, and even then there were some technical developments that subsequently lay at the basis of the hydrogen bomb. When I recently looked through some of Sakharov's old reports from the Lebedev Physical Institute, one of them caught my attention (the report of Tamm's group from January 1949). In it, Sakharov wrote that they were working out an idea expressed earlier by Zeldovich. He meant the beginning of the work — the 'rough sketches' of the hydrogen bomb, made back in 1948.

In 1949, we conducted the first tests of Soviet atomic bombs, made in accordance with the American design, but our scientists were already working on more modern and refined constructions, which were also tested before long. The use of the tested design for our first bomb avoided unnecessary risk. At that time, it was important to show the world as soon as possible that we also had nuclear weapons. However even then, it was clear to us that this was only the beginning, and the initial creation is not the last word in weapons technology.

We continued our work on improving the nuclear charges and developing hydrogen bombs. Sakharov and Zeldovich (each of whom had a group of young theoreticians) were already working on the development of a hydrogen bomb. In 1951, the Americans performed several physics experiments in which it was possible, using an atomic bomb, to 'ignite' a small quantity of a liquid mixture of deuterium and tritium. In November of 1952, they first created a thermonuclear explosion with a power of about 10 megatons. But that was an explosion of a huge ground structure the size of a two-storey building, which was, of course, not the same as a bomb. This was only an intermediate stage on the road to the creation of a bomb. The first true hydrogen charge, with a power of about 400 kilotons, prepared in the form of a transportable bomb without any liquid, low-temperature component, was tested by the Soviet Union in August of 1953. This test became a prominent, high-priority achievement for our physicists, first and foremost for A.D. Sakharov and V.L. Ginzburg. This test should not be identified with the American tests using a small quantity of tritium and deuterium in 1951, nor with the explosion in 1952, which used thermonuclear fuel in a liquified state at a temperature near absolute zero.

The *Voenno-Istoricheskii Zhurnal*[26] asserted that, with the aid of Fuchs, it seemed that we had obtained all our information about the hydrogen bomb from the Americans. H. York, in his well known book *The Advisors: Oppenheimer, Teller, and the Superbomb*,[27] describes in detail how it all happened in reality. H. Bethe, who directed the Los Alamos theoretical department, noted that, from October 1950 through January 1951, Teller — the father of the American hydrogen bomb — was in complete despair: the Polish

[26] *Journal of Military History*. [Translator]

[27] Stanford University Press, 1989.

With Yu.B. Khariton at their joint 150-year jubilee (the 80th birthday of YuB and the 70th birthday of YaB). Moscow, 1984.

mathematician Ulam, who was participating in the project, had found serious errors, which nullified all results that had been obtained up to that time. And Fuchs at that time was already in prison, and was not capable of transferring anything. So the claim that the Soviet programme had adopted the American design was completely unfounded.

Recently, in the West, it has been asserted that when the Americans conducted their first thermonuclear explosion, we were probably able to collect secondary products of the explosion contained in atmospheric precipitates, and analyze them to recreate an entire sketch of the explosion process. In fact, we were fundamentally incapable of doing this, since the capture and analysis of radioactive fragments was not sufficiently developed in the USSR at that time. Thus, this claim likewise has no grounds.

Two groups worked in parallel on the hydrogen bomb — under Sakharov and Zeldovich. The research was conducted with the groups in close contact, without a sharp division between them. At one point, during our journey to the 'object,'[28] Tamm complained to me that he was so immersed in the project that he had become cut off from modern physics. He then commented that Zeldovich was somehow able to be fully up-to-date on all the latest scientific news — it must be that he works at night, since he is occupied all day with this main work. How did he manage it? I explained this very simply — he was a unique, absolutely incredible, person. (If you look at the two-volume selected works of Zeldovich, which includes a list of his publications, it immediately becomes clear how much prominent research he carried out.)

[28] The highly classified establishment near Arzamas where work was being conducted for the Soviet atomic weapons program. [Translator]

However at some point he decided to leave work at the 'object.' I could see that he was full of ideas, and the project didn't give him enough room to develop them. On the other hand, a generation of strong scientists had already grown up, so that his departure would not engender any special tragedy. I couldn't object to his decision; I had no moral right to do so. It would have been a sin to try to keep him there.

There is something I'd like to say in conclusion. Recently, there has been a tendency to contrast Sakharov and Zeldovich, not as scientists, but according to their social activities. It seems to me that it is important to understand that some people have a bent for such activities and others don't. Yakov Borisovich was infinitely captivated by physics, absorbed by it. He understood very well that under the circumstances, there was no reason to expect political activity on his part to lead to sensible results, while it would unquestionably disturb his work and take his attention away from science. It was clear from the example of Sakharov that political activity would not pass without leaving its mark. In general, Zeldovich chose his path in both life and science.

For me, the years spent in close contact with him, the friendship that bound us for many years, remains a time of enormous happiness. When solving some complex problem, torturing myself over it, I always knew deep inside that there was Zeldovich. It was always worth turning to him, and he was always able to find the solution to even the most complex problem — and what's more, he could do so beautifully and elegantly. I remember very well one particular case. Having arrived in Sarov for a visit, Kurchatov conducted a meeting on an acute scientific–technical question. Yakov Borisovich energetically participated in the discussion. After long discussions, we finally arrived at an agreement, and people went off their own ways. I remained for some time with Kurchatov. For a while, he sat in silence, and then sighed, hit his fist in the palm of his other hand, and said, 'Yes, Yashka is a genius ... .' His was a completely fantastic intellect. I bow before him — as a scientist and a person.

It was like magic

V.A. Tsukerman

He held out to me a grey tube with a diameter of about 12 mm and a length of 50 cm, that looked rather like a piece of spaghetti and said, 'It would be very interesting to measure the speed with which this thing burns in the middle and at the edges. We are studying the rate of burning of such powders at pressures of tens of atmospheres. It seems logical that the central part of the tube should burn more slowly than the edges. You've already obtained X-ray pictures of bullets in flight. Do you think it might be possible to use your technique to study phenomena during the combustion of powders?'

That is how I remember one of my first encounters with Yakov Borisovich Zeldovich. That was in Kazan, in January or February of 1943, in the middle of the Second World War, while the fight for Stalingrad raged, and the great battle at the Kursk Arc was just being prepared for. At that time, we were conducting successful experiments on X-ray photography of phenomena occurring during the explosion of small charges. Yakov Borisovich followed the development of this work attentively.

Three years later, we met at the time when work on the creation of Soviet nuclear weapons was under way. Yakov Borisovich was appointed head of the theoretical department. With time, it became clear how sound that decision had been. A broadly educated physicist, in excellent command of gas dynamics and the physics of explosions, mercurial, active and energetic, Zeldovich was the soul and the symbol of our department over two decades. He was able to capably resolve not only theoretical questions associated with the creation of weapons, but also the overwhelming majority of experimental works, many of which had been begun on his initiative. He posed current unsolved problems to researchers, and often himself proposed means and methods to solve them. He actively worked on the selection of qualified theoretical physicists for the scientific staff.

Zeldovich turned out to be responsible for the arrival of another excellent theoretical physicist with high inventive potential — Evgenii Ivanovich Zababakhin. The story of his arrival is as follows. In 1944, having graduated from the Zhukovskii Air Force Academy with distinction, Zababakhin was enrolled as a postgraduate student at the Academy. The supervisor of his dissertation, Prof. D.A. Ventsel, suggested a theme to him: Studies of processes in converging detonation waves. After Zababakhin completed work on the dissertation, Ventsel recommended that he be sent to the Institute of Chemical Physics. There he fell into the hands of Zeldovich, who immediately understood that the Zababakhin's research was quite close to our own tasks. In the spring of 1948, Captain E.I. Zababakhin arrived at the institute, and soon became one of the leading researchers in the theoretical department. He has made an enormous contribution to the development of applied nuclear physics and to the solution of a number of physical problems associated with the creation of various types of nuclear weapons. In 1968, he was elected as an Acting Member of the USSR Academy of Sciences.

At that time, there was a great dispute between theoreticians and experimentalists. According to the estimates of the theoreticians, the bulk velocity of the products of the explosion should be 2 km/s, while the experimental estimates yielded a smaller value. The experimentalists lost and found the error in their experiments themselves. It was then necessary for them to 'pay up.' Since most of the researchers in the department lived in Moscow at that time, the wine was drunk at the Institute of Chemical Physics on Vorobiev Road.

Our clashes in connection with the question of the true bulk velocity of the products of the explosion did not end with this episode. When we began to measure this parameter, it turned out to be 25 to 30% less than expected. We went to Yakov Borisovich, who said he would think about the possible miscalculations in the experiment or theory overnight. The telephone call came in the early hours of the morning: 'It seems that I've found the reason for the discrepancy. Come over to the hotel.'

We went. In Zeldovich's opinion, the origin for the problem was the short length of the charges used. If the charge length were increased to a metre — or even better, to two — the initially spherical wave front would become nearly planar. Then, the measured velocity should grow. During the pointlike detonation of explosive material, a diverging wave with an infinite derivative at its front forms; consequently, a very thin pressure peak develops, which rapidly decays toward the barrier. In order to obtain accurate results, experiments must be done using charges whose length is no less than ten times greater than their diameter.

But why should processes occurring in the detonation wave front be affected by phenomena occurring tens or hundreds of microseconds before the formation of the front? We couldn't understand this. However, to our surprise, Zeldovich was right. The pressure in the wave front grows appreciably when the length of the charge is increased.

What was it? Intuition, knowledge, talent .... Without these three components, creativity in science is not possible. Two decades of work with Yakov Borisovich convinced us that he had in full measure those rare qualities that enable a scientist to search for and find solutions for complex physical or physico–chemical problems.

With the permission of Yu.B. Khariton, I would like to illustrate this using a particular incident. At a meeting dedicated to the solution of a complex problem, we spent a long time searching for the correct solution. Suggestions were disproved one after the other. And then Zeldovich enters the scene. A chain of formulae appears on the blackboard, and after a matter of minutes, the problem was solved. It was just like magic. After the participants had gone their separate ways, and only Yulii Borisovich and Igor Vasil'evich were left in the room, Kurchatov said, 'Yes, Yashka is a genius ....'

It is difficult to define the boundary between talent and genius. It could be that it doesn't exist. However I myself was the witness of such an episode. At the beginning of the 1950s, after the conduction of an important experiment, a large group of scientists were waiting while oscillograms and the output of other instruments were being processed. Igor Vasil'evich lamented: 'It's a shame that Yashka isn't here. He would take his little slide rule from his pocket, move the frame a few times to the right and to the left, and give us all the necessary numbers.' (In those distant years, the slide rule was the main instrument for both theoreticians and experimentalists.)

The creative style that remained in the institute from its first days was due in large part to Yakov Borisovich. However, he was intolerant of any type of pseudo-science. We kept his open letter from January 18, 1963, addressed to the experimentalists at the institute. (At the beginning of the 1960s, the so-called Dean machine, which was used for its work on gravitational forces, was widely discussed. The letter was composed in connection with the announcement of a seminar dedicated to the Dean machine.)

'I have deep respect for the collective of experimentalists ... . This is an independent institute. In our common area of physics, it has made a large, inestimable contribution. Therefore, it is not merely funny, but painful for me to see the advertisement for the incompetent 'invention' of Dean.

It is clear to anyone who understands the fundamentals of mechanics that this machine is nonsense. One can make guesses as to how they granted a patent for this nonsense, or how it appeared on the pages of a journal, but, indeed, that isn't so interesting.

Sometimes people say, 'Of course, Dean's experiment contradicts theory, but it represents an experiment, facts — a concrete thing, and so forth.' Here, we must understand that theory — in this case, classical mechanics — is the concentrated result of an enormous number of experiments and results, which are published, verified, discussed many times, and which are consistent.

Do you remember the fuss about the transformation of time into energy which was 'discovered' by one of the Leningrad physicists? Several years ago, no matter what topic any public lecturer talked about, he was always asked about that. And where is this 'theory' (if one can call it that) now? We all understand that if it had been correct, it would not have been silenced and suppressed.'

And much later, at the joint session of the USSR Academy of Sciences and Academy of Medical Sciences in November of 1980, a short but memorable presentation by Zeldovich was directed against extrasensory perception, a concept that is still popular now. He pointed out that even the 'magician and sorcerer' Grigorii Rasputin, who enjoyed the patronage of the tsar's family, was not admitted to the sessions of the pre-Revolution Imperial Academy of Sciences. And there are cases where 'philosophers' who are not competent in the natural sciences, using academic titles as a cover, organize public demonstrations of the 'tricks' of the not unknown Juna.[29]

Yakov Borisovich wrote a number of excellent books, which have been and will be used to educate more than one generation of physicists. His ability to open up the essence of complex problems and his bright, vivid language are delightful. Many pages in his books form a genuine paean to science and inspired creativity. It is not possible to set aside a book with a foreword written by Zeldovich ending in the words: 'You can read this book and study it, but you can also use it as a source of inspiration. It may be that this is the best compliment that a book with such a specialized title can receive.'[30]

---

[29] This refers to Juna Davitashvili, who supposedly possessed powers of extrasensory perception, and was widely known in the Soviet Union in the late 1970s and early 1980s, owing to her frequent appearances on television. She was rumoured to be a personal healer to Brezhnev himself. [Translator]

[30] Barenblatt, G.I., *Similarity, Self-similarity, and the Intermediate Asymptotic*, Leningrad: Gidrometeoizdat, 1982, p. 10.

In our memory live his brilliant review articles published in *Uspekhi Fizicheskikh Nauk*, written in a free and informal style, sometimes with elements of mischievousness. The oral presentations of Zeldovich always produced genuine satisfaction and delight in the audience, no matter who they were.

Yakov Borisovich was a multifaceted, well-educated man, and the circle of his interests was exceedingly wide. Lines of poetry are often cited in his scientific articles. He defended his literary and artistic sympathies as energetically and consistently as he did his scientific convictions. He hotly and passionately stood up in favour of the nomination of the writer Chingiz Aitmatov for a full membership of the USSR Academy of Sciences.

The sudden death of Yakov Borisovich on December 2, 1987 from a heart attack astounded all those who knew him. We saw each other for the last time at the funeral of Khariton's daughter Tatiana Yulievna. I stood not far from the coffin, and suddenly had the impression that someone was staring at me. At that time, my vision had already deteriorated, but the sharp feeling that someone was looking at me caused me to turn around. It was Zeldovich. We embraced and kissed each other. Who would have thought that meeting would be our last? He never complained about his heart, never sought medical care ... .

A person's life passes rapidly, very rapidly. Who would have thought that such a robust and optimistic fellow as Yakov Borisovich would leave our world so soon? He was only a year younger than I, but succeeded in doing much more in science and life.

At the civil funeral ceremony on December 7, Andrei Dmitrievich Sakharov made a splendid speech about Yakov Borisovich Zeldovich, to which each who worked with him adds his voice.[31]

From the eulogy of A.D. Sakharov[32]

Yakov Borisovich Zeldovich has left life. This is difficult to believe, since thoughts of death do not accord with his figure. It is unbearably bitter to realize that he is no longer with us.

Yakov Borisovich's indefatigable scientific activity and staggering versatility and intuition were always striking. He began working early, and continued to the last day of his life, having accomplished an incredible amount in a whole series of very different fields of physics. In the November issue of *Uspekhi Fizicheskikh Nauk*, which appeared after his death, we saw an article that made a bridge back to the beginning of his scientific work — chemical physics, surface phenomena, combustion and detonation, chemical

[31] See the next contribution.

[32] Given on December 7, 1987 in the hall of the Presidium of the USSR Academy of Sciences, where the valedictory speeches for Ya.B. Zeldovich were made.

and nuclear chain reactions. Then came jet propulsion technology, and the years of participation in the development of Soviet atomic and thermonuclear weapons. His role in this last area was exceptional, and can now be spoken of freely. He is the author of several eminent works on elementary particle physics, including the beginnings of 'current algebra,' the prediction of the existence of the Z^0 boson and some of its properties, and a formulation of the problem of the cosmological constant. His last 25 years were devoted primarily to astrophysics and cosmology. He was always at the leading edge, and always surrounded by people. Everyone who had contact with him learned lasting invaluable lessons, both in connection with specific scientific questions and in terms of examples of how to work in science and modern technology.

I was honoured to spend many years side by side with Yakov Borisovich. Thinking back to this time, I feel just how much I owe him. In the extremely intense circumstances of those years, we had a simple, friendly, extremely benevolent relationship. This was in spite of the fact that Igor Evgen'evich Tamm and I, in a manner of speaking, invaded his territory; it took outstanding objectivity on Yakov Borisovich's part in order not to react negatively and with offence. In place of such negativity, Yakov Borisovich and his colleagues provided inestimable help and collaboration for the sake of the common cause. And then we worked shoulder to shoulder in the general charge of 1954 and 1955. Those who participated in these events together with us (many are now in this room) remember what that meant. In the area of fundamental physics, much of my research arose from my contacts with him, under the influence of his work and ideas.

In science, Yakov Borisovich was a person of enormous avidity (in the best sense of this word), and at the same time, he was absolutely honest, self-critical, prepared to acknowledge his mistakes, and also the correctness or authorship of others. He expressed an almost childish glee when he was able to accomplish something important or overcome some methodological difficulty using a beautiful technique, and he suffered deeply over unsuccessful endeavours and mistakes. Overall, he was a modest person in relation to science. It often seemed to him that he was a dilettante and was not sufficiently professional in regard to some questions, and he made enormous efforts in order to overcome his deficiencies (though others were not always aware of these).

Our forty years of acquaintance included some thorns, offence, and cool periods. Now, that all looks like no more than foam on the sea of life; but, so to say, what happened, happened. Once, many years ago, Yakov Borisovich telephoned me and said, 'There are words that are not to be repeated every day, but which sometimes must be spoken.' Today, parting with Yakov Borisovich, I want to say what a huge role he played in my life, as he did in the life and work of very many people, how much we loved him, how I loved him, how much he will be missed, how much we needed him!

The beginning of the physics of extreme states

L.V. Altshuler

In his valedictory speech, A.D. Sakharov described Ya.B. Zeldovich a person of universal interests. According to the calendar, Zeldovich lived a normal human lifespan. However, this lifespan encompassed several scientific biographies of enormous volume and lasting importance. It turned out that these biographies were devoted to explosions of increasing power (detonations) and chemical explosions, chain reactions and nuclear reactions, and the Big Bang that formed our universe 15 billion years ago. Questions concerning combustion and detonation were a first and lasting love of Yakov Borisovich, to which he remained faithful to his last days.

The beginning of his 'second biography' is marked by the publication of three papers written with Yu.B. Khariton, about the development of chain reactions in uranium. This research, which demonstrated the fundamental possibility of releasing atomic energy, were published on the eve of the Second World War. Both authors of those papers participated in the Soviet atomic project from the time of its official birth in the spring of 1943.[33] Later, a large number of scientists became involved in this global problem, among them V.A. Tsukerman, the head of the X-ray laboratory of the Institute of Machine Science of the USSR Academy of Sciences, and myself, his former colleague.

Having been friends since our schooldays, we worked together for about 15 years, with a two-year gap when I served in the army. V.A. Tsukerman was not able to serve in the army and at the front due to his progressive loss of sight, which led over the course of several years to complete blindness. We were distinguished from many of our colleagues by our frequent — and, in the opinion of those around us, somewhat excessive — activity. We often behaved like conquistadors of science, setting out to discover still unknown continents. I have carefully preserved a monograph from Zeldovich, given to us in 1953, with the meaningful inscription: 'To my brothers and rogues Altshuler and Tsukerman, from an author who has not yet become one of their victims.'

In Kazan, Tsukerman installed X-ray equipment in hospitals, developed a launcher for hurling Molotov cocktails over large distances, and, due to an unfortunate accident, himself ended up playing the role of a burning tank. Having returned to Kazan after his hospitalization, he worked on the realization of his main idea — instantaneous X-ray photography of processes occurring over millionths of a second in explosive discharges. This new method made it possible to understand the almost mystical mechanism

[33] Golovin, I.I. and Smirnov, Yu.N., *Eto nachinalos' v Zamoskvorechye* (*It Began in Zamoskvorechye*), Preprint No. 19263, Moscow: Kurchatov Atomic Energy Institute, 1989.

for the 'armour-burning' Faust patrons[34] used by the Germans against our tanks. In numerous discussions of this and other related problems, an acquaintance that quickly grew into friendship arose between him and Yulii Borisovich Khariton, as well as Yakov Borisovich Zeldovich.

In 1946, Khariton, in rather unclear terms, came to Veniamin Aronovich [Tsukerman] with the proposal that we both take part in a very interesting and complex problem: 'You can begin the work in Moscow, but it will be necessary to spend one and a half or two years outside Moscow in order to finish it up.' This timescale was extended in our case to ten years. The formalities of our departure were rapidly dealt with, since we were both included in the governmental directive.

My meetings with Zeldovich began while I was still in Moscow, at the Institute of Chemical Physics of the USSR Academy of Sciences. Our discussions were conducted in a small room at a blackboard, to which a torn rubber boot had been nailed to hold chalk and an eraser rag. The tone of our discussions was extremely free and easy, and often terms that were not standard for scientific publications were used. This surprised me a bit, but it felt quite natural. Subsequently, many scientists used similar slang when it was necessary; it was to the point and expressive, and this was especially true of the test sites just before experiments. I was told that there were special instructions for the drivers of the cars used to transport the researchers — in order to preserve state secrets, they were forbidden to repeat words and phrases spoken by their scientific passengers. The result was unexpected — the drivers stopped swearing.

During one meeting at the Institute of Chemical Physics, Yakov Borisovich, having greatly simplified ways to obtain supercritical states, proposed that I analytically investigate them and compare their advantages according to a simple criterion. It turned out to be clear that one of the methods was highly preferable. We were able to realize this method in 1951, and it demanded a great deal of creativity and intense work on the part of several scientific collectives. The first, more primitive construction of the 'item'[35] was tested, as is known, in August of 1949. For this purpose, a large institute fenced off from the outside world by barbed wire was built outside Moscow, on the site of the former Sarov Monastery, in early 1947. This was one of numerous islands of the 'white archipelago,' where the Soviet atomic project was developed under the closest secrecy. In order to ultimately obtain an explosion 'brighter than 1000 suns,'[36] it was necessary to study the properties of matter at high and ultrahigh temperatures and pressures, and to create and develop a new scientific discipline — the physics of high

[34] A cumulative anti-armor munition. [Translator]

[35] 'Item' — *izdelie* in Russian — was the word used for an atomic (and later hydrogen) bomb. [Translator]

[36] Jungk, R., *Brighter than a Thousand Suns*, Harcourt Brace Jovanovich: New York, London, 1958.

energy densities.[37] Brilliant chapters in this work were written by Ya.B. Zeldovich, A.D. Sakharov, D.A. Frank-Kamenetskii, and the experimentalists working at the 'object.'

The living and working conditions for the researchers were excellent, especially for the experimentalists. It was possible to prepare parts for experiments very quickly in the factories and workshops, and then carry the tests out in wooded areas. The rapid progress in solving the problems at hand was also facilitated by the nearly infinite faith of the young researchers in their supervisors, none of whom were more than 30 to 35 years old. Close, almost daily contact with the theoreticians was also very important. In the early years, there were very few of them. Apart from Yakov Borisovich, the experimentalists associated most of all with Evgenii Ivanovich Zababakhin and Grigorii Mikhailovich Gandelman. At the end of 1947, Zababakhin was a postgraduate student at the Zhukovskii Air Force Academy, and defended a dissertation dedicated to converging detonation waves. The dissertation ended up at the Institute of Chemical Physics, where it greatly interested Zeldovich — and it interested the workers in the security clearance department even more. 'Where do you keep your notebooks and manuscripts?' they asked Evgenii Ivanovich sternly. 'In the drawer of my desk,' he answered innocently. This was followed by an anxious silence, the calm before the storm that broke out and carried Evgenii Ivanovich, together with his dissertation, to the theoretical department of our 'sacred abode.' This was a fortunate turn of events for the 'object,' and for the atomic project as a whole.

For nearly two years, the leading scientific groups at the 'object' determined the pressures for detonations of powerful explosives. In the structures created, the gaseous products of the explosion formed during detonation played the role of the 'working body,' as does water vapour in tubes and other thermal machines. Their pressure and other properties determined the efficiency of the developed models. At that time, there were no unambiguous theoretical predictions about these phenomena. According to the estimates of German scientists, the detonation pressure of TNT was 120,000 atmospheres, while Landau and Stanyukovich estimated it to be 180,000 atmospheres. In order to determine the true value, first experiments were conducted in the laboratory of Tsukerman. There, instantaneous X-ray photographs were obtained of the propagation of detonations and of millimetre steel spheres located behind the detonation front — which, to our amazement, remained stationary. This result called forth a strong reaction from N.N. Semenov, the future Nobel Prize winner, who happened to be visiting the 'object' at that time: 'If your method doesn't record the bulk

[37] Zeldovich, Ya.B. and Raizer, Yu.P., *Fizika udarnykh voln i vysokotemperaturnye yavleniya* (*Physics of Shock Waves and High-temperature Phenomena*), Moscow, 1963; Altshuler, L.V., *Primenenie udarnykh voln v fizike vysokikh davlenii* (*Application of Shock Waves in the Physics of High Pressures*), UFN, 1965, vol. 85, pp. 197–258; *The Physics of High Energy Densities*, New York: Academic, 1971.

velocity of the products of the explosion, that means that it is not good for a damn thing.' Fundamentally, however, the method was very effective. We only had to increase the strength of the charge and replace the spheres with thin foils. After this, we obtained results that were close to the values predicted by Landau and Stanyukovich.

In my group, the detonation pressure was derived from the velocities that were acquired by thin plates of various metals attached to the ends of cylindrical charges during an explosion. Again, the first experiments were discouraging, and suggested low pressures. I told Khariton and Zeldovich about this late one evening in December 1947. Everybody left in low spirits. At eight the next morning, Yakov Borisovich telephoned me and asked me to come to see him at the hotel (by the way, this was a hotel constructed at the beginning of the twentieth century for the Tsar's visit to the Sarov monastery). I don't know if it was during late-night study or in his sleep, but Yakov Borisovich came to understand that night that, in our experiments, a diverging detonation wave forms with an 'infinitely thin' peak pressure, which rapidly decays in the plates used. In order to broaden the wave front and obtain the correct results, he suggested that we conduct the experiment using a longer discharge container, with a length of about a metre. All of us, even Yulii Borisovich, were surprised by this recommendation, and reacted to it with doubt, and even thought it was a little silly. However, experiments using the proposed scheme were conducted, and fully confirmed the correctness of Zeldovich's analysis. It seemed that everything had become clear.

Then, unexpectedly, the validity of the project was again placed in doubt. This occurred at the beginning of 1949, when the prominent Soviet physicist E.K. Zavoiskii, who was working at the 'object,' reported his latest results. In his method, a charge was placed in a uniform magnetic field, and the measured quantity was the electromotive force in conductors placed in the charge, which was proportional to their velocity. The basis of his method was irreproachable, as in the two other examples mentioned above. Attempts to arrive at consistent conclusions in the commission formed for this purpose were unsuccessful.

The two laboratories 'opposing' Zavoiskii had to try to reproduce the rather complex apparatus used in the electromagnetic approach. In a short while, a small methodical error was discovered, which had lowered the velocity of the explosion products. The green light was given for testing with the first nuclear charge.

However, it is not right to consider the opinion of Yakov Borisovich to be indisputable — to treat it like one of the laws of thermodynamics. For a long time, he didn't believe in the electrical conduction of the explosion products discovered by the experimentalists, and even rashly placed a bet of several bottles of cognac on the matter. He lost, the cognac was amiably consumed, and the article of Brish, Tarasov, and Tsukerman reporting the conductivity was published. To this day, this pioneering work remains frequently cited.

Our discussions were not always conducted in a gentlemanly way. For example, one of the researchers under Zavoiskii, now prospering, maintained that the Russian language was almost never heard at the object, and the experiments of Tsukerman contradict Marxist dialectic. On our side, our drafts invariably went to the factory under the code name 'DZ,' which stood for 'Death to Zavoiskii.' When the dispute had almost concluded in our favour, during a break in one high-level meeting in the presence of Kurchatov and Khariton's assistant, K.I. Shchelkin, Yakov Borisovich for some reason began to tell a story about children pretending to be a car. The oldest of the boys said, 'You'll be the right wheel, you'll be the left wheel, you're the motor, and you're the steering wheel.' 'What about me?' asked the youngest in a voice filled with tears. 'You're going to run along behind us and foul the air with petrol fumes.' 'Who, in your opinion, is fouling the air here?' asked Shchelkin, frowning gloomily. YaB's answer was clear to all: 'In any case, it's not Altshuler and Tsukerman.' On this note the scientific disputes effectively ended. I should note in fairness that a modified form of Zavoiskii's electromagnetic method later became one of the main methods for studying detonations and their development in transient regimes, both in the USSR and abroad.

The investigation of the properties of metals under ultrahigh pressures became as urgent as the study of explosive materials had been. It was necessary to know a realistic equation of state for them, which would make it possible to calculate their pressure from the density and temperature. At that time, these relations were completely unknown; therefore, the calculations were carried out in two somewhat arbitrary versions — the hard 'K' and soft 'D' cases. Even at high-level meetings, these were referred to in these terms:

> Case K was plucked from thin air
> Case D from a full beard of hair.

('The Beard' was at that time an informal nickname for I.V. Kurchatov.)

The physical instrument enabling extreme states of matter to be attained and studied was strong shock waves. Therefore, first and foremost, it was necessary to measure the pressure and density in shock waves with various amplitudes. Shchelkin told us in 1948 that our laboratory collective would have to 'move' to the megabar pressure region and report information about the compression of metals under pressures no less than three megabars. This task was fulfilled by the experimentalists with a wide margin, and the ceiling for pressure measurements became 10 Mbars. The results obtained were published in papers in 1958 to 1963. In 1988, American scientists wrote that these results were obtained on apparati not described anywhere in the literature, and were not exceeded by any other group.[38]

The main role in these accomplishments was played by the founder members of our departments: K.K. Krupnikov, A.A. Bakanova,

[38] Nellis, W.J., Moriarty, J.A., *et al.*, 'Metals Physics at Ultrahigh Pressure: Aluminium, Copper, and Lead as Prototypes,' *Phys. Rev. Let.*, 1988, vol. 60, no. 14, pp. 1414–17.

M.I. Brazhnik, and their colleague R.F. Trunin. Zeldovich assigned the theoretical calculations of the metal compression to G.M. Gandelman. Through constant contact with him, we were able to correctly understand the properties of compressed transition and rare-earth metals that were being discovered in the experiments. Much later, on the basis of new and more complex ideas, American scientists conducted their own calculations, obtained the same results, and announced — in my view, rather tactlessly — that they considered their own results to be genuinely correct, and the results of Gandelman to have turned out correctly just by accident. Poor Grigorii Mikhailovich was extremely offended, and lamented the fact that he didn't live at the time of D'Artagnan, when he could have defended his honour in a sword duel. It took a great effort from Yakov Borisovich to calm him.

Apart from Zeldovich, few people in those distant years realised that consideration of a single shock adiabat was a completely inadequate way of fully understanding the properties of metals in extreme states. On the phase diagram, one could liken the shock adiabat to a path surrounded by unexplored jungle. In accordance with an expanded programme set by Zeldovich in 1948 and published in 1957,[39] the efforts of Soviet researchers were for many years directed toward acquiring the additional information needed. With this goal in mind, the group of S.B. Kormer measured the velocity of sound behind the fronts of strong shock waves. Another novel research direction was provided by studies of the shock compression of powder-like metals in the collectives of Krupniov and Kormer. Much later, another of Zeldovich's experimental ideas was realized jointly with the laboratory of V.E. Fortov — the detection of isentropic contours for the expansion of metals.

The year 1956 was marked by the reorganization of our laboratory. Several very highly qualified researchers from my department became heads of new laboratories. They were all younger than both me and Yakov Borisovich. As he told me, direct contact with them became for him simpler and more comfortable. This did not affect our friendly relationship.

Especially close contact developed between him and S.B. Kormer. At the suggestion of Zeldovich, the temperatures of shock compression of ionic crystals and their indices of refraction were measured for the first time in Kormer's laboratory, and the smoothness of the shock fronts was estimated. He discussed the results with the experimentalists for hours.

His contact with our department continued when he, Yu.M. Styazhkin, and I got the idea for a new method for determining the smooth compression of metals under ultrahigh pressures of the order of 100 Mbars. Unfortunately, Yakov Borisovich didn't live to see the glorious publication of the results of these interesting and labour-intensive investigations.

Over the course of many years, Yakov Borisovich and A.D. Sakharov

[39] Zeldovich, Ya.B., *Issledovaniya uravnenii sostoyaniya s pomoshch'yu mekhanicheskikh izmerenii* (*Investigations of the Equation of State Using Mechanical Measurements*), *ZhETF*, 1957, vol. 32, pp. 1957–58.

were the heart and mind of our 'object.' Now, after the untimely death of Yakov Borisovich, he is often accused of conformism, and of not wanting to take part in the battle with ugly social phenomena. Yakov Borisovich firmly believed that a person who had dedicated himself to science, and had made science the main business of his life, should not — did not have the right to — split himself in two, to waste his energy on other endeavours. In his opinion, the dissidence of scientists was as useless as titling at windmills. He felt concrete, personalized help to get individuals involved in science was more important. He helped me a lot in this way. Yakov Borisovich did not participate in official campaigns to condemn Sakharov, in contrast to a number of other academicians. He viewed Sakharov as a phenomenon of nature, unique in all respects.

In their memoirs, V.A. Tsukerman and Z.M. Azarkh refer to Yu.B. Khariton, A.D. Sakharov, and Ya.B. Zeldovich as spiritual giants.[40] Yulii Borisovich considered and considers his most important task to have been the creation of a nuclear shield for our country. Andrei Dmitrievich, an extraordinarily prominent scientist and the spiritual leader of our time, like Pushkin's prophet, 'inflamed the hearts of the people with a word.' The greatness of Yakov Borisovich consisted in his enormous scientific potential and absolute devotion to science. He defined the goals and paths for studies of special states of matter for decades to come. In this field, he will always remain a leader.

A Knight of Science

Yu.N. Smirnov

On one autumn evening in 1968, when work on preparing a memorial for Igor Vasil'evich Kurchatov was in full swing, I had the privilege to be the guest of his widow, Marina Dmitrievna. We sat at a table set for tea, on which bright, cheerful flowers stood in a round vase. The cordiality of my hostess, her cultured simplicity, and the atmosphere of calm and remembrance led us into conversation. Gradually, our discussions of museum affairs shifted to the past, and Marina Dmitrievna began to reminisce about life in Leningrad and the intense period of self-deprivation during Igor Vasil'evich's activities in Moscow. At one point, we began to talk about the book by P. Astashenkov — *Kurchatov*. Marina Dmitrievna did not favour the book, which must have dismayed rather than gladdened her, and expressed regret that she had shown the author letters from her husband, which she had kept. However, when I remarked that the best book about Igor Vasil'evich was probably still to be written, Marina Dmitrievna sadly disagreed: 'No ... Another time has

[40] Tsukerman, V.A. and Azarkh, Z.M., 'Lyudi i vzryvy' ('People and Explosions'), *Zvezda*, 1990, no. 11, pp. 93–122.

come. Now the cosmonauts get all the glory. The creators of nuclear weapons have passed into shadow ... .'

Nearly a quarter of a century has gone by. Delight and feverishness over each new space launch are a thing of the past. And few remember the offensive claim that we learned how to make rockets on a conveyer belt, like big sausages. When speaking of defence, we have long used the balanced phrase 'nuclear rocket shield for our country,' paying due homage to the heroic contributions of both physicists and rocketeers.

I see some analogy in this situation when, in the year of the widely celebrated 70th birthday of Andrei Dmitrievich Sakharov, right on the heels of my remembrances about him,[41] I was called upon to write about Yakov Borisovich Zeldovich. It is easy for a society that has awakened from numbness or ignorance, possibly feeling penitence, to take up one name, unintentionally pushing another into shadow.

The greatness of Andrei Dmitrievich will not suffer if I dare to assert that there was another of our countrymen whose talent as a physicist matched his own — Yakov Borisovich Zeldovich. The issue here is not the identical official distinctions they received for their work, though each of them was honoured with the name of Hero of Socialist Labour three times over and was awarded a Lenin Prize while still far from his jubilee years, long before the 'star-spangled epoch of stagnation.' The essence here goes much deeper. It is not by chance that the patriarch of Soviet nuclear physics, Yu.B. Khariton — who, by the way, was awarded these same distinctions and was well acquainted with both the 'open' and 'closed' work of both his celebrated colleagues — describes one of them as a 'completely unique phenomenon in our science' and the other as an 'absolutely exceptional scientist.'

The rich theme of the equally great talents of the two leading figures of physical science, Sakharov and Zeldovich, will attract more than one researcher. However, it seems appropriate to me to draw attention to two or three noteworthy circumstances. Indeed, while the fundamental ideas of thermal isolation of a hot plasma using a strong magnetic field and of a magnetic thermonuclear reactor belong to A.D. Sakharov and I.E. Tamm, Ya.B. Zeldovich and Yu.B. Khariton had the honour of laying the foundation of the modern physics of atomic reactors and nuclear energetics, and of carrying out in this field ground-breaking work of enormous fundamental importance. While the original idea of creating explosive magnetic (magneto-cumulative) generators, making it possible to obtain record ultra-strong impulsive magnetic fields belongs to Sakharov, Zeldovich, for example, proposed a means of confining very slow neutrons in a cavity (the 'Zeldovich nuclear bottle'), which marked the beginning of a new field of neutron physics. Yakov Borisovich also proposed the use of colliding particle beams in physics experiments, an idea that has been realized in modern accelerator technology. Andrei Dmitrievich and Yakov Borisovich independently

[41] Smirnov, Yu.N., 'This man has done more than all of us ...' (*Andrei Sakharov. Facets of Life*, France: Editions Frontieres, 1991, pp. 591–619).

expressed pioneering ideas about the possibility of muon catalysis nuclear reactions in deuterium, and then jointly worked out these ideas.

The contribution of Andrei Dmitrievich to the development of the Soviet thermonuclear bomb was decisive. However, once again it is appropriate to turn to the authoritative opinion of Khariton: '... Here, the roles of many others were also rather important. In general, it was a collective effort. In one of the reports from the very beginning of the period, Andrei Dmitrievich says that he is developing some ideas expressed by Zeldovich. So it is difficult to say if he would have had those decisive thoughts if the earlier works of Zeldovich had not been there.'[42]

These same thoughts about the close creative collaboration between Yakov Borisovich and Andrei Dmitrievich have been emphasized by other veteran colleagues. Yu.A. Romanov, speaking of A.D. Sakharov's work after the first test of a Soviet hydrogen bomb in August 1953, draws attention to the fact that 'in the early spring of 1954, in discussions with Zeldovich, the same ideas were born at which Ulam and Teller had arrived in 1951.'[43] Talking about this period with V.B. Adamskii, I learned that it was even difficult to understand who made the decisive breakthrough in their splendid creative interaction; his memory had been stamped with the boyish spontaneity of happy Yakov Borisovich, jumping for joy, announcing his understanding achieved during a discussion with Sakharov of 'how it should be done ... .'

Yakov Borisovich was among the main creators of the first Soviet atomic bomb, whose successful test on August 29, 1949 liquidated the USA's atomic monopoly. For participation in this work, in accordance with a Directive of the USSR Council of Ministers signed by Stalin on October 29, 1949, Yakov Borisovich and several colleagues were presented with the title of Hero of Socialist Labour, awarded a large sum of money and a *Pobeda* ['Victory'] car, became a holder of a First-degree Stalin Prize, and received a country home (dacha) built at government expense.

Finally, it goes without saying that Andrei Dmitrievich, having obtained a number of eminent results in astrophysics and theoretical physics, was arguably the most refined, successful, and unique at revolutionizing engineering applications of his brilliant physical ideas, which were so important for the defence of our country and the development of science. At the same time, Yakov Borisovich, having made an enormous contribution to the atomic defence of our country and laid the physical foundations for the internal ballistics of solid-fuel rockets, left behind not only an immensely rich scientific legacy, but also gave birth to scientific schools in the fields of chemical physics, hydrodynamics, the theory of combustion, nuclear physics, the physics of elementary particles, and astrophysics.

---

42 Khariton, Yu.B., 'Radi yadernogo pariteta' ('For the Sake of Nuclear Parity'), Dosier, *Literaturnaya Gazeta*, 1990, pp. 17–19.

43 Romanov, Yu. A., 'Otets sovetskoi vodorodnoi bomby' ('The Father of the Soviet Hydrogen Bomb'), *Priroda*, 1990, No. 8, pp. 20–24.

If we speak of the creative traits of Yakov Borisovich and Andrei Dmitrievich, it would be fair to say that they enriched and complemented each other, and, in a certain sense, moved toward each other from opposite directions. Indeed, Zeldovich, when entering a field of research which was new to him, preferred to set and solve a problem, and came to understand problems and their deep interconnections through generalization. For Andrei Dmitrievich, with his broad erudition, in the words of Tamm, 'physical laws and the connections between phenomena ... are directly visible and perceptible in all their inner simplicity'[44] which enabled him to achieve genuine wonders of technical invention.

I cannot resist citing a few words from Andrei Dmitrievich: 'Yakov Borisovich's indefatigable scientific activity was always striking, his lively interest in everything new, his amazing multi-facetedness and intuition ... He was always at the leading edge, always surrounded by people ... The influence of Yakov Borisovich on students and those around him was immense. In them, fruitful creativity was often discovered, which would not have been realized without this influence, or would only have been realized partially. I feel just how much I owe him. In the extremely intense circumstances of those years — our friendship was simple, friendly, and extremely benevolent. In the area of fundamental physics, much of my research arose out of my association with him, under the influence of his work and ideas. In science, Yakov Borisovich was a person of enormous avidity (in the best sense of this word), and at the same time, he was absolutely honest, self-critical, prepared to acknowledge his mistakes, and also the correctness or authorship of others. He expressed an almost childish glee when he was able to accomplish something important.'[45]

Having arrived at the 'object' in A.D. Sakharov's group in August 1960, I immediately felt the hypnotic, bewitching charm of Zeldovich. He was such a vivid individual that it seemed to me, having read *Atoms at Home* by Laura Fermi, that if Yakov Borisovich could be compared with any of the famous participants of the American atomic project, then it was with the legendary Italian Enrico Fermi. And I felt a 'quiet joy' when I learned much later that L.D. Landau had remarked that he had not known any other physicist, apart from Fermi, with such a rich store of new ideas as Zeldovich.

The talent of Yakov Borisovich first appeared unusually early. In 1938, when he still had only a Candidate of Science degree, the Science Council of the Institute of Chemical Physics unanimously nominated him for membership of the Academy, although this bid was unsuccessful. 'Having succeeded at the age of 24 to already do so much for science, and continuing to work and grow more intensively, he will undoubtedly further enrich science with

[44] From the report of I.E. Tamm on the scientific activity of A.D. Sakharov, *Priroda*, 1990, no. 8, p. 12.

[45] Tsukerman, V.A. and Azarkh, Z.M., 'Lyudi i vzryvy' ('People and Explosions'), *Zvezda*, 1990, no. 11, pp. 118. [See also the partial reproduction of Sakharov's presentation at Zeldovich's memorial service on p. 98. [Translator]

Ya.B. Zeldovich, A.D. Sakharov, and D.A. Frank-Kamenetskii during their years at the 'object' (1950s).

even more valuable results. However, even now, independent of his age, the quality and quantity of the work accomplished by him unquestionably makes him worthy of the title of Corresponding Member of the USSR Academy of Sciences.'[46]

The talent and quick wit of Zeldovich delighted Semenov and Kurchatov. At a time when electronic computers did not yet exist, one of the more vivid and gifted colleagues of Yakov Borisovich at the 'object,' N.A. Dmitriev, compared him with ten working calculators. At that time, this was the highest praise imaginable!

Everyone understood that Zeldovich was an eminent personality. The ease with which he periodically switched from one field of physics to a completely different one, very quickly getting up to speed and almost immediately becoming an acknowledged leader, was striking. And he did not desert his earlier interests, as often happens with other scientists. He even joked about it, describing his research in the field of combustion and detonation, to which he returned again and again, as 'evergreen.'

In June 1980, in the conference hall of the Institute of Chemical Physics, a day was being celebrated in memory of D.A. Frank-Kamenetskii — a man of encyclopedic knowledge and unusually broad scientific interests. I had been acquainted with Yakov Borisovich for some time already, and had seen

[46] 'Fiziki o sebe' ('Physicists about Themselves'), Leningrad, Nauka, 1990, p. 391.

him in various situations. However, I was very fortunate that this event gave me a chance to observe this academician — if I can put it that way — as never before, in all his polyphonic brilliance. During various specialists' discussions of the contributions of David Albertovich to one field or another of physics, the unifying centre and distinctive 'translator' was invariably Yakov Borisovich. Without the least effort, he could at any time turn from the depths of atomic nuclei and astrophysics to refined questions on the theory of combustion or chemical physics, surprising us with his dazzling displays of details, old and new references, and names. When speaking on that same day about his friend, he unintentionally gave a characterisation of himself as well: 'David Albertovich was very light on his feet. I remember once, when we were still in Leningrad, in our youth, we went walking outside the city ... We came across a station ... A train went by, and we got on it and ended up in Novgorod, looked at some ancient churches ... Now, looking back, I can see that in science, I have more than once walked in the footsteps of David Albertovich. I started working in combustion later than he did. Many years later, when he turned to the stars, some time later I also began to work in astrophysics ...'

The leaders of our country's atomic department valued Zeldovich highly. When he was leaving defence work at the end of 1963, the question of whether or not to let him go was painfully decided by the Central Committee.[47] His departure was viewed as a great loss. This is echoed in the opinion of Khariton, who felt 'deep certainty that if Sakharov and Zeldovich had continued their activities in the field of defence themes, they would have struck upon something important.'[48]

Quite recently, when talking with E.P. Slavskii, who has headed the Ministry of Medium Machinery for three decades, I asked his thoughts about Yakov Borisovich. Efim Pavlovich was transformed and responded with delight: 'Zeldovich is a supertalent! Now that's ta-a-alent! What a smart guy, what a great man!'

In the years of collaborative work, there were various situations which made it possible to 'sense people deeply, including supervisors, under unusual or critical circumstances'. For example, at the time of the first Soviet atom bomb test, when the smallest mistake could end tragically for its creators, there were people who were predicting failure, who didn't hide their gloomy prognoses from the high-level administration. One of the supervisors of the atomic project met with a group of such people, and, having heard from them again that it wasn't going to work, suggested that they set out their thoughts in written form.

They did this, and only then they were told that the tests had already been carried out and had gone successfully.

---

[47] Of the Communist Party of the USSR, de facto the highest power in the country. [Translator]

[48] Khariton, Yu.B., 'Radi yadernogo pariteta' ('For the Sake of Nuclear Parity'), Dosier, *Literaturnaya Gazeta*, 1990, pp. 17–19.

Let me describe a significant incident. When, during a presentation given by a minister, L.V. Altshuler (in a sharp manner) didn't agree with one of his controversial statements, Zeldovich couldn't control himself: 'Lev Vladimirovich! You've acted tactlessly. Efim Pavlovich is a very fine fellow, and you're making such remarks ...' However, soon Yakov Borisovich telephoned his colleague and relieved the tension in his own style: 'Lev Vladimirovich, please understand, with age, I'm losing my sense of humour ...'

But Yakov Borisovich was never short of humour! Sparkling, sometimes with a thin veneer of causticity and cockiness, humour was inseparable from Zeldovich, it was a part of his nature. And God help you if you came under the lash of his tongue!

He might come up to the board and, with the words 'now we'll draw a top-secret circle,' would decisively sweep his hand, to the approving buzz of those present. If someone at a seminar got carried away with statistical computations, he loved to say, 'There are lies, ugly lies, and ... statistics!' He would reproach a speaker who kept going round and round without getting to his point: 'Here you are, like in the Indian love epic — flowers, birds, sun ... Come on and tell us about the sweetheart!'

Yakov Borisovich knew a lot about poetry and literature, and kept track of new works that came out. When quoting (as he did rather frequently), the literary passages he 'borrowed' were distinguished by their freshness, unexpectedness, and pertinence. One November or December evening in 1962, I saw him in the city reading hall, rapturously absorbing 'One Day in the Life of Ivan Denisovich'[49] in the new issue of *Novy Mir*.[50] He had a weakness for Ilf and Petrov,[51] and used their images and expressions very successfully and opportunely.

To the joy of the local wits, from his lips also flew 'salty' phrases: 'Doing secret work is the same as cocking a snook at someone with your hand in your pocket,' 'it's cheekiness without satisfaction,' and so forth. He even made mischievous jokes about his birthday when he turned 49: 'I'm now seven squared!' Having been born on March 8,[52] he never missed a chance to remark that this meant he was a gift to women. At a time, when March 8 was still a working day, he once breached all security barriers and prohibitions to bring a crate of champagne to work for the women in the neighbouring department.

When it came to mischievousness, stunts, and jokes, as in his sharp reactions, everything happened naturally and suddenly with Yakov Borisovich. When a former colleague who had a neighbouring dacha turned to him

[49] A story by A.I. Solzhenitsyn that brought the author into the limelight. [Translator]

[50] *New World*, a renowned 'heavy' literary journal. [Translator]

[51] The writers I. Ilf and A. Petrov were the authors of the very popular comic novels *The Twelve Chairs* and *A Golden Cow*. [Translator]

[52] International Women's Day, a holiday that became popular during the Soviet era. [Translator]

for advice, Zeldovich, having already left the 'object' and secret work, promptly countered, 'For God's sake, spare me that! I'd like so much to 'sterilise myself' of it all! What the hell good to me are these bombs — Chinese, American ... Damn them all!'

Zeldovich's cheerful disposition and wittiness made a favourable impression on Kurchatov. Once, in dramatic circumstances, he himself became an involuntary participant in a risky prank. Here is how the tale is told by the eyewitness V.I. Alferov:

'During the preparation for the test of our first atomic discharge, the final operations were carried out strictly in accordance with technical instructions: one man read what was to be done, and others (as a rule, two or three men) acted accordingly. It was usually Khariton who did the reading, sitting on a chair. The instructions were very detailed: the right hand should take this, left hand take that, approach in such-and-such a way, join in such-and-such a way, and so forth. It was time for the most sensitive operation — the assembly of the plutonium hemispheres; Kurchatov and Zeldovich came for this. The two hemispheres were assembled and sent for technical examination using the corresponding drafts. And then it was found that the edges of the plutonium hemispheres were bevelled, and this wasn't indicated on the drawings! It was clear that the part did not correspond to the drawing. And under such pressure and responsibility ... Igor Vasil'evich, though he understood that the bevelling didn't pose any danger to the operation of the item, spoke up: 'Why was this bevelling done, when it isn't indicated in the drawings?' Explanations were found. And at that moment, as bad luck would have it, Beriya[53] arrived, surrounded by his retinue. In order to defuse the situation, Kurchatov turned to Zeldovich: 'Yakov Borisovich, calculate the effect of this bevelling. How long will you need?'

'About fifteen minutes.'

'Good. Go and do the calculation.'

Not even ten minutes had passed when Zeldovich returned.

'Well, have you done the calculation?' asked Kurchatov.

'Yes, everything will be okay.'

'Show me!'

I happened to be standing behind Zeldovich. He opened his folder, and inside was a blank sheet of paper. Igor Vasil'evich, instantly picking up on the game, clapped Zeldovich on the shoulder: 'Right! Great work, Yasha!' So that decided the matter, and Beriya, fortunately standing aside from the two, didn't understand what had happened.'

There is another version of the last part of this story, described by I.N. Golovin based on the words of V.A. Davidenko, who also participated in the assembly of our first atomic bomb for testing:

'Kurchatov calls Zeldovich, and names a committee to answer the question of whether the bevelling will disrupt the compression. The

[53] Head of the secret service (MGB/KGB) under Stalin. At the time, he was the government official overseeing the atomic weapons programme. [Translator]

members of the committee work in different rooms, so that their arguments will be independent. Their conclusion is that the bevelling isn't a problem, the symmetry of the compression will be preserved, and they write this into the log for the assembly of the "item." The assembly continues.'[54]

Despite the apparent differences between them, it seems to me that both versions describe real events in different proportions and possibly at different times, and therefore cannot exclude each other. Whatever actually did happen, the resourcefulness and quick-wittedness of Zeldovich could not have been absent from such a dramatic situation.

His instantaneous reactions to words and actions during discussions with colleagues called forth surprise and delight. And also a certain desire to ... pay him back in kind. When this occasionally happened, the case became legendary. For example, the story was passed from mouth to mouth of how, at a seminar at the Keldysh Institute, in response to a question from Zeldovich, the speaker — a well-known mathematician — brushed him aside: 'Well, Yakov Borisovich, explaining this to you would be like explaining to General So-and-so how to ...'

Jokes were made about his constant travels between Moscow and the 'object,' where he supervised groups of physicists:

> Uncle Yakov has a cart of great fame —
> It can go backward and forward the same!

And since the pressure in one group was reduced somewhat with the departure of Yakov Borisovich, a poet remarked:

> Let the other half have a good scratch,
> This one will rest and relax for a patch ...

Yakov Borisovich valued beauty in life. But he explained that, for him, no less important was the sense of beauty that arises when it is possible to understand something in nature and its laws. Frequently, however, colleagues played on only the first part of this statement. One of them even gave Yakov Borisovich an epigram:

> ... To conquer ladies' hearts,
> YaB isn't sparing with his arts:
> To acquire a style genteel,
> YaB bought an automobile.
> So ladies, never you mind —
> If he's passing, you won't be left behind!

This same theme was present in the 'Departure aria of the theoretic,' from one of the skits done at the 'object':

> We are going, going, going,
> Called by Khariton.
> And each a theoretician —

[54] Golovin, I.N., *Kul'minatsiya* (*The Culmination*), Prep. No. 4932/3, Moscow, Inst. atomnoi energii im. I.V. Kurchatova, p. 15.

Almost a Newton.
We are rushing to our work
Our cares don't worry us a hoot.
And our left hands are kept busy
Calculating square roots.
What beauty! O what beauty!
To Gandelman to do our duty (one voice)
Zhen'ka Zababakha,[55] (two voices)
Also the bully Yashka,[56] (women)
Dmitriev Nikolai —
What a great bunch of guys!
What a great bunch of guys!

I've already had occasion to describe how, for us young people, Yakov Borisovich and Andrei Dmitrievich were a new and surprising type of academician.[57] With their democratic attitude, devoid of any hint of grandeur or overbearing, and their openness and accessibility, they destroyed the traditional picture of academicians as objects of universal reverence and worship. And this delighted us.

Their creative 'kitchen' was open to all. Both discussions between them and heated disputes at the board with any of us were the norm, as were opportunities to become acquainted, for example, at the suggestion of Yakov Borisovich, with his calculations or writings in his notebook.

It is difficult even to imagine Zeldovich or Sakharov giving one of the younger researchers an overbearing dressing down. The amicable atmosphere in the collective was their doing, a consequence of their attention to each of us. We knew that both of them had refused the usual financial bonuses granted for reaching key milestones in their research, and always redirected them, as we would say now, to charitable purposes.

In our small city, it was possible to run into Yakov Borisovich in the cinema or the library, when skiing or skating, at a cozy little 'general' canteen for the local elite, and even in the grocery store. Once, returning from a sports practice, I ran into him at a bus stop. He had spent the evening skating. We talked about sports. It turned out that he had experience of just about any kind of sporting activity you could think of. 'You can't have done boxing as well?' 'Boxing, too,' he affirmed, and then added with a chuckle, 'But only until the first knockout!'

He participated with us in noisy gatherings for New Year, responded lightly to the invitation and arrived at the dormitory to share in the joy of the occasion. He participated in our festive meal, and we drank some Blood Marys, which were popular at the time.

---

[55] The future academician Evgeny I. Zababakhin (1917–1984).

[56] *Yashka* is a nickname for Yakov (Zeldovich).

[57] Smirnov, Yu.N., 'Etot chelovek sdelal bol'she, chem my vse ...' ('This Man has Done More than All of Us...'), in memory of A.D. Sakharov.

He habitually disturbed scientific tranquillity, and supplied the latest news from extremely varied fields of physics and the physics groups of our country. Zeldovich's creative potential was unlimited. He was constantly drawn toward fundamental science, work which was held in high esteem at the 'object.' Yakov Borisovich tried to be at the leading edge, to create something new, and didn't allow us to relax and turn into commonplace 'techies.' He took pains that he himself should be in fine scientific shape, and was a genuine live source of information for the collective, for, in the words of Adamskii, Zeldovich didn't keep to himself either his knowledge or his ideas, and rapidly disseminated them. His skill was to generate scientific ideas, add flesh to the tones of physical conjecture, ignite and inspire those around him, especially young scientists, with the process of scientific creativity.

Even on our first encounter, when I ran into him in a hallway, and he immediately extended his hand and introduced himself — 'Zeldovich. And who would you be?' — I didn't escape without an invitation to stop into his office for a seminar that was about to begin. By the way, he was the speaker at that particular seminar, talking about something in the field of elementary particles. He caught himself making an error, and straight away corrected the mistake at the board. In conclusion, he remarked with satisfaction: 'You see how useful it can be to present seminars!' Even then, I noted the atmosphere of directness, benevolence, and merry, inexhaustible wittiness. This tone was given, of course, by Yakov Borisovich.

It could also happen like this. Zeldovich has only just begun his talk, when Kolya Dmitriev jumps up and, as if seriously concerned, suggests, 'No, wait. Let's call Andrei Dmitrievich! As an inspector.' Sakharov arrives and sits down. Zeldovich continues his report. Of course, there is no question of a real inspection. But the goal has been achieved: the correct mood for the seminar has been set.

I should also point out that such anecdotes have acquired in various tellings and remembrances a life of their own, and have sometimes merged and acquired different accents. For example, we shouldn't understand in absolute terms (outside the mischievous atmosphere of a seminar) the following remark by Yakov Borisovich: 'Andrei, of course, it is all clear to you! Therefore, you can go for a walk for a half hour or so, until we get to the main point. Then, we'll call you back.'

However, it is true that Yakov Borisovich always singled out Andrei Dmitrievich, to see how he would react. In general, they made a splendid scientific duet, and often acted in concert with one another. I learned a noteworthy story from V.D. Shafranov. Nearly 35 years ago, when still a young theoretician, Vitalii Dmitrievich did some research on the structure of shock waves in plasma, and sent an article to a scientific journal. It turned out that at almost the same time, Yakov Borisovich was carrying out quite similar work, and his manuscript was sent to the same journal. Each paper had its good points, but in one place, there was a fairly close correspondence between the two. Doubts probably arose among the editors as to how to proceed with Shafranov's manuscript, given the existence of the article by Zeldovich. Therefore, they decided to organize a meeting between the authors, giving them a chance to discuss the situation. The meeting took place. However, to Shafranov's surprise, Yakov Borisovich came not alone, but ... with Andrei Dmitrievich. V.D. Shafranov remembers: 'Yakov Borisovich, in the presence of Andrei Dmitrievich, gave me a sort of examination. Since to me all the issues were already clear and well thought-out, I was able to answer all his questions, including what seemed to me a tricky question about the role of the electric field. The atmosphere changed sharply, and he ended our conversation in a very amicable fashion. Andrei Dmitrievich, who seemed to me to be playing the role of an impartial arbiter, sat silently nearby, and didn't ask a single question. Our articles were both published, either in the same issue of the journal, or one just after the other.

The flow of scientific publications by Zeldovich in extremely varied fields of physics never ran dry. We even used to make joking remarks, with the feigned bravado of youth: 'Well, have you read about YaB's latest university homework problem in the journal?...' However, even then, we understood very well both the value of these 'problems,' their depth, and the daring of Yakov Borisovich's scientific imagination.

Perhaps one of his most fleeting ideas, as if stumbled on by chance, was the concept of colliding beams. With similar lightness, Andrei Dmitrievich expressed the idea of magnetic collimation, when, in the words of Yu.A. Romanov, this 'proposal was born literally before my eyes.'[58] And I can't resist the pleasure of reminding the reader of the elegant refinement with which the 'culprits' responsible for the new direction in accelerator physics later wrote about the idea of colliding beams and its realization.

A.M. Budker: 'The idea of colliding beams is not new, and is a trivial consequence of the theory of relativity. To my knowledge, the first to express this idea was Academician Zeldovich, though with a rather pessimistic tone. This pessimism is quite understandable. In this case, the target was a second beam whose density is 17 orders of magnitude lower than the density of condensed media — the target of an ordinary accelerator.'[59]

Ya.B. Zeldovich: 'To characterize the punctiliousness of A.M. Budker, I will remind you of one occasion: in a presentation to the General Assembly of the USSR Academy of Sciences, he made reference to my comment that colliding beams are energetically very favourable, however, he also remarked, quite validly, that I had considered colliding beams to be unattainable in practice due to difficulties with their focusing. It is very characteristic of the scientific daring of Andrei Mikhailovich that he took in the positive part of my remarks. At the same time, my simple (and also naive) pessimistic

[58] Golovin, I.N., *Kul'minatsiya* (*The Culmination*), Prep. No. 4932/3, Moscow, Inst. atomnoi energii im. Kurchatova, p. 15.

[59] Budker, G.I., *Uskoriteli so vstrechnymi puchkami chastits* (*Accelerators with Oncoming Particle Beams*), *Uspekhi Fizicheskikh Nauk*, 1966, vol. 89, iss. 4, p. 534.

estimates didn't frighten him: he found a way of surmounting the difficulties.'[60]

If I had to give a brief characterisation of Yakov Borisovich, I would say that he had a reckless, youthful keenness for science and inconceivable energy. In science, Zeldovich was attracted by (to use his own words) not so much the cascade of previous discoveries, but most of all the manifest, gaping incompleteness of theory. In creative associations with him, there invariably arose an indescribably excellent impression of having had contact with the leading edge of the investigations, behind which dominate not so much answers, as an avalanche of questions. He possessed an unerring and mysterious talent — to feel the 'new buds' of science.

Even Zeldovich's arrival at work was an event. He drove dashingly up to the building in his white Volga, made an energetic turn, and, putting his car into reverse, parked his car against the wall. Just as energetically, he bounded up to our floor, where, once in the corridor, he set off for his office, giving instructions to somebody, inviting someone else to see him. And all this invariably with some merry joke. He might appear in a well-cut suit, even sometimes with one of his gold stars (but never with two or all three!) on the lapel of the jacket, which, once he was in his office, would end up hanging on the back of his chair.

On a hot day, Yakov Borisovich might get several colleagues together and set out in his car for a pond to take a swim during work time. After an hour or so, Yakov Borisovich would command: 'Now for work!'

His mischievous directness was also striking. On my way to lunch, I drove up to the 'general' canteen on my motorcycle just as Yakov Borisovich was coming out. Once he saw me on my motorcycle, he shouted excitedly, 'Yura! Let me have a ride!' The motorcycle began to thrum, and Yakov Borisovich, not terribly sure of himself (after all, many years had gone by since he had switched from driving a motorcycle to sitting behind the wheel of a car), but with gradually increasing speed, disappeared around a bend. As more and more time went by and he didn't return, I became increasingly anxious. After 15 or 20 minutes, he appeared, smiling broadly, happy, already self-assuredly mounted on the bike. Pointing to his Volga, he cried out, 'Let's swap!'

And here is a story told by another participant in an extraordinary episode, V.S. Komel'kov: 'Zeldovich's directness attracted me. He could talk completely freely on any topic, and our time together passed rather merrily. However, once he unexpectedly suggested that we ... wrestle. I took in his short height, our difference in weight, and felt bewildered — why did he want to do this? But nonetheless, I agreed! We 'messed about' for some time, and then I pinned him down in the dust. Yakov Borisovich took this all in stride, quietly got up, shook himself, and we went on, continuing our conversation as though nothing had happened.'

---

[60] Academician G.I. Budker, 'Ocherki i vospominaniya' ('Essays and Remembrances'), Novosibirsk, Nauka, p. 85.

G.L. Shnirman remembers the commotion that broke out on the test site when it was suddenly discovered that the Corresponding Member of the USSR Academy of Sciences Ya.B. Zeldovich entered the local guarded hotel not through the entryway, but rather (for which were not at all clear reasons) by crawling through the window of his room on the ground floor. 'He romped like a person who had not yet reached adulthood.' And he very much loved to romp. On another occasion, several of the colleagues, including Yakov Borisovich, unsuccessfully tried to convince D.A. Frank-Kamenetskii to go for a walk. He was obstinate and insisted he didn't want to go. Then, led by Yakov Borisovich, four of them took the corners of the blanket on which David Albertovich was resting on his bed, lifted him up, and went out of the hotel building with their unusual burden, past a group of bewildered generals from the local authorities.

Yakov Borisovich never missed a chance to fool around, and even the two bodyguards once assigned to him could not avoid his pranks, right up to his sudden 'mysterious' disappearance from the premises. One of these guards was plump, middle-aged, and didn't know how to swim (apparently, he was serving out his term toward his pension). During his trips to the test site, Yakov Borisovich loved to get limbered up and dive into the waters of the Irtysh. You have to see the funny side of the situation, with this poor bodyguard, waiting on the shore 'in the course of duty,' helplessly fretting, while his ward, a tireless practical joker, played around in the river, his head nowhere to be seen, hidden under the water.

However, Yakov Borisovich's passion for science dominated all others. He was never parted from his little logarithmic slide rule, in the use of which he was a virtuoso, nor from the thick notebook in which he carried out his calculations and computations. He never passed by a chance to discuss scientific questions of interest to him. Once, in the winter of 1961 or 1962, we — young theoretical physicists who had recently graduated from Leningrad University and were working in the groups of Zeldovich and Sakharov — were pleasantly surprised when we suddenly saw Yakov Borisovich on our floor in the company of a professor from that university, O.A. Ladyzhenskaya. At that time, Yakov Borisovich's office was being reorganized, so we then discussed various scientific issues in the office of Andrei Dmitrievich, who was absent. Olga Aleksandrovna remembered her unusual trip to get there: 'For me, the trip to the 'object' was very exotic. Without any clearance documents, without reading or signing any papers. They brought me there in a blindfold, and took me away after a week, also in a blindfold ... There were no other contacts — it was pure chance. I didn't even suspect that our discussion took place in the office of Andrei Dmitrievich, though I heard his name. The trip was preceded by the following events.

After I got my results on the Navier–Stokes equations (the global unique solubility of two-dimensional hydrodynamics problems), it was not mathematicians, but physicists who invited me to present my work at the Division of Physical and Mathematical Sciences of the Academy, which was unified at that time. The hall was full, and there were celebrities present.

Somehow, I was touchingly greeted by Leontovich, Landau; Yulii Borisovich Khariton also came up to me ... He introduced himself and asked if I would be willing to come and make a presentation for his colleagues — my former mathematics students. I agreed, gave him my address and telephone number, and returned to Leningrad.

Events unfolded further within the next month. They telephoned me, and I arrived in Moscow. There, I was met by a man I had never seen before, who then escorted me everywhere. We went to the 'object' by train at night. I understood that outsiders were not allowed, and that it was not without reason that this man escorted me. (We also returned by train.)

At the 'object,' I read the mathematicians a sort of series of lectures. Yakov Borisovich invited me to meet the theoretical physicists. He was a very vivid personality, mercurial. On the next day, he even organized a ski trip together with the theoreticians. By the way, afterwards, I myself walked through that unusual forest and, of course, came across a ruined church. I didn't consider it proper to ask Yulii Borisovich about the 'object' or the church. Since they don't say anything, that means they can't. And I didn't want to put anyone in an uncomfortable position. After I returned to Leningrad, Vladimir Ivanovich Smirnov and I figured out that I must have been at the Sarov monastery.'

Indeed, Olga Aleksandrov had been at the top-secret 'estate' of Khariton, Sakharov, and Zeldovich on the territory of the once famous Sarov monastery, about which one can read in the *Encyclopaedic Dictionary* of Brokgauz and Efron: '...Founded in the seventeenth century at the site of the Tatar town Sarakly. The first church was constructed in 1706. The monastery acquired a reputation for strictness; especially revered is the memory of the hermits Mark and Serafim.' In accordance with the bizarre circumstances, the Sarov monastery became the centre for the creation of Soviet nuclear weapons.

Professional discussions with Yakov Borisovich were distinguished by their dynamism and expressiveness. After acquainting myself with his article 'Initial Stages in the Evolution of the Universe,' which had just been published in the journal *Atomnaya Energiya*,[61] and finding several inaccuracies, I stopped by his office. Without delay, we went to the board, and I could barely manage to follow the dancing chalk in Yakov Borisovich's hand. A remark, a calculation, another remark, an objection, a counter-objection ... And suddenly, with a contented smile: 'That's it — you've nailed the Academician!' Then, he went to his desk, opened his general notebook and proposed that I read the manuscript for his latest work — if it was interesting for me. It was interesting ... Having left for his traditional two weeks in Moscow without seeing me, he left me a note and a task. He supported me patiently and with interest until the task was done.

When the work was completed and I proposed that we publish it jointly,

---

[61] Zeldovich, Ya.B., *Nachal'nye stadii evolyutsii Vselennoi, Atomnaya Energiya*, 1963, no. 1, pp. 92–99.

Yakov Borisovich categorically refused: 'No! It's your article!' Once he had become acquainted with my manuscript, he made well-founded criticism and gave me (in written form!) several pieces of advice about how to improve the text. This was an unforgettable lesson from the master, which made a great impression on me.'[62]

Zeldovich provided his young colleagues with an important example. It was precisely then, at the end of 1961 and beginning of 1962, that he made a sharp turn away from the physics of elementary particles toward relativity theory and cosmology. In the process, he decisively reoriented the themes discussed in his seminar, bringing the young researchers with him. Our initial positions became equal: we began together to get acquainted with the 'Field Theory' of Landau and Lifshitz. From one seminar to another, we studied and made chapter-by-chapter reports about this text, which Zeldovich liked very much. (Moreover, he once remarked that, having thumbed through the monograph of V.A. Fok's *Theory of Space, Time, and Gravitation*, he thought it was overloaded with mathematics, and saw it as an example of 'how not to write a book'.) Our discussions gave rise to more questions, which were transformed into physics problems, and inconspicuously led to publications by participants of the seminar. Yakov Borisovich actively got up to speed, rapidly and surely becoming an authority in this field of physics, which was still new for him.

It wasn't possible, however, to avoid some misfires. I have already mentioned Zeldovich's article in *Atomnaya Energiya*. In that article, he proposes a 'cold' model for the Universe that contrasted with the Big Bang in Gamow's model. He placed great hope in his hypothesis. It was not by chance that he published the article in a special issue of the journal dedicated to the memory of I.V. Kurchatov, and ended it with the words: 'In both practical and theoretical research, we will remember what Igor Vasil'evich taught us — to take up the basic, fundamental questions, to be honest, brave, and passionate in both work and life.'

Yakov Borisovich developed his cold model of the Universe on an exceptionally large scale. He popularized it among his colleagues and at all possible public lectures, which he always gave with enviable mastery; he had an inexhaustible supply of engaging problems, and some of his young colleagues — participants in his seminar — developed the model proposed by the master with enthusiasm.

Andrei Dmitrievich was also interested and caught up by it, and published at least two articles in the framework of Zeldovich's hypothesis. As he was absorbed by his work at the 'object,' Sakharov's name had been absent from the pages of open general physical journals during previous years. Turning to Zeldovich's hypothesis was for him a sort of 'awakening.' Talking about one of his works from that period, he remarked in his

[62] Smirnov, Yu.N., *Obrazovanie vodoroda i ^{4}He vo Vselennoi na dozvezdnoi stadii (v modeli Gamova) (Formation of Hydrogen and ^{4}He in the Universe Before the Stellar Stage (in the model of Gamow))*, *Astronomicheskii Zhurnal*, 1964, vol. 12, no. 6, pp. 1084–89.

'Memoirs': 'I again came to believe in my strength as a theoretical physicist. There was a certain psychology of "escape," which made my subsequent research of those years possible.'[63]

This continued for two or three years. But, with the Nobel Prize-winning discovery in 1965 of cosmic microwave background radiation by A. Penzias and P. Wilson, the cold model in its original form collapsed. However, the fact remains that Yakov Borisovich, in his work on his cold model and simultaneously in comparisons with the Big Bang model, was at the very peak of gripping fundamental research, and one step away from triumph. That time, as happens in science, luck was not on his side, and the decisive successes were left to others.

Yakov Borisovich suffered over this. L.V. Altshuler remembers: 'Once he arrived at our home very dispirited. He said that, from general physical considerations, there had been a good basis to push forward his cold Universe model. "But it was my responsibility to say," went on Yakov Borisovich, "that, if the universe was hot, there should be relict radiation! In this way, I could have indicated to the experimentalists what and how they should observe."'

Serving truth in science, he accepted defeat with dignity. (One amusing thing was that he later appreciated Fok, noting his 'deep and brilliant mathematical technique.') And once he had done this, he passionately proclaimed the epochal importance of the discovery of background radiation for our understanding of the development of the universe. I remember how he gave splendid reports at the Institute of Physical Problems, in a conference hall overflowing with attendees, with B.P. Konstantinov the seminar chairman. At an international conference on shock waves in plasma in the 'Akademgorodok'[64] at Novosibirsk, he gave presentations to mathematicians, attracting their attention to the problems of cosmology. More than one academic and non-academic auditorium listened to this lively man of modest height giving lectures about the universe. A picture was engraved in the memory of a friend of mine who was present at one of these lectures: 'Zeldovich appeared in the auditorium, in simple clothing, a plaid shirt with an unbuttoned collar. Without any board, he began to speak very simply and intelligibly. His emotion caught up the audience, and the contact with the listeners was complete. Before us was a man all caught up, lively, like a little bouncing ball ...'

And again the growing flood of publications!

Zeldovich avoided political conversations, although sometimes he too was subject to a certain feeling of inevitability engendered by our life at the 'object.' However, as observed by V.B. Adamskii, he didn't remain indifferent in the presence of rivalry between scientific schools. And if a heated

[63] Sakharov, A.D., *Vospominaniya* ('Memoirs'), cited from *Nauka i Zhizn*, 1991, no. 4, p. 14.

[64] Academic city. [Translator]

scientific argument had already arisen, he sometimes merrily exclaimed in the heat of the discussion: 'I'll put my head on the block, but I won't bet ten roubles!'

He was one of only a few physicists who dared to argue with Landau himself, and was very proud when he was able to perform some work that corrected or developed the results of the teacher.

His great talent doesn't fit in the Procrustean bed of ordinary everyday concepts. Its manifestations gave rise to a variety of reactions in everyday life, as, by the way, does the surrounding world. However, in the confusion and the chaos besetting us, we are saved by a clear and wise thought, contained in the poetical lines by Alexander Blok that were so loved by Yakov Borisovich:

> Wipe away the chance features —
> And you'll see: the world is wonderful.

As a physicist, Yakov Borisovich possessed great talent. And, like any very gifted person, his style was not for everyone and was not always simple, 'correct,' and comfortable. However, remembering this eminent personality, we come to understand that time, having swallowed details and random traits, preserves his true and majestic scale.

... Before my eyes is a warm, sunny autumn day with a bright palette of multi-coloured leaves frozen in the windlessness of the trees. A meeting between Khariton and us theorists has just ended. Zeldovich has gone to see him out. From the window of my office, on the top floor of the building, I see golden autumn, the empty asphalt entryway leading to the building, and Yulii Borisovich and Yakov Borisovich slowly appearing there, pacing shoulder to shoulder.

They pass along the building, turn around and come back, caught up in their conversation. The soft, broad gesticulation of Yakov Borisovich — Zeldovich is speaking. The complete peace of the pacing conversationalists, and Zeldovich's hands folded behind his back — Khariton is talking. They walk back and forth, while Khariton's attentive guard waits for him to one side, in a car. And one can be sure that this conversation would be followed by action and, without fail, results.

And my glance slides from the conversationalists along the shrivelled grass of the lawn, where here and there young cherry groves are gathering their strength. It passes through the guarded strip of harrowed earth and the protective fence of our area, with sentries in towers at the corners, slides along the crowns of the trees. And far in the distance, kilometres from the vigilantly and attentively guarded external perimeter of the 'closed' secret zone enclosing the city and the 'objects,' I can make out the barely visible smoky curve of the Big Land — our immense Motherland. In that moment, it was impossible not to think that her safety to a large degree rested on the shoulders of three giants, the splendid scientists Yakov Borisovich Zeldovich, Andrei Dmitrievich Sakharov, and Yulii Borisovich Khariton.

A born leader

L.P. Feoktistov

At the very beginning of 1951, after my graduation from Moscow State University, I was sent to the highly secret place. The degree of secrecy can be judged from the fact that during the processing of my transfer in Moscow (in a gloomy office on Tsvetnoi Boulevard), when I casually asked how long it took to fly to the place, I promptly received a half-hour dressing down — it turned out that I shouldn't have said the word 'aeroplane' aloud.

Upon my arrival, after a short examination — for which I was very well prepared, since I had precise information about its content from classmates who had arrived a little earlier — I ended up in the theoretical group of D.A. Frank-Kamenentskii. This group was, in turn, part of the department headed by the Corresponding Member of the USSR Academy of Science Yakov Borisovich Zeldovich. At that same time, I also became acquainted with other famous people: I.E. Tamm, N.N. Bogolyubov, and A.D. Sakharov, who headed other theoretical sections.

The level of secrecy at the 'object' was extremely high: in principle, the different groups, each working on their own problem, were not supposed to know what their neighbours were doing, though, of course, everyone knew about everything. Young people like me ended up in a closed space, behind barbed wire, without the right to leave the object, even during vacations. This restriction didn't apply to the big bosses.

I began to work under the touching guardianship of D.A. Frank-Kamenetskii, who taught me all sorts of difficult and specialized science not from books. Suddenly a rumour began to spread — 'the Corresponding Member is coming.' David Albertovich (or DA — that was the way we named all the bosses, according to the initial letters of their first name and patronymic — YaB, YuB, AD, and so forth) took me to be introduced, and I believe was more nervous than I was. I saw a mobile man of modest height with an intelligent and mocking expression. YaB began a free and easy conversation, which developed into a professional discussion, and, as if in passing, asked me to write the equation of hydrodynamics on the board. 'It has begun,' sounded in my head, but out loud only the pitiful utterance, 'We didn't cover that.' But it all turned out to be not so frightful, since one or two hints made it clear that all of mechanics is built on the laws of conservation of mass, momentum, and energy.

Nowadays, it seems that many people love to start things, but hardly anyone is willing or able to finish things. They get stretched out indefinitely, tediously, and uselessly. It was not at all like that at that dynamic time, and, as always, it was leaders who defined the situation.

When I was beginning my research, Zeldovich's department was working on questions connected with the thermonuclear burning (detonation) of deuterium in a tube. This was far from a simple problem, and, in fact, was nearly unsolvable. The reason for this is that the radius of the tube had both lower and upper limits. This upper limit was especially surprising, since, as a rule, the larger the size, the lower the losses, and the better the detonation. In the situation being studied, however, a phenomenon developed that later acquired the name 'Comptonization': soft Bremsstrahlung photons scattering on hot plasma electrons did not lose energy, as in the known Compton effect, but instead gained it, and tended toward a Planck (Wien) equilibrium. An additional 'outflow' of energy from the matter to radiation arises. In order to decrease this energy flow, it is necessary to make a medium that is as transparent as possible, so that the photons will escape from the medium without acquiring extra energy.

We spent four years on work in this direction, which was very fine and engaging from the physical point of view; however, suddenly, at some point, having decided that the work was not proving to be competitive, the administration of our department reoriented the research of our entire collective literally within a few days. Because of this type of decision, a year later we had a device that has not gone out of date even now. I believe that much research on thermonuclear detonation carried out under Zeldovich that retains large theoretical and practical interest should be declassified and published.

As it turned out, a tritium–deuterium mixture burns so much faster than a purely deuterium medium that the radiation does not have time to come to equilibrium with the medium, and the temperature of the material rises sharply — a fact that I learned from Zeldovich, and which had an extremely strong influence on my subsequent research. Later, the Doppler broadening of the energy of the 14-MeV neutrons was used to determine the temperature, which exceeded a billion degrees — a feat worthy of inclusion in the Guinness book of records.

Zeldovich's simple remark that neutrons with energies up to 28 MeV could occasionally be observed in the neutron spectra of DT reactions (due to collisions between neutrons) initiated a large series of explosion experiments to create far transuranic elements.

In spite of how busy he was, YaB gave a lot of time to our education. Like any big boss, he came in a little later than us in the morning, with his thick notebook, and began to talk to us. It would turn out that he had already managed that morning to write some article or perform some calculation. At that time, for example, I learned the elements of quantum electrodynamics. I remember that we enthusiastically calculated the formulae for the Compton effect using Feynman diagrams. And what is more surprising, our answers were sometimes correct.

Whenever YaB arrived from the capital, he immediately got us all together and reported the latest scientific news. At one of these arrival gatherings one evening, he was telling us about something as usual, but we were tired, or perhaps it was too complicated for us to absorb quickly. He apparently sensed this, because the following morning, he tried to find out who had understood what. I answered his questions more or less clearly, and even was deemed worthy of some praise. It was only years later in a chance conversation with YaB that I confessed that I'd outwitted him: that morning, I'd read the relevant section in Landau's textbook, having sensed that a dressing-down was coming.

The chief was a born leader and he knew this, but he never emphasized it; in fact, there was no need, it was quite obvious. And all the same ... Yakov Borisovich loved and highly respected David Albertovich Frank-Kamenetskii (Dodik). However, once I was present during an argument between them: DA announced that he could read a 300-page book in two hours. It became clear that YaB didn't believe that he himself was unable to do this, and his authority was shaken. YaB began to get agitated, and they finally argued. DA shut himself up in his room, and two hours later, the verification began. YaB opened the book at an arbitrary place and read a line, and DA continued it nearly word-for-word. The extremely kind DA's memory was extraordinary!

Another episode has also been imprinted in my memory, probably because it is the only one in which I was able to correct an eminent scientist. He growled with dissatisfaction: 'One–nil to you,' and continued our conversation until I had made some mistake, which, you understand, was not terribly difficult to do. 'One–one,' said my teacher contentedly.

Probably each of us has experienced a feeling of euphoria, a sense that we have reached an apex, that we know everything. Usually, this occurs after passing some exam, graduating from school, or from a university. Of course, soon afterwards, a sad sobering-up sets in. In that distant time, we occasionally went out to the Semipalatinsk test site. There also, YaB contrived to write 'formulae.' We supposed that our work would begin after the 'phenomenon' (we never used the words 'explosion,' 'bomb,' 'plutonium,' 'tritium,' saying instead 'phenomenon,' 'item,' and so forth), so we openly loafed around. At some point, the chief couldn't stand it any more and initiated a conversation: 'During the explosion, a powerful electromagnetic signal arises ...' and then assigned G.M. Gandelman and me the job of discovering its nature. This task proved to be unusually captivating. After several days, we had no doubt that we had managed to find the right approach. The point is that instantaneous fission gamma-rays escaping from the explosion point push electrons in the direction of their flight and polarize the air. We rapidly made a 'report' (calculation and description) and, satisfied, came to YaB. It was clear to us that he was pleased, which made his conclusion sound all the more strange: 'In your formulation of the problem, the amplitude of the radio signal is strictly equal to zero.' 'Why?' we piped up together. 'Because the electric vector is directed radially, and where is the magnetic vector directed — to the right? To the left? Why should either be preferable? A spherically symmetrical system does not radiate.' It was not difficult for us to correct our result taking into account the asymmetry of the problem, but our bitter resentment at our own illiteracy remained for a long time.

In our lives, which were saturated and anxious due to living under the pressure of secrecy, various pranks were usual. One time, I was riding a bicycle while, to one side, YaB was hurrying somewhere. He saw me, stopped, and began to talk. Further, it was all like in a fable. 'How well you ride your bicycle! You know, I had a poor childhood, I never had a bicycle ...' 'What? You don't know how to ride? But it's so simple.' I had a real chance to demonstrate my superiority, at least at something. YaB clumsily got himself into the seat, squeaking, 'Oh, hold me!' And then for several seconds more I watched with consternation the rapidly fleeing figure, who had decided to solve his transportation problem at my expense.

In the time of Beriya, there was a mysterious military personality at the object — a representative of the Sovnarkom.[65] Nobody knew clearly what he did. We had to deal with him once a year, if we wanted to apply for a pass to go outside the zone. Here, each relied on his own ingenuity; for example, one friend of mine went out each year to sell a goat in response to inquiries from a village soviet; I — to get married. Several people who had graduated from military academies worked among us. We took the military uniform of one of them, dressed one of our friends in it (the one with the 'goat'), disguised him as this 'representative,' and began to let through to him one by one our colleagues for 'interrogation.' God, how much news — good and bad — we learned about our bosses and the Soviet system.

This 'execution' came to an end when I ran into YaB in the hallway. 'What's all the noise?' he asked (this was the 'victims' exchanging their impressions). I enthusiastically explained. YaB first laughed, then became sombre and said: 'For such jokes some organs[66] will cut out certain of your organs, and in that case, I can't help a bit.'

Not long before his death, I met Yakov Borisovich by chance on a walk in the Lenin Hills. We talked a bit. He was full of impressions from a visit to Greece, and enthusiastically told me about his astrophysical successes. This made the words he uttered upon his departure all the more strange: 'Can you guess what was the brightest time for me? Yes, yes, that one ... I still have a dream: to write one more book about detonation.'

I count the years backward, and also bless that time. Not only because it includes the best period of my youth, but because fate led me to very wise and talented first teachers.

A man of universal interests

Andrei Sakharov

On 2 December, 1987, Academician Yakov Borisovich Zeldovich died suddenly from a heart attack. He was an outstanding physicist, who made

---

[65] Council of People's Commissars, the equivalent of a governmental cabinet. [Translator]

[66] A common Soviet term referring to the secret service (KGB). [Translator]

enormous contributions to many branches of science and technology. I am not a specialist in every sphere of his activity (indeed, he was probably unique in the breadth of his interests), and hence I will touch on certain sides of his work only in broad outline. From 1948 to 1968, however, the two of us worked closely together, and I know a great deal of about that period of his life.

In his memoirs, Zeldovich states that in 1931 (when he was a 17-year-old laboratory technician in the Institute of Processing of Useful Ores) he went on an excursion to the laboratory of chemical physics of the Leningrad Physical Technical Institute and got into discussion with some of the staff on the forms of crystallization of nitroglycerine. He was invited to work in the laboratory in his free time, and soon afterwards he transferred there officially. So began his path in science, which was to last for 56 years. In 1936 he defended his dissertation for the degree of Candidate of Sciences; later he wrote of the 'happy times when permission to defend [a Candidate's dissertation] was granted to people who had no higher education.'

In the 12 to 15 years following his move to the Physical Technical Institute, Zeldovich carried out outstanding work on the theory of combustion and detonation, adsorption and catalysis, the fixation of nitrogen, chemical chain reactions, and (before, during and after the war) propulsion technology. His interest in chemical physics lasted his entire life, and his last work now seems to have been a return to this first love.

The discovery of uranium fission changed Zeldovich's scientific destiny, as it did that of many other scientists, years before — on a far wider scale — it changed the destinies of us all. His pioneering researches, in collaboration with Yu.B. Khariton, on the theory of explosive and controlled fission chain reactions, were simultaneously the last to be published in the open literature until the veil of secrecy was lifted from the subject. They had a great influence on everyone working in this field. From the very beginning of Soviet work on the atomic (and later the thermonuclear) problem, Zeldovich was at the very epicentre of events. His role there was completely exceptional.

In the middle of the 1950s, Zeldovich found himself a new area of activity — first of all, the theory of elementary particles (in which, soon after the end of the Second World War, there had been a breakthrough, and which has developed up to the present time), and then, in the 1960s, the no less dynamic and captivating field of astrophysics and cosmology.

In 1955, Zeldovich and S.S. Gershtein together put forward the hypothesis of the conservation of the weak charged vector current. This idea (which was discovered independently by Feynman and Gell-Mann) played an important part in the formulation of the theory of weak interactions and the unified theory of weak and electromagnetic interactions, and also in what became known as 'current algebra.' In another paper of the same period, Zeldovich predicted the existence and certain properties of the Z^0 boson.

For the last 25 years of his life, astrophysics and cosmology had a central place in the thinking of Zeldovich and his pupils. He was universally acknowledged as a world leader in the field — for the exceptional clarity and concreteness of his physical thinking; for his intellectual daring as a theoretical physicist, which was applied with equal facility to physical laws and theoretical methods, to the formation of Liesegang rings in test-tubes, to the grandiose processes of the explosion of a supernova with the formation of a neutron star or black hole, and to even more extreme processes of the cosmology of the early universe; and for his closeness to observations. I shall enumerate in no particular order a few of his contributions to astrophysics and cosmology.

In Zeldovich's work of 1967, there is a formulation of the problem of the cosmological constant. According to Zeldovich, it followed from the theory of elementary particles that this constant is small or equal to zero. At the present time, the cosmological constant is one of the central problems in attempts to construct a unified theory of all fields and interactions. In publications with Ya.A. Smorodinskii and S.S. Gershtein, Zeldovich considered the cosmological limits on the masses of the electron and muon neutrinos. These contributions are examples of the new directions in science that arose during the 1960s, that lie at the junction of cosmology, astrophysics, and the theory of elementary particles, and that to a considerable extent are associated with Zeldovich. Here, the entire universe plays the role of a gigantic laboratory. Zeldovich's works on the generation of particles in a gravitational field are of great importance, and include a joint paper with Pitaevskii which contains a remarkable discussion with S. Hawking. The effects of polarization of a vacuum make possible the generation of particles by a classical field. Soon after this discussion, Hawking himself published his famous theory of radiation evaporation of black holes. The effects of polarization of a vacuum (quadratic conformal anomaly), according to the theory of Zeldovich's colleague, A.A. Starobinskii, led to cosmological solutions without singularity. (I don't give this formulation my wholehearted support and would propose instead that the main causes of inflation are 'false vacuum' effects.)

Closer to classical astrophysics, but nonetheless important, are Zeldovich's publications on neutron stars and black holes, and accretion and radiation in solitary objects and binary stars. In his first paper on black holes (1964), he put forward the idea of observing black holes by the radiation from material moving in its gravitational field (simultaneously and independently, a similar proposal was published by Salpeter). Soon afterwards, there followed his work in collaboration with O.Kh. Gusseinov on radiation in binary systems, one component of which is a black hole. In a work of 1964 (in collaboration with M.A. Podurets), and in a number of subsequent papers, he considered the dynamics of neutron emission during the formation of black holes. As a result, black holes became accepted as really observable objects, and appropriate observational programmes were worked out and began to be implemented. These programmes, as is well known, have already yielded very interesting results.

Another important aspect of Zeldovich's interests (in which A.G. Doroshkevich, I.D. Novikov, R.A. Sunyaev, S.F. Shandarin, and others also took part) was the formation of galaxies and galaxy clusters. He established the nature of the spectrum on initial perturbations of density, which have observable consequences, and he predicted singularities of the large-scale structure of the universe — according to that theory, there are gigantic 'black regions' free from galaxies and filled with hot low-density gas of 'pre-galactic' composition. There are data suggesting that this is really the case, although the question must still be considered open.

Zeldovich and his colleagues analysed cosmic electromagnetic radiation and proposed experiments to be carried out in a range suitable for the discovery and investigation of relict radiation. These ideas were not generally known abroad and were not properly exploited by those who did know them. Immediately after the discovery of relict radiation, Zeldovich recognized it to be of enormous importance, not only as a confirmation of the hot model of the universe but also as a powerful means of investigating many other important questions in cosmology and astrophysics. In a number of publications (in collaboration with R.A. Sunyaev and others) he examined the effect of various cosmological factors on the anisotropy of relict radiation. As is well-known, this line of investigation has acquired great significance.

In his last years Zeldovich published works in which there was an especially profound reflection on the interconnection of the theory of elementary particles and cosmology, considering cosmic domains, and cosmic structures, the astronomical consequences of the rest mass of the neutrino and other postulates of the theory of elementary particles. He attempted to indicate the outlines of what he called 'full cosmological theory' (corresponding to the question of the character of the pre-classical, that is, the quantum-gravitational, stage of development of the universe, and the origin of the qualitative and quantitative characteristics of the classical stage, including the polytropy spectrum of 'initial perturbations').

All his life Zeldovich was at the leading edge of science, and he was always surrounded by people. His effect on his pupils was remarkable; he often discovered in them a capacity for scientific creativity which without him would not have been realized or could have been realized only in part and with great difficulty. An essential factor here was his scientific style and personality — his immense energy, his sensitivity to what was new, his intuition, his striving for theoretical simplicity and elegance, his scientific honesty, and his readiness to admit his own error or acknowledge the priority and correctness of another. In science, Zeldovich was a humble man (although the manner in which he sometimes took part in discussions, defending what he considered to be the scientifically irrefutable, could give a somewhat different impression). He was almost childishly delighted when he had managed to achieve some important piece of work, or had overcome a methodological difficulty by an elegant method, and felt failures and errors keenly.

To Zeldovich, it often seemed that he was not professional enough in this or that field, and he summoned up greater efforts to fill in these gaps. Here, too, his approach was creative. He often found new, more comprehensive ways of describing and dealing with a problem. So there came into existence numerous papers and articles of a pedagogic nature, and monographs and books (more than 20 of them) which always included much that was original. His books *Relativistic Astrophysics*, *Theory of Gravitation and the Evolution of Stars*, and *The Structure and Evolution of the Universe* (jointly with I.D. Novikov) acquired great renown. It is impossible to overestimate the importance of this side of his work, which helped a great many people to come to science by the most direct route.

Special note should be made of his book *Higher Mathematics for Beginners*. In one of his papers, Zeldovich wrote: 'The so-called "strict" proofs and definitions are far more complicated than the intuitive approach to derivatives and integrals. As a result, the mathematical ideas necessary for an understanding of physics reach school-pupils too late. It is like serving the salt and pepper, not for lunch, but later — for afternoon tea.' I wholeheartedly agree with him on this point.

As I have already said, for many years I worked closely with Zeldovich. This was a relationship of comradely goodwill, mutual readiness to help, and strenuous work towards a common aim. There was no negativism, hostility or signs of unhealthy competition (this, moreover, although the group of I.E. Tamm, to which I belonged, had 'fallen in' to an already well-established team from outside). In the 1950s and 1960s, our offices and homes were next to one another, and several times a day we would get together to consider a basic theme; we would also talk about general scientific problems. Often we discussed interesting physical and mathematical trifles (what I call 'amateur problems'). Sometimes we played games, as it were, competing in the speed and elegance of solutions (the one who solved it first ran to tell the other at any hour of the day or night).

Much of my own work on fundamental science had its beginning in these contacts. Here is one example. Once Zeldovich rang me up and said that he had been lecturing to one of the Moscow scientific seminars on his work on the cosmological constant (described above) and had met with puzzlement. I immediately appreciated the importance of the problem, and a few days later I rang him up with a further development, the idea of induced gravitation. Zeldovich received the idea enthusiastically, and, in his turn, wrote a paper which treated electrodynamics in an analogous way. I, too, understood the role played in the problem of the cosmological constant by the compensation of the boson and fermion contributions — unfortunately, neither of us managed to think the problem through as far as supersymmetry.

I should also like to put it on record that Zeldovich helped a great number of people in a purely personal way, even in matters of everyday life. Of course, one should not assume that in every case Zeldovich appeared in the best possible light. He was no angel.

My relations with Zeldovich were not always unclouded. In the 1970s and 1980s, especially in the Gor'kii phase of my life, hurt feelings and mutual coldness crept in. Zeldovich strongly disapproved of my social work, which irritated and even frightened him. He once said 'People like Hawking are devoted to science. Nothing can distract them.' I did not understand why he could not give me the help which, given our relationship, I considered myself justified in asking for. I know that all this tormented Zeldovich. It tormented me too, as described in my memoirs. Today, the events of those years seem like foam, carried away on the stream of life. Unfortunately, after my return from Gor'kii I met Zeldovich only once or twice, and then in company, and was hardly able to communicate with him in human terms. Yet another lesson is that it is not always possible to put things off to another day.

Now, when Yakov Borisovich Zeldovich has departed from us, we, his friends and colleagues in science, understand how much he himself did, and how much he gave to those who had the chance to share his life and work.

Thus he remains forever in my memory

T.A. Sakharova

My father Andrei Dmitrievich Sakharov had warm and kind relations with many of his acquaintances and colleagues, but valued only a few specially. First and foremost among these were Igor Evgen'evich Tamm and Yakov Borisovich Zeldovich.

In spite of the difference in their characters, temperaments, and some views on life, Papa and Yakov Borisovich were made similar by something higher — which was the main pivot point for their mutual affection — God-given talent and a reckless devotion to physics. For them, physics was the main content of and reason for life. They very much valued this talent in each other. It may be that they, in some sense, supplemented each other. In the writings of Yakov Borisovich, there are sometimes remarks such as 'the very deep idea of ADS!'

Papa didn't hide his delight with the unusual scientific productivity of Yakov Borisovich and the astounding organization of his work, and when I was a child he often told me that Yakov Borisovich got up at five every morning, exercised energetically, and then worked for many hours until evening, allowing himself to take only short breaks. But then, Yakov Borisovich dedicated the evening to rest. Papa's style of work was completely different. A problem on which he was working and that tormented him would not let him go for a minute, often sending him into a half-conscious state. The unexpected solution would arise at a completely inappropriate time — during a distracting conversation with Mama, with me, or with friends. Then, Papa would suddenly 'turn off' and not hear a single word of the conversation, which had become unimportant.

Certain bright memories of childhood have remained forever in my memory: Yakov Borisovich comes up to the gate of the area around our cottage at the 'object,' crying loudly: 'AD! Are you home?' This frequently happened at dinner time at our home. However, to Mama's dismay, Papa would promptly forget about dinner. His face beamed with joy and liveliness. And I remember as if it were yesterday how long they walked along our street from end to end, infinitely repeating their path and conversing about something very important to them, two men outwardly so different — Papa, who was tall and slow, and short Yakov Borisovich, mobile as mercury. Yakov Borisovich often stopped by our home at the 'object,' and Papa loved to spend time at his place.

One strong impression of my childhood is a ride with Yakov Borisovich on a motorbike and sidecar. I, of course, sat in the sidecar. The wind whistled in my ears, and it was fun and a little frightening. I remember from that time that he valued every minute, and his speech was always rapid and to the point. His questions were along the lines of 'Do you do sports?,' 'What are you reading?,' 'How are your studies?' He liked short and accurate answers.

Papa himself was not very athletic, and was always delighted with the athletic successes of Yakov Borisovich. More than once, I was able to convince myself of the justness of Papa's opinion. One time, Yakov Borisovich was waiting for Papa in the yard of our home. I was playing nearby, next to the stone wing of the side entrance to our cottage. 'What do you think — can I jump on this wing?' asked Yakov Borisovich, 'Do you want to see?' I thought it was impossible to jump onto such a high wing, but wanted to see Yakov Borisovich try to do it. However, Yakov Borisovich came running, jumped up, and, to my amazement, ended up on the wing. That is how he remained forever in my memory, literally radiating energy.

As fate had it, throughout my life, the people closest to me worked and had close contact with Yakov Borisovich. When I got married, it turned out that my husband, essentially a student of Yakov Borisovich, felt great love and respect for him. Again, I heard words familiar from my childhood about how Yakov Borisovich really can work, how he gets up every day at five o'clock, and about his enormous talent. When Yakov Borisovich later became the chairman of the theoretical department of the Institute of Physical Problems, where my husband worked, nearly every day, I heard stories about what an unusual person Yakov Borisovich was. Each day at the theoretical department, a queue of people formed — from students to academicians — wishing to discuss their problems with YaB. It is impossible to imagine a field of physics in which, according to his colleagues, YaB hadn't made an important contribution. My husband said that the strikingly rapid and deep YaB was able to delve into problems that were far removed from him. For YaB, important conversations with a venerable scientist and with a student were equally — and YaB never forgot to ask the student his first name and patronymic.

Soon, other stories began to be added to those of my husband — told by my daughter Marina, who was often at the Institute of Physical Problems about Yakov Borisovich's human charm, wittiness, and optimism.

I always felt that he was somewhere nearby, and I knew that, in any difficult situation, I could turn to him and he would always help.

The unexpected death of Yakov Borisovich shook all of those who loved him. Allow me to cite what Papa wrote in his book *Vospominaniya* [*Memoirs*] after Yakov Borisovich's death: 'Everything that is superficial and trifling has fallen away, and the results of his truly immense work remain. And those who have gone into science with his help. I sometimes catch myself thinking I'm having a dialogue with YaB on scientific themes.'

Two years after the death of Yakov Borisovich, in the same month, my father died ...

'For me, they were happy years'[67]

V.S. Pinaev

In February 1956, I defended my undergraduate thesis at the Laboratory of Measuring Instruments of the Academy of Sciences (currently the Russian Scientific Centre of the Kurchatov Institute) and, according to my assignment, was supposed to go to the Ural atomic 'object.' Work related to the production of fissioning materials didn't suit me very well. While still at school, I expressed interest in the physical sciences and research.

I shared these thoughts while taking a walk in the pine forest near Pokrovskoe-Streshnevo with my school friend Anatolii Larkin,[68] who had just defended his undergraduate thesis with A.D. Sakharov at the little-known 'Privolzhskii Bureau.'

There are probably turns in a person's fate that can decide the course of the rest of one's life. For me, one such turn was that bitterly cold February day. After hearing my laments about my future work, Anatolii suggested that he have a word with his bosses. The 'Bureau' was growing and needed young physics specialists.

We made the decision and immediately began to act. The first telephone call was to Yu.N. Babaev,[69] a colleague of A.D. Sakharov. Yurii Nikolaevich immediately understood the situation, and proposed that he, in turn, talk about me with Ya.B. Zeldovich, who (as fate would have it) was in Moscow at that time. By the evening there was already some concrete development: after listening to Yu.N. Babaev, Yakov Borisovich (YaB, as he was called by his colleagues at the 'Bureau') asked Yurii Nikolaevich to ask me to get in touch with him by telephone the next morning.

[67] This article was not included in the original Russian version of this book, published in 1993. [Translator]

[68] Anatolii Ivanovich Larkin, a theoretical physicist, Academician, and researcher at the Landau Theoretical Physics Institute, currently working at the University of Minnesota in the USA.

[69] Yurii Nikolaevich Babaev (1927–1986), a theoretical physicist and Corresponding Member of the Russian Academy of Sciences.

Barely containing my anxiety, I called and introduced myself. There was a muffled voice at the other end. YaB invited me to his flat on Vorobiev Road at eight on Monday morning.

I remember this first meeting very distinctly. Anatolii and I made our way to Vorobiev from Kaluzhskaya by bus. The frost had abated and a thick snow was falling. The door was opened by Yakov Borisovich himself, his face covered with foamy soap. We went into a room with a prominently placed small blackboard. I introduced myself. Yakov Borisovich apologized for the fact that he would be shaving while asking me questions. That was how our acquaintance began.

After rapidly asking where I had studied, the theme of my undergraduate thesis and what interested me in physics, YaB began to examine me for real. In connection with my thesis topic, which was something like 'Dependence of the gas-separation coefficient in a tube with porous walls on turbularization parameters,' YaB asked me to derive the logarithmic profile of the velocity distribution in a turbulent flow. Fortunately, I remembered the main postulates of Prandtl theory and so was able to manage with this test. The next question was connected with a course we had at the Institute on the theory of charged-particle accelerators: derive the operating conditions for a betatron inductive electron accelerator. Setting the Lorentz force equal to the centrifugal force, I was able to derive the dependence of the magnetic field on the electron velocity and radius of its orbit. But I had to sweat over the condition for stability of the orbit, namely, that the mean magnetic induction field should be twice its value at the orbit. Further, there was a question about Fermi–Dirac and Bose–Einstein statistics. But Anatolii Larkin and I had spent all of Sunday in the dormitory on Zatsepa 2a going over the newly issued tutorial on *Theoretical Physics* by A.S. Kompaneets, and this had not been in vain — it now came in handy. Finally, there followed, as I later learned from others, Yakov Borisovich's classic textbook question: compute the integral $\int x \ln x\,dx$. This was simple; I rapidly took the integral and waited for further questions. No, no further questions.

The freshly shaved YaB, beaming with blue cheeks, turned to Larkin with a smile — 'Well, shall we take him?' — and told me that he would be in the ministry today and would talk to the necessary officials, finally asking me to call him tomorrow morning.

At the entryway, an official car[70] was already waiting for Yakov Borisovich. YaB kindly offered to let us out at the nearest metro, and about ten minutes later Tolya and I parted company with him at Kaluzhskaya.

The next day, I called as requested. In my anticipation, it seemed to me that I could hear my heart beating. And then there was the voice of YaB: 'You're cleared to work with us. Get in touch with Tishkina at the ministry, she'll explain what you have to do. See you later.'

My fate was decided. I was going to work at the 'Privolzhskii Bureau,'

[70] A ZIM, a luxury Soviet-built sedan produced in small numbers for top government officials only. [Translator]

where Zeldovich and Sakharov worked, along with many of the best scientists, whose names were still unknown to me then. I was going to labour side by side with my friend Larkin, who shared my schoolboy enthusiasm for physics and mathematics, on very secret scientific problems of importance to the government. But all this would begin in a month, after my holiday, and now my spirit was ringing with joy and gratitude toward Yakov Borisovich, who had decided my fate.

I'll try to call up memories about my first impressions of Yakov Borisovich. His outward appearance was something of a disappointment. I had prepared myself to see a solid, noble member of the Academy of Sciences, whose enormous merits had been tersely described to me by Anatolii Larkin. But Yakov Borisovich was relatively short, wore round glasses, was not handsome, and was not distinguished by his apparently elegant suit jacket. I remembered this because later, when conversing with Yakov Borisovich, I suddenly realised that I feast my eyes on him when he is speaking, presenting a talk or discussion at a seminar, sitting at the wheel of an automobile, dancing, wherever. I noted the same in the eyes of others around him. Yes, the first impression had been mistaken: in life, Yakov Borisovich was very charming, vivid, with inexhaustible energy and humour. His eyes shone with a youthful ardour, and sometimes with mischievousness. He was 'eternally young.' From the first minute of our meeting, I felt at ease in interacting with YaB. There was no commanding, arrogant tone, he never pushed forward his scientific merits. The most important thing for him was the subject of a problem, a question, an issue. One thing that was striking even then was the wide scientific erudition that was felt in his questions and commentary. This feeling only increased with each new interaction, and grew into marvel and awe. The breadth of the fields of physics and astrophysics to which YaB made contributions is such that some foreign scientists (in particular, Steven Hawking has been mentioned in this context) got the impression that an entire collective of authors were working under the name of Ya.B. Zeldovich, as in the case of N. Bourbaki.

When I arrived at the 'Privolzhkii Bureau' in early April 1956 (it turned out that it was located in Sarov, a historical place of Russia associated with the holy Seraphim Sarovskii), a little over 20 people worked in the two theoretical sectors, headed by Ya.B. Zeldovich and A.D. Sakharov. These were primarily young theoreticians without higher degrees who had worked there for between two and four years, some of whom had nevertheless already received Stalin Prizes or government awards. There were also eminent scientists in the sector, such as D.A. Frank-Kamenetskii, G.M. Gandel'man, and N.A. Dmitriev. This modestly sized collective worked on computational–theoretical studies of nuclear charges. Here, new, freshly born ideas were translated into theoretical estimates of structural parameters for charges. On the basis of this, specifications were formulated for applied mathematicians on the execution of complex, detailed computations; technical specifications were formulated for engineers and technicians, who transformed these ideas into blueprints and models; and finally, technical specifications were formulated for the gas-dynamical experts and experimental physicists on physical measurements and experiments with the charges. The main aim was to make nuclear charges more powerful, lighter, smaller, and more economical in use of special materials, etc.

This was the essence of the work of a theoretical physicist, specific to this new field of military technology, and was explained to us, a small group of 'recruits' to the ranks of nuclear arms workers, by Ya.B. Zeldovich. He listed the areas of physics with which one needed to be familiar for the work, and marked out the first steps with which to begin. He did not avoid the question of the importance of work on nuclear weapons, the enhanced requirements and responsibilities connected with this work, the high price of errors. Later, all this would become part of our world view, and Yulii Borisovich Khariton would constantly and demandingly remind us that we must know ten times more than what was needed for any specific task.

A high spirit of creative enthusiasm soared in all departments of the 'object': only a few months had passed since the testing of a thermonuclear bomb based on a two-stage design in November 1955. Ya.B. Zeldovich was one of those whose contributions to the development of the thermonuclear bomb were especially weighty and lasting. The theoreticians were inspired by the opening perspectives for further success with the development of the new design. The spirit of confidence and optimism, and also the satisfaction gained from solving a major problem, reached new heights. There were many more ideas than possibilities of testing them quickly.

Toward the end of April, we were given access to all classified documents, including the secrets of the new weapon, and 'weapons' physics began to permeate our lives.

1956 was not a successful year for YaB. The attempt to appreciably improve the technical and military characteristics of the device tested in 1955 ended in two 'failed' tests (as tests with appreciably lower power output were called). Here, it was not that any unforgivable mistakes had been made. It was just that Nature was uncooperative when our knowledge of certain physical parameters was insufficient for us to choose the correct construction design. More than a year passed before the complex quantum-mechanical calculations and various physics experiments led to better understanding and success with this work.

YaB was very anxious because of the lack of success with these experiments. In a fit of temper, he told the theoretician supervising work on these devices, 'I have exhausted the limit of failures, and cannot support your plans.'

YaB's theoretical sector had three departments. However, YaB and ADS (Andrei Dmitrievich Sakharov) did not express any administrative fervor in acquiring assignments for members of their departments to work on the development of charges. In practice, the development of charges was carried out by groups of workers from various departments. The fact that the collectives organised themselves in the best way possible supported the creative atmosphere. All our offices were on a single floor of the building. The offices of ADS and YaB were next to each other, on one side of the corridor. These rooms are now occupied by computer repair services; they are filled with test benches and fumes from soldering irons, and nothing is there to remind one of the past. It is very sad that time is so ruthless.

I remember the first task assigned to me by YaB: to determine the time after which it was possible to explode a nuclear bomb carried by a bomber subject to the action of neutrons from the nuclear explosion of an antimissile rocket. The problem reduced to an estimation of the number of delayed neutrons that would lead to a premature onset of the chain reaction when the bomb was ignited. There were no computers at that time, and these types of problems were solved by reducing them to simple models that could be solved analytically. I learned many lessons from this early work.

In applied studies, before beginning massive computations, it was important to be able to rapidly obtain numerical answers that could be trusted to be correct to order of magnitude (i.e., to be able to give a correct qualitative estimate of the effect). YaB and ADS were great masters in this art. It seems that there was no problem to which they could not obtain a numerical answer after some time calculating at the blackboard, chalk in hand. For us, who were students just yesterday, these were clear lessons in the use of dimensional, similarity, and symmetry arguments, and most of all a demonstration of physical intuition that was able to distinguish in physical phenomena the main factors determining their fate. Problems with self-similar solutions were very popular at that time, especially those associated with non-linear heat-conduction and hydrodynamics. Some of these have a direct relation to nuclear physics. Ya.B. Zeldovich and A.S. Kompaneets published results on a thermal-wave regime as early as 1950 (the American publication by R. Marshak appeared in 1958). The monograph *Physics of Shock Waves and High-temperature Hydrodynamical Phenomena*, written by YaB together with Yu.P. Raizer (YuPR), became known worldwide. Written by one of the creators of nuclear weapons (YaB) and a specialist in the physics of nuclear explosions (YuPR), this book became a text for all whose work was related to regimes with high temperatures, pressures and densities.

For Yakov Borisovich, 1957 was marked by the successful practical realization of his old idea, developed together with L.P. Feoktistov, of increasing the efficiency of the explosion of an atomic charge via a so-called booster regime (to use the American terminology). YaB assigned the young theoretical physicist Vitalii Morozov to this work. The computational theoretical work was carried out over more than a year, and ended in December 1957 with a successful test. This was a momentous event not only for the theoreticians, but also for the designers and technicians.

The charge activity entered its maturity, and new ideas waited for their hour to be tested. For YaB, there was the further development and broadening of the booster regime, and also the realization, together with L.V. Altshuler and Yu.M. Styazhkinii, of the idea of a fission regime in which a chain reaction that has just begun is damped. In the future, the conduction of such 'incomplete' explosions was important for scientific investigations of a number of questions in weapons physics.

As the head of a theoretical sector, Yakov Borisovich was responsible for the entire 'production' coming from his subordinates (i.e., he was called upon to confirm their reports, to have a large number of results from computational and analytical investigations passed by him). Of course, this was a big load, and for YaB a simply enormous one, given his scientific interests beyond weapons. He never lost a minute, and sometimes solved problems on the move.

Apart from theoreticians, YaB had many interactions with experimentalists: L.V. Altshuler, V.A. Davidenko, S.B. Kormer, V.A. Tsukerman, and their co-workers. The sharing of ideas bore fruit: in Sarov was born one of the world centres for applied and fundamental studies in the new fields of physics. It was not by chance that, at a conference dedicated to the physics of high energy densities (1969, the Fermi School in Varenna, Italy) Edward Teller said:

'We have two reasons to be sorry. One is the absence of Altshuler and Zeldovich, two people who have, perhaps more than anyone, helped facilitate the discovery of this new field of investigation. The second is the absence among us of scientists from the excellent laboratory in Los Alamos.'

The moratorium on nuclear tests (from the end of 1958 to September 1961) did not really affect the work of theoreticians, although a number of colleagues switched their interests partially to non-classified topics.

One noteworthy event was the organization of excursions to Moscow to hear lectures on quantum electrodynamics, which were given by L.D. Landau for two semesters in 1959 and 1960 at the Physics Department of Moscow State University. YaB and ADS managed to 'push through' permission with the head of the 'object,' B.G. Muzrukov, for about 15 people to go on the trips.

L.D. Landau's lectures made big impressions on us, with their depth and the manner in which they were presented. Yakov Borisovich also attended these lectures. 'As a theoretical physicist, I consider myself a student of Lev Davidovich Landau,' wrote YaB in his autobiographical afterword to his *Selected Works*. This deep admiration of YaB for the teacher resounded in everything that had to do with Landau.

Yakov Borisovich played a leading role in creating the scientific atmosphere at the object. Being often in Moscow, and having contacts at seminars and personal meetings with the leading scientists of the country, YaB was up-to-date on all the latest accomplishments in physics. When he arrived at the 'object,' YaB organized seminars to share the news. Often, a talk contained further developments of a question posed by YaB himself, or the formulation of new problems. The seminars ranged from small ones, intended for the theoreticians, to those that filled large auditoria to bursting, attended by all who were interested. Everyone loved to listen to Yakov Borisovich. The seminars were attended by both heads of large divisions and laboratory assistants. His keenness for science, thirst for understanding, sense of the beauty of an effect resounded in his presentations and was conveyed to the audience. Here is a brief list of some talks given by YaB at seminars: creation of a theory for superconductivity; discovery of non-conservation of parity; storage of cold neutrons; light nuclei oversaturated in neutrons; hot and cold models for the universe; the most rigid equation of state. This list does not include the seminars on hydrodynamics, processes in shock waves, optical phenomena, and self-similar solutions.

One could only be amazed by how YaB was able to get to know this mass of new information. He himself said that the best way to study a new area of physics was to write a scientific review about its current state. How many such reviews YaB wrote for the journal *Uspekhi Fizicheskikh Nauk*!

YaB was always overflowing with problems and ideas that he generously shared with the audience at seminars. At times, having formulated a problem only the day before, the morning of the following day, YaB would 'call all hands on deck' to report the solution.

One of Yakov Borisovich's seminars played an important role in my choice of a theme for my future dissertation work. In the beginning of January 1959, having returned from Moscow, YaB told us about the hypothesis of B. Pontecorvo that the weak interaction of electrons and neutrinos with the Hamiltonian in the form proposed by Feynman and Gell-Mann could lead to the bremsstrahlung emission of neutrino pairs by an electron in the field of a nucleus. According to this mechanism, the hot cores of stars should lose energy unhindered in the form of neutrino radiation, since the absorption of neutrinos by the stellar material is negligible. Would someone like to check if this is indeed the case?

This problem interested me. Armed with my modest experience in computing Feynman diagrams, for which I owed the young Masters student and future Minister of Atomic Energy, Viktor Mikhailov, I decided to give it a try. My interest was supported by G.M. Gandel'man, and together we showed that, indeed, in the latest stages of evolution of a star, the radiation of neutrinos due to electron–neutrino interactions could exceed the radiation of photons.

Having learned of these results, YaB organized a meeting with B. Pontecorvo, which took place at the apartment of D.A. Frank-Kamenetskii in Moscow. The result was a letter from B. Pontecorvo to the *Zhurnal Eksperimentalnoi i Teoreticheskoi Fiziki* and my article with G.M. Gandel'man. The publication of these papers provoked a wide response, and a flow of articles on other, more efficient processes for neutrino radiation followed. Yakov Borisovich was one of the first to consider neutrino losses during gravitational collapse. The detection of neutrinos from Supernova 1987A in the distant Magellanic Cloud confirmed that neutrinos carry the bulk of the energy of the star in a supernova explosion.

In September 1961, the nuclear tests were reinitiated. During the years of the moratorium, theoreticians had carried out a lot of work on analysing experimental designs for charges and had determined directions for their further refinement, having composed an impressive list for the tests. An exciting time began, when some devices would be on their way to the test range, others were being prepared to take their places on the railroad ramp, others, regardless of the time of day, were being assembled in the factory workshops, and still others had just been laid out on blueprints or pressed out in punch cards for computer calculations.

For Yakov Borisovich, the series of nuclear tests of 1961 to 1962 was the last. This was work that was stressful, demanded great responsibility, and, it seemed to me, was not bringing YaB satisfaction even then. One could sense that his interests were pulled toward the discovery of unknown mysteries of nature, that there was not enough room for him in the framework of applied weapons physics. 'Cockroach races' — that was how YaB humorously described his work in the last year before his departure from the 'object.'

YaB acquired a deep attraction to the general theory of relativity, cosmology, and astrophysics. This could be evidence of the effect noted by someone that, in many theoretical physicists entering the second half of their lives, there appears a burning interest in questions of the nature of the universe. Of course, his acquaintance with the physics of processes occurring in stars and thermonuclear explosions played a not insignificant role in this choice.

After a short while, YaB had mastered the theory of general relativity, using the *Theory of Fields* theoretical text by L.D. Landau and E.M. Lifshitz. This was no simple self-education. Participants in the seminar organized by him (1960 to 1963) remember how YaB presented and discussed unsolved problems in general relativity, in particular, the question of the evolution of massive stars and gravitational collapse. YaB named the result of catastrophic compression of a star the 'gravitational grave' (a term that may be no less rich in content than 'black hole'). The seminar became more lively when A.D. Sakharov was present. The opinions of academicians did not always coincide, and discussions of details sometimes went far beyond the topic of the talk. One of those present at the seminar once remarked, 'Isn't this rather too many models of the universe for a single design bureau?' YaB's enthusiasm for general relativity didn't prevent him from remaining true to the physics of elementary particles, nuclear and atomic physics, chemical kinetics, the physics of continuous media, etc.

In July 1962, the Theoretical workshop on the most important problems of astrophysics took place in Tartu, for which YaB prepared his very comprehensive review lecture 'Modern physics and astronomy.' Yakov Borisovich assigned me to give a talk about neutrino processes in stars. The seminar in Tartu was very memorable. This was the first time I saw YaB anxious about presenting a talk in front of an audience of astronomers and astrophysicists. The seminar was very representative: its participants included I.S. Shklovskii, B.M. Pontecorvo, D.A. Frank-Kamenetskii, S.B. Pikelner, A.G. Masevich, R.Z. Sagdeev, S.S. Gershtein, G.S. Saakyan, and other well-known scientists. Perhaps the reason for YaB's anxiety was that he, the theoretical physicist, specialist on combustion and detonation, hydrodynamics and shock waves, one of the pillars of the Soviet atomic project, had burst in upon the avant-garde of astrophysicists. The conclusion of YaB's presentation resounded like a credo: 'There is enormous pathos in the problem of describing all the variety of observed phenomena and general laws of the universe on the basis of the existing laws of physics, established by laboratory experiments and theoretical analyses.' Several years later, with the same success, YaB would use data obtained from observations of the universe to establish limits in the physics of elementary particles that were inaccessible to experiments on Earth.

One other event is also memorable. One day at the Tartu University, where the workshop took place, there was a campaign to collect signatures in support of a protest against the USSR conducting atmospheric nuclear tests. YaB, not waiting to be asked for his support, left, having said that he could not sign documents on the development of charges for nuclear tests and a petition against testing with the same hand.

In 1964, Yakov Borisovich moved to Moscow and headed a small theoretical department at the Institute of Applied Mathematics. A new stage in his life began, completely dedicated to his beloved science. He had given 20 years to the development of atomic and hydrogen weapons. 'For me they were happy years,' Yakov Borisovich would write in his autobiographical afterword to the *Selected Works*. Ahead lay a little more time, set aside by fate to be dedicated to the untiring investigation of the mysteries of space, time, and matter. World recognition was imminent.

> Archimedes and Galileo were not the last to derive a forward push in their science from the defence needs of the larger community. The Wilsons and Zeldoviches of our day — through their mathematical predictions and the test, which those predictions have survived — have created a standard for their scientific colleagues the world over.

These were the words spoken in 1982 at a symposium in the USA dedicated to the 60th birthday of D.R. Wilson by John A. Wheeler, the author (together with N. Bohr) of the theory of uranium fission, one of the developers of nuclear weapons in the USA, and an astrophysicist and cosmologist.

Part IV

Elementary Particles and Nuclear Physics

On the path towards universal weak interaction

S.S. Gershtein

I will always remember my first meeting with YaB. In the autumn of 1951, after my graduation from Moscow State University, I ended up as a teacher of physics and mathematics in a high school in a village called Belousovo in the Kaluga Region, 105 km from Moscow. Sitting in the evenings at dinner with the old peasant woman who kindly took care of me, and in whose hut[71] I lived, I looked at my university notes and, listening to the rain beating on the window, beyond which was pitch darkness, I consciously bade farewell to the scientific activity for which I had prepared at university.

However, my friends who graduated in the same year didn't leave me to my fate. One of them, Sergei Repin, asked, 'Why not try to pass Landau's minimum?'[72] and gave me LD's [Landau's] telephone number. I called 'Dau'[73] and began to take his exams. At the beginning of 1953, when I had passed the last of them, LD said, 'Think of a theme for your work and try to go to the seminar. Unfortunately, I can't accept you in my group. I've already looked into it, and it's impossible.'

It is easy to say 'think of a theme.' I was able to read fresh journals only in fits and starts when I came to Moscow, and at that time, we didn't study nuclear physics and elementary-particle physics at all at university (these subjects were taught only at the secret nuclear section of the Physics Department). Nonetheless, I tried to set aside one free day at the school — Thursday — to go to LD's seminar (this was far from simple with a workload of 44 hours per week in two shifts), and I came to my first seminar two weeks after my conversation with LD. Before the beginning of the seminar, I sat somewhere in the middle of the hall and began to look through a fresh issue of *Physical Review*, which I had been loaned for a couple of hours. Unexpectedly,

---

[71] *Izba* in Russian. [Translator]

[72] Landau's 'theoretical minimum' or simply 'minimum' was a sort of qualifying exam, which he insisted that anyone who was to have professional contact with him must pass. [Translator]

[73] A widespread nickname for Landau used among Soviet physicists. [Translator]

a strange man sat down in the free seat next to mine; he was of modest height, wearing round glasses, and struck me with a sort of unusual internal energy. This energy could be felt in his unusually lively eyes and impatient motions. He seemed to find it hard to stay in his place, looking around at the people gathering for the seminar and searching for a glance from someone he knew. In passing, he unceremoniously looked at my open journal and turned away disappointed, having convinced himself that he had already seen that issue. Soon, the seminar began. I don't remember who the speaker was. I think the talk was about some article on kinetics. Unexpectedly, my neighbour got up and started arguing with the speaker. LD also got involved in the argument, which immediately turned into a discussion between Dau and my neighbour, who rapidly moved to a place in the front row. 'Who in the world is that?' I thought, knowing that it was rare that someone could hold their own for very long in a discussion with LD. I was even more surprised by Dau's words: 'Yes, I think you're right.' That was an extremely rare event. None of my acquaintances were sitting near me, so I couldn't ask who my neighbour had been.

After the seminar, Dau went out into the corridor, then, returning to the hall, came up to me, asked me to follow him, led me to that same stranger who had so interested me, and presented me with the following words: 'Here's the young man I told you about. Take him on, he is in every way thirsty for activity.' His short companion said, 'Let's go,' and we went to his home together with two other people, whom he had already invited. From the conversation, I understood that they were both from Leningrad (they were I.M. Shmushkevich and L.S. Sliv, who later became my friends). Entering the flat, our host briefly gave a command to the woman who opened the door — 'Sasha, something to eat' — and, turning to us, said, 'All discussions after lunch.' When we had eaten, we went into the small room that served as his office, and he began a scientific conversation with my two new acquaintances, and extended to me some typewritten pages with the words, 'You have a read for now.' I read the name of the author of the article, and only then realized that I was talking with Zeldovich. I suddenly felt shy. At that time, YaB's name was surrounded by a blanket of secrecy. Studying at Moscow State University, we heard about the pre-war work of Zeldovich and Khariton on uranium-fission chain reactions, about his research on combustion and detonation; therefore, we didn't doubt that he had become involved in work on the atomic bomb. This was confirmed by a story told by one of the older students, of how he had seen YaB make a presentation in defence of N.N. Semenov at a discussion in which Semenov had been accused of forcing through idealism in the theory of combustion (this type of discussion was common at that time). YaB arrived at the gathering with the star of a Hero of Socialist Labour on his chest. It was clear that he had not put it on by chance. The appearance on the podium of a speaker with a piece of regalia so rare at that time exerted a significant influence on the results of a 'scientific' discussion of this sort, and the fact that the decree ordering the decoration had not been published lent even more importance to the presentation. Knowing all this, I was struck by the simplicity of YaB's relations with us, and his democratic attitude.

Having finished his conversation with the two Leningraders, YaB called me to his desk and began to tell us about his new theory of β decay. It was clear that he was very enthused by and satisfied with this result. When we parted, he gave me several typewritten copies of his papers and said that I should read them and come again in two weeks, when he would be in Moscow again. In this way, I first became acquainted with YaB and, having been to his home, wholeheartedly took a liking to that home and its inhabitants.

When I next appeared at Landau's seminar, he asked, 'Well, how do you like YaB?' and, without waiting for an answer, said, 'He won't let you be idle. I don't know of anyone else with so many ideas in his head. Maybe only Fermi. In addition, YaB has an iron grip, and he'll force you to work at full steam.' This coincided with my impressions. The articles that YaB gave me were dedicated to several very different questions. There was a paper on baryon charge (which YaB rediscovered, not knowing about the earlier work of Wigner and Stuckelberg, and which subsequently served as an example for his scheme for lepton charge), another on the Compton effect in polarized electrons (enabling the measurement of circular polarization of γ rays), and finally, one on various types of β decay. This last article interested me most, and I decided to work in this area.

Later, I reflected more than once on how lucky I was: I became acquainted with YaB precisely when he became attracted to elementary-particle physics. It is well known that, during his years of scientific activity, YaB changed the main area of his interest several times: from chemical physics (catalysis, combustion, detonation) to nuclear physics, the theory of elementary particles, and then to astrophysics, gravitation, and cosmology. These shifts in interest were not undertaken by chance. Possessing broad views and an amazing intuition, YaB chose the most interesting and pressing problems of his time. In the beginning of the 1950s, one such problem was the physics of weak interactions. After the discovery at the end of the 1940s of the $\pi \rightarrow \mu\nu$, $\mu \rightarrow e$ decays and of μ capture, it became clear that these processes occur under the action of forces that have the same order of magnitude as that for the known β decay. In this way, the idea arose that all these processes are due to a special 'weak' four-fermion interaction bearing a universal character. It is evident that the hypothesis of the existence of a new elementary interaction (together with the strong, gravitational and electromagnetic interactions) was of fundamental importance; however, verifying it required first and foremost a good understanding of the weak, four-fermion interactions for all known processes. And these had not been reliably established experimentally, even for the long-studied β decay. Of the five possible types of four-fermion interaction (scalar S, vector V, tensor T, axial vector A, and pseudoscalar P) and β decay, preference was given to type T, and it was also supposed that one of the so-called Fermi types (V or S) existed.

Beginning to work in a new research field, YaB tried first of all to distinguish the main problems in this field and search for general principles for solving them. In doing this, he often revealed new problems that had not previously been known to specialists in the field, but which subsequently became fundamentally important. This was the case, for example, with the problem of the observed limit to the magnitude of the cosmological constant. YaB decided that the general principle that should be used as a basis in searches for types of β decay was whether the theory could be renormalized. As is known, this principle subsequently played an important role in the creation of a theory for the electroweak interaction. However, this was done in a different form to that which YaB proposed. At the beginning of the 1950s, it was known that, in addition to electrodynamics, theories with scalar and pseudoscalar particles were renormalizable. Based on this, YaB proposed that the intermediary particles in the β decay were precisely these particles, and obtained a $(V+T)$ form of β decay, which, as it then became clear, was in full agreement with the experimental data. Therefore, it is easy to understand his joy. Striving to derive further experimentally observable consequences of his theory, which could either confirm or disprove it, YaB noted that, in his scheme, taking into account the Coulomb field of the nucleus should lead to a mixture of various types of β interactions, which should be visible in the β-decay spectra of electrons. He assigned to me the calculation of this phenomenon.

We usually met as follows: YaB, coming to Moscow for a short time, would send me a telegram to Kaluga Region and give the time when I should come to his place (as a rule, at six o'clock in the morning). Managing with difficulty to get to Vorobiev Road on time (public transport was only beginning to operate at that hour), I would find YaB half-dressed, but already sitting at his desk. Occupied primarily with his 'special business,' YaB set aside rare hours for doing purely scientific research. Sometimes our meetings occurred only a few hours before YaB's departure for the 'object,' and in that case, the door was opened by the 'secretaries,' who, as I guessed, were his escorts. After our discussions, he insisted that I join his family for breakfast, and I, at first shy, was struck by the atmosphere of kindness and naturalness that ruled in the Zeldovich home. It was clear that this was a deeply work-loving family, where each conscientiously went about his business, be it scientific work or studying at school. I didn't see any trace of the ambitions or feelings of exclusivity that were so characteristic of the wives and children in many highly placed families. There was no lacquered furniture, nor expensive things set up on display, and the housekeeper Sasha was essentially a member of the family. I will never forget the kindness and sympathy exhibited by Varvara Pavlovna.

For many people who came into contact with YaB, it was a mystery how, when he changed his field of interest, he managed to both get oriented very quickly in the new area and remain up-to-date on developments in former fields of work, from time to time returning to them and obtaining new and spectacular results. Through my work with YaB, I came to understand that this was no mystery. Apart from his talent (of the sort that can be said to be God-given), this ability was connected with YaB's working methods. When he began to work in a new area of physics, he first and foremost sought out young specialists; talking with them 'as equals,' he could rapidly get acquainted with the most important results obtained in that area.[74] In many ways, the learning process was aided by YaB's democratic nature and lack of professorial conceit. He was never too shy to ask 'stupid' questions in any situation, and often got condescending answers from the specialists (I remember a joke that certain 'serious' young people liked to repeat: 'YaB wants to apply physico–chemical methods to elementary particle theory.'). However, it usually soon turned out that some of these 'stupid' questions left the specialists stumped — that they, indeed, represented new formulations of a problem or completely new problems that had not been considered previously.

In addition, an active approach to the 'learning' process was characteristic of YaB. When he came across unclear questions or new problems, he educated himself in the very process of solving them. I remember his words when he learned that I was studying the renormalization theory: 'What use it is to study abstractly? You should find and solve some problem, that way you'll really learn.' As he accumulated new problems, YaB assembled around him young physicists who were just starting their careers, and, working together with them, rapidly became familiar with areas that were new to him. In this way, a new school was also formed, which continued its investigations after the interests of the teacher had shifted to another field. The personal contact, attention, and concern demonstrated by YaB toward his former students meant that he was always up-to-date on developments connected with his former research. Young researchers constantly came to him for advice or to discuss new results. He got drawn into discussions that frequently led to new ideas in some 'old' field.

YaB experienced the need to discuss his new research with various people, listen to their criticism, find counterarguments, and thereby test the soundness of his own ideas. These discussions were very fruitful and educational for his students. We involuntarily became infected with the joy he experienced when he discovered some new, unknown effect, solved some new problem, or understood the possibility of a simple, 'first principles' explanation for some phenomenon. He usually gave his students the manuscripts of his own papers to read and asked them to make comments, for which he unfailingly expressed thanks in his articles. This stimulated deeper study of a subject, and increased one's confidence in one's own abilities.

YaB usually discussed the contents of his scientific talks with his students. I remember one episode in particular in connection with this. In 1954 to 1955, I.V. Kurchatov decided to organize a cycle of lectures on the most topical problems of physics, in order to expand the horizons of his researchers and their possibilities for choosing new directions of scientific activity. He asked YaB to read several lectures on β decay and weak

---

[74] For example, Landau recommended to him one of his most talented students — V.V. Sudakov — when he wanted to 'study' β decay.

interactions. YaB's attitude toward this request was serious, since he quite correctly believed that physicists, especially experimentalists, could make large contributions to these problems. (These hopes were shown to be fully justified only a few years later, when the brilliant research of P.E. Spivak's group on β decay of neutron and of I.I. Gurevich's group on $\mu \to e^-$ decay appeared.) Therefore, he tried to make the material in his talks as accessible as possible to experimentalists who hadn't worked in these fields before. I should say that YaB could master this art like no one else. Preparing for lectures, he invited Sudakov and myself to help him, and we had long discussions together, not only about the contents, but about the best way to present individual questions. Our sort of reward for this was that YaB took us to his lectures, and, after they were finished, presented us to Kurchatov as his assistants.[75]

By that time, I was already living in Moscow again. Having worked through my required three years at the village school, I moved to Moscow in order to be closer to science; I hoped to find a position as a teacher at one of the universities or, if I was lucky, at a scientific institute. YaB and Dau recommended me for several places, but each time, when the question got to the personnel department, they would drag it out for a month or two, and I would then receive a rejection on some more or less plausible pretext. I made ends meet by giving lessons. I didn't want to go back to work at a school, all the more so because hope was dawning ahead. Times gradually changed, and at some point, Landau told me, 'There is talk that Kapitsa will be re-appointed director of the Institute of Physical Problems. In that case, I'll be able to take you on as a postgraduate student.' However, weeks and then months passed, and my situation didn't change. I don't know what I would have done without the help of my friends.

YaB, interested in my affairs, offered me money on many occasions, which I, naturally, refused, saying that I didn't need it. The circumstances highlighted the delicacy with which YaB could insist on his own way (subsequently, I often encountered similar actions toward both myself and other people). He once asked if I would agree to give private lessons to the daughter of a friend, a schoolgirl who had survived a serious disease in her earlier childhood. I accepted this offer, and worked enthusiastically with the capable girl, who was keen to learn in spite of the serious after-effects of her disease. After a while, YaB, asking about my activities, asked, 'Since you're giving lessons, I'd like to ask you to work a bit with my girls.' I began to refuse, saying that this was quite unnecessary for them, their studies were going fine, they got straight As in school, and would only be embarrassed if they were assigned a tutor (at that time, it was considered shameful).

75 As science becomes more bureaucratic, there are fewer and fewer high-ranking scientists who are willing to learn in public, thereby acting to motivate lower-ranking 'bosses' in their own investigations. Therefore, I often remember sadly the over-full hall, and Kurchatov sitting in the front row with the notebook in which he wrote YaB's lectures, asking questions from time to time when something wasn't clear to him.

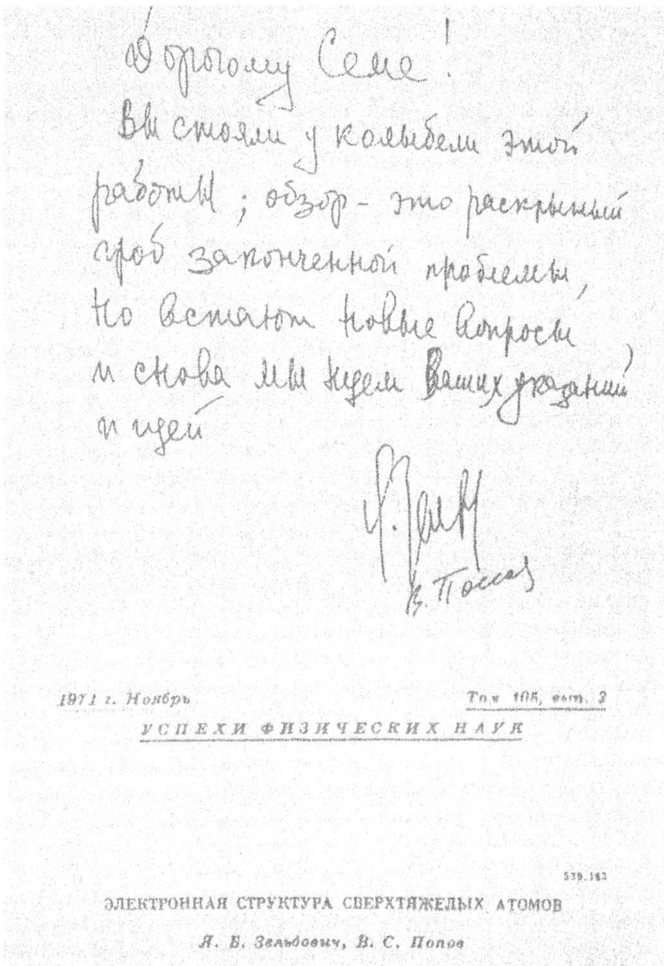
Дорогому Семе!
Вы стояли у колыбели этой
работы; обзор - это раскрытый
гроб законченной проблемы,
но встают новые вопросы
и снова мы ждем Ваших указаний
и идей

1971 г. Ноябрь Том 105, вып. 3

УСПЕХИ ФИЗИЧЕСКИХ НАУК

539.183

ЭЛЕКТРОННАЯ СТРУКТУРА СВЕРХТЯЖЕЛЫХ АТОМОВ

Я. Б. Зельдович, В. С. Попов

Inscription written by Ya.B. Zeldovich to S.S. Gershtein on a copy of an article entitled 'The Electronic Structure of Superheavy Atoms' in *Uspekhi Fizicheskikh Nauk*, November 1971. The inscription reads:

To my dear Syoma!
You stood near the cradle of this work; the review is the open grave of a finished problem, but new questions are posed, and again we are seeking your suggestions and ideas.

Ya. Zeldovich

However, YaB found the following objection: 'At school, they don't teach experimental skills. Buy some instruments, and let them do some experiments.' I had no choice but to agree (I had some experience in this area, having tried to establish a similar physics laboratory at the village school, so that each student could carry out experiments). I was able to obtain some rather good equipment in educational supply stores — an optical bench, lenses, galvanometers, and so forth. I also found a small Wilson chamber. And so Olya, Marina, and I began to set up various experiments and solve problems connected with them. At times, these experiments ended in embarrassment for me — several times, I burned out sensitive electronic instruments. YaB gently made fun of me; sometimes he himself took part in the experiments, and I, together with my students, listened with interest to his commentaries about the applications of the phenomena being studied in various technical devices. I was surprised by YaB's experimental skills and his ability to work with his hands. 'After all, I began as an experimentalist,' he told me, 'and before that I worked as a lab assistant' (at that time, I didn't yet know that YaB had not continued his formal education beyond high school). I don't know if my lessons (thought up by YaB, as I understood, as a way of helping me) were of any use to my students; but they taught me many things, and helped me support myself until May 1955, when Landau was finally able to realize his plan and take me on as his postgraduate student at the Institute of Physical Problems.

In the meantime, my scientific discussions with YaB continued. The calculation of the effect that I carried out proved to be useless, since the experimentalists had by that time established with certainty that the β interaction was a sum of the scalar and tensor types (and not the vector and tensor types, as followed from YaB's theory). I could feel that he was sorry that my work had turned out to be for nothing, and was looking for a new problem for me to work on. Once, when discussing the experimental data, he told me: 'Experiments show that the ratio of the Gamow–Teller and Fermi constants does not greatly exceed unity. It would be very beautiful and important for our understanding of the nature of weak interactions if these two constants were the same for a "naked" nucleon, and the small difference between them arose because the nucleon was immersed in a pion mantle. Let's calculate this effect.'[76] Remembering YaB's comment that the best way to learn was by working on problems, I proposed that we perform calculations for all possible versions of the interaction. YaB agreed to this. When we independently concluded our calculations and began to check over our results, we discovered that there was a significant divergence between our results for the vector case. I, working as a student, had carefully used the isotopic spin matrices, while YaB, trying to obtain his results more quickly, had acted

[76] I'll mention in connection with this that an important element of YaB's natural philosophy was a belief in the existence of beauty and internal connections between various phenomena. In addition to this guess, which turned out to be completely correct, I remember that even back then, he told me more than once, 'Note the following: the constant for weak interactions of fermions with intermediate bosons has units of electrical charge. It would be very beautiful and natural if it were simply equal to the electric charge. This could be used to estimate the mass of the intermediate bosons.' Having forgotten that it was necessary to use Heaviside units for the charge, he obtained a mass of about 30 GeV for the intermediate bosons. If this is multiplied by the omitted factor $\sqrt{4\pi}$, the resulting value is very close to the experimentally measured mass for the W boson.

somewhat primitively, using the same constant for interactions of a neutral pion with protons and with neutrons (while, in fact, the constant has different signs for these two cases). When he discovered the reason for the discrepancy between our results, YaB exclaimed: 'Now I understand how a symmetrical theory differs from an isotopic invariant theory!'

However, unexpected results awaited us in the future. Taking into account in the vector case the possibility of β decay of a charged pion (which YaB earlier calculated based on the Fermi–Young model, using, in essence, only the isotopic invariance of this model), we discovered that pion corrections didn't change the constant for the weak vector interaction. This result immediately interested us, although it seemed not to be related to the real behaviour of the β interaction established in experiments. First, we thought that the result was associated with a low correction of the perturbation theory with which we were working. However, after some discussions, we understood that it was an analogy of the conservation of electric charge of the proton, which does not change during strong interactions with pions, and we even saw some the possibility of rigorously proving this using the Word identity. We were very excited about our discovery, and regretted very much that it remained a simple mental game, and did not exist in Nature.

We understood that, in the vector case, possibilities would open for the precise experimental verification of the behaviour of the universal weak interaction. From a pragmatic point of view, we were able to gain satisfaction only in the confirmation of YaB's guess. Our result for the scalar and tensor cases showed that the observed ratio of the Gamow–Teller and Fermi constants was completely consistent with the proposal that these constants were equal for a 'naked' nucleon. 'We should write a paper,' said YaB. However, a new hindrance arose. When I arrived at YaB's home at the appointed time, I found a note warning me that he had been delayed and asking me to wait for him. The note began with the words, 'Woe to us ...,' and beneath it lay a new issue of *Physical Review* with an article by Finkelstein and Moshkovsky, where they considered the same problem for the scalar and tensor cases and came to the same conclusions. The novelty of our work was lost. Furthermore, their calculation had been based on the model of Chu and Lowe, which, it stands to reason, took into account the strong interaction more accurately than our perturbation theory, from which we expected only a qualitative confirmation of the hypothesis.

I was upset, and thought that YaB would also be distressed, and to an appreciable extent because of me. But YaB wasn't depressed. When he returned home, he proposed that we work through Finkelstein and Moshkovsky's paper together, and we discovered several inaccuracies. They did not change the main conclusions, but were important for comparisons of β and μ decays. 'This is the subject for a short note,' said YaB, 'At the same time, we'll note the properties of the vector case. It's important that your first published work comes out soon.' By the time of my next visit, he already had a draft of the article. After reading it through and making several corrections, I asked, 'Won't Dau criticize our treatment of the vector case? After all, he doesn't like abstract discussions about things that don't actually exist.' — 'It's impossible just to walk by such a beautiful possibility,' answered YaB, 'And if you're afraid of Dau, let's add the words "does not have practical importance, but is methodically very interesting" before the paragraph on the properties of the vector case.' And we sent the article off to the publisher. A few years later, the law of conservation of vector current was rediscovered by R. Feynman and M. Gell-Mann in their famous work, in which they proposed almost simultaneously with Marshak and Sudarshan their theory of the universal ($V-A$) weak interaction. As later noted by Feynman, neither he nor Gell-Mann knew about our work (but in the future always referenced it). Numerous experimental studies performed after the work of Feynman and Gell-Mann confirmed that precisely the vector and axial-vector types of weak interaction exist in nature.

Returning to our work many years later, YaB always expressed satisfaction that, in spite of the experimental situation at that time, we had noted the properties of the vector interaction, which played an important role in the experimental verification of the universal character of weak interactions, and stimulated the development of gauge theories. This, in my view, showed YaB's characteristic striving for a thorough analysis of any question he considered. His scientific uninhibitedness and the fact that he never worried about looking silly enabled him to widely discuss effects that would seem to be unrealizable in nature.[77] Many such comments, made as if in passing, later proved to be very important. I'd like to present two examples in connection with this. A.M. Budker recalled that his desire to work on colliding beams was initiated by something YaB said during one of his talks: 'This question could be elucidated in experiments with colliding beams, however it is probably impossible to create them.' Budker said that the word 'impossible' cut him to the quick, and he began to search for ways to realize the idea technically. Subsequently, when this became possible and prominent discoveries were made in colliding-beam experiments, Budker frequently referred to this comment by Zeldovich.

Another example is a comment made in the classical pre-war work of Zeldovich and Khariton on fission chain reactions. Discussing the possible

[77] It is characteristic that Landau also loved to talk about such things within a narrow circle of colleagues, but tried not to refer to them in his published work. In my opinion, this sometimes led to certain losses to science. There are few now who know that, in the well known work by Landau, A.A. Abrikosov, and I.M. Khalatnikov, there was initially a sign error in the formula relating the physical and seed charges. The situation that arose because of this was very pleasing to Landau, and he fully developed a philosophy of asymptotic freedom, which now has a place in quantum chromodynamics. However, when A.D. Galanin and B.L. Ioffe pointed out the error to Landau and he was able to correct the article before its publication, this 'philosophy' was omitted in the new version (which led to zero charge). Not many people know about this. If it had been known, this would undoubtedly have helped subsequent researchers in the field of quantum chromodynamics. (In any case, Yu.B. Khriplovich, who changed the sign in quantum chromodynamics told me that he was not aware of this event.)

realization of fission reactions in natural uranium, the authors noted that approximately a billion years ago, when the percentage content of ^{235}U was substantially higher, this would have been comparatively easy. The importance of this comment was demonstrated after the discovery of a natural reactor that operated in Africa about two billion years ago. YaB was very distressed that, having made this comment, he hadn't gone on to predict the existence of natural reactors in the past.

By the way, I should note that YaB, a very sincere and direct person, never hid his annoyance with himself in situations when, having embarked on the right path to solving some problems he had set himself, he bypassed some consequences or the final solution for a question. However, this did not prevent him, like a true scientist, from being delighted with the results obtained by other investigators, and from praising them in every possible way. I remember how YaB criticized himself for the fact that, having considered the creation of pairs in the field of a rotating black hole, he didn't think of the general effect of black-hole evaporation, and was delighted with the results of Stephen Hawking. I also recall the following words of Dau at the end of the 1950s: 'YaB has amazing flair and luck.' This is no longer a secret, so I can tell you about it. When YaB first raised the question of thermonuclear reactions, it became clear that it would not be possible to bring them about in deuterium, since the effective cross section for *dd* reactions measured experimentally was too small. Then, YaB began to present, as it then seemed to me, a completely unfounded suggestion that the cross section for the reaction of deuterium with tritium should be substantially larger. When it was measured, this cross section, indeed, turned out to be 100 times higher than the *dd* cross section, due to the resonance of the ^{5}He nucleus, which we hadn't considered. However, even before this, YaB's confidence in the large cross section for deuterium–tritium reactions was so great that they had already begun to build factories.[78] Later, when Sakharov was able to find an

---

[78] In reality, a narrow circle of people involved in the hydrogen bomb project already knew about the large deuterium–tritium cross section from intelligence data. The fact that YaB did not have the right to tell about it even to Landau, characterises the situation in which they worked. By the way, V.L. Ginzburg, who suggested to use solid LiD in the hydrogen bomb (the second idea, in the terminology of A.D. Sakharov), did not know about the large deuterium–tritium cross section, either.

Initially, V.L. Ginzburg assumed that the reaction between neutrons occurring in an atomic bomb explosion and lithium-6 isotope nuclei (in which reaction tritium is formed!) could be used only as an additional source of energy (but not to produce tritium reacting to deuterium).

1. History of Soviet Atomic Energy Project, *Proc. Int. Symp. HISAP-96*, Dubna, 1996; Moscow, 1999.

2. Goncharov, G.A. Thermonuclear Milestones, *Physics Today*, **49 (11)**, 44 (1996); Goncharov, G.A. American and Soviet H-bomb development programmes: historical background. *Physics-Uspekhi* **39 (10)**, 1033–1044 (1996); Goncharov, G.A. On the history of creation of the Soviet hydrogen bomb. *Physics-Uspekhi* **40 (8)**, 859–867 (1997); Adamskii, V.B. and Smirnov, Yu.N. Once again on the creation of the Soviet hydrogen bomb. *Physics-Uspekhi* **40 (8)**, 855–858 (1997).

exceedingly efficient technical solution to the problem, YaB told everyone: 'What have I done? But look at Andrei!'

My collaboration with YaB continued after my matriculation as a post-graduate student with Landau. One new push for our collaborative activities was the discovery in 1957 by Alvarez *et al.* of muon catalysis nuclear reactions in hydrogen, and the establishment of the universal weak interaction. We saw that these two areas were connected, since various meso-atomic and mesomolecular processes in hydrogen that determine the behaviour of μ catalysis also substantially influence the capture of muons by protons. It was essential to study this phenomenon experimentally in order to comprehensively verify the theory of universal weak interactions. In the field of mesomolecular processes, we were able to discover a number of unknown mechanisms that explained the existing experimental data and predicted new phenomena. In June of 1958, when YaB was elected as a full member of the USSR Academy of Sciences, he chose precisely this theme for a talk at the Physical Mathematical Division of the Academy (since it was forbidden to speak about the results of the classified research to which YaB had devoted his energy and talent over 15 years). I remember that the morning before the talk, YaB asked me to visit, and we discussed some details of his presentation. And then I, having listened to his talk, spent almost the whole day worrying at the Institute of Physical Problems, waiting for the results of the vote. There was some cause for such worry. Landau, stopping into the office of his post-graduate students, had talked about various pre-election academic intrigues. He feared that YaB's independent character, for whom work always occupied first place, might arouse discontent among certain academicians. In this way, I first came up against the 'wrong side' of the Academy, and suffered deeply for YaB. Happily, it all ended favourably, and YaB was elected by the majority as an academician. (Dau described how Kurchatov's presentation in support of YaB had made a large impression on those present.)

Under the influence of YaB, muon physics determined the area of my scientific interests for many years. With his approval, I moved to work in Dubna, where several experimental groups of the Laboratory of Nuclear Problems (including those of B.M. Pontekorvo and V.P. Dzhelepov) had begun to study μ capture and μ catalysis, while the group of Yu.D. Prokoshkin had begun to design experiments for the detection of pion β decay. YaB's idea about the possibility of resonance amplification of the formation of mesomolecules (when they have a level with a small binding energy)[79] served for us as a guiding thread for the interpretation of the unexpected results from the groups of Dzhelepov and Ermolov, who detected an increase in the output of μ catalysis during deuterium fusion reactions as the temperature increased. The possibility I noted for the

[79] This idea was contained in work published by YaB in 1954 (before the discovery of Alvarez's group), with the first estimates testifying to the possibility of observing μ catalysis in a mixture of hydrogen isotopes.

existence of a vibrational–rotational level of deuterium mesomolecules with a binding energy of several electron volts led the Estonian physicist E. Vesman, then my postgraduate student, to the discovery of a peculiar mechanism for the resonance formation of deuterium mesomolecules; subsequent very refined calculations by L.I. Ponomarev's group not only confirmed the existence of this level in the deuterium mesomolecule, but also established the presence of an analogous level in the deuterium–tritium mesomolecule. This last result proved to be very important, and led to the prediction in 1977 that a single muon could give rise to about 100 nuclear fusion reactions in a deuterium–tritium mixture. Subsequent experiments conducted in the USSR (Dubna), USA, and Switzerland confirmed these predictions and put forth (in accordance with the idea of the Leningrad Institute for Nuclear Physics researcher Yu.V. Petrov) μ catalysis as an alternative method for the processing of nuclear fuel. In the course of this work, it was elucidated that the most important restriction on the efficiency of μ catalysis was the 'sticking' of muons to helium nuclei indicated by YaB, which arises as a result of nuclear fusion reactions. Currently, more than fifty groups in the world are working in the area of μ catalysis, international conferences regularly take place, and a specialized international journal is published.

In 1960, YaB and I wrote a review dedicated to μ catalysis for *Uspekhi Fizicheskikh Nauk*. During our work on this, YaB gave me his notebooks from 1957, after Alvarez's discovery. Reading these notebooks allowed me to penetrate more deeply into YaB's creative laboratory. Judging from their contents, the notebooks were written during his time off work at the 'object' and in Moscow. They demonstrated how YaB worked and thought: he set himself a problem, made estimates, rechecked them using various methods. He didn't shun routine calculations, computations of integrals, extrapolations, or numerical estimates. The notebooks include brief summaries of Alvarez's work (together with a derivation of his results done by YaB himself), and also Frank's work, published in *Nature* in 1947, and Sakharov's report, published in 1948 (YaB learned about these last two papers only in 1957). One also encounters traces of discussions between YaB and Sakharov from that time: 'ADS proposes ...,' 'the very deep idea of ADS,' and so forth. One of the notebooks contains a manuscript for some joint work by YaB and Sakharov.

In our review, I tried to develop some of the ideas expressed in YaB's notebooks, guided by certain beliefs about how, in his opinion, a review should be. 'In a review,' said YaB, 'there must be, first of all, a good introduction, which presents a clear formulation of the problem that is accessible to any physicist, even if he hasn't worked in this area. Furthermore, it must be complete. Don't be afraid to explain general questions when this is necessary in order for the main content to be useful both to specialists and to those wishing to enter work on the problem in question.' I tried always to be guided by these instructions, and when now, after 40 years, I still encounter references to our review, I remember these instructions with gratitude. I

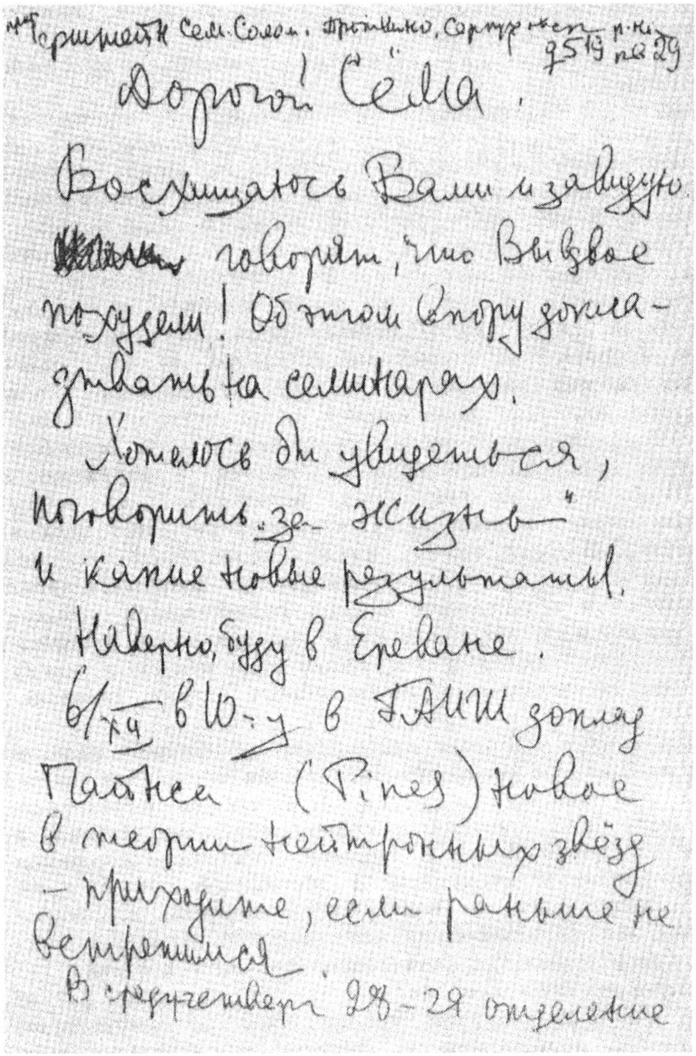
Дорогой Сёма.
Восхищаюсь Вами и завидую.
говорят, что Вы вдвое похудели! Об этом вопросу докладывать на семинарах.
Хотелось бы увидеться, поговорить „за жизнь" и какие новые результаты.
Наверно, буду в Ереване.
6/XII в 10-у в ГАИШ доклад Пайнса (Pines) новое в теории нейтронных звёзд — приходите, если раньше не встретимся.
В среду-четверг 28–29 отделение

presented L.I. Ponomarev with YaB's notebooks, for use in the education of his students, who, in essence, are YaB's scientific 'great grandchildren.'

One of the most important and acclaimed accomplishments of modern physics is undoubtedly the understanding of the deep internal connection between the theory of elementary particles and cosmology. YaB was at the source of this flow of research. In the 1960s, when his interests had shifted toward astrophysics and cosmology, he entered these fields enriched with knowledge of the latest problems of elementary-particle physics, and immediately began to obtain interesting results at this junction of different scientific disciplines. Since those years were marked by major discoveries in each of these fields, YaB worked in two directions simultaneously. He enthusiastically adopted the idea of quarks and estimated (together with L.B. Okun'

PS: Берегитесь!
Вот сообщу Вашей жене,
что Вы можете получить
открытие и тугрики и
не делаете этого —
тут и конец мирному
сосуществованию

A note from Ya.B. Zeldovich to S.S. Gershtein:

Dear Syoma!

I am both proud of you and envious. I hear you've lost half your own body weight. It's time to hear you give a talk about this at the seminars.

It would be nice to see you, to talk about life and new results.

I'll probably be in Erevan.

On December 6 at 10.00 there will be a talk by Pines at the Sternberg Institute on a new theory of neutron stars — come along, if we don't meet before then.

On Wednesday–Thursday 28–29 there will be the meeting of the section.

P.S. Watch out! I'll tell your wife that you have the opportunity to make discoveries and get bonuses and that you don't do it — that will be the end of peaceful domestic bliss.

and S.B. Pikelner) the possible concentration of free quarks in the surrounding medium in the framework of a hot theory of the universe. After this, he initiated experiments to search for free quarks in matter. I remember that he took me to the laboratory of V.B. Braginsky in order to acquaint me with such experiments, and to discuss them. To a large degree, the negative results of these experiments allowed modern concepts on the impossibility of the existence of free quarks to be established, and gave a push to the search for mechanisms for quark confinement.

In 1966, YaB and I were invited to Hungary by Professor Marx. It was the first trip to a foreign country for both of us. YaB was, as they say, a hit. Many people who hadn't known him earlier were enchanted not only by the depth and breadth of his knowledge, but also by his energy, brilliant wit, and resourcefulness in discussions. It was during discussions on Lake Balaton that our idea about the possibility of obtaining restrictions on the total mass of all types of stable (or quasi-stable) neutrinos from cosmological data was born.

YaB loved Hungary very much, and visited it again several times. He much regretted that he was not able to go to the 'Neutrino-72' conference in Balaton, which brought together many eminent Western physicists, including R. Feynman. I had the pleasure of talking with Feynman for several hours. He was interested in work on supercritical charge begun by YaB and me and then continued by YaB with V.S. Popov, who discovered one rather important error in our earlier work. Feynman wrote the problem for $Z > 137$ on a special card, which he placed in his pocket. He said that a desire to meet YaB had been one incentive for him to come to the conference, since he learned from V. Telegdi that Zeldovich often attended neutrino conferences in Hungary. The possibility of travelling outside the 'socialist camp' for contact with major Western physicists (which could have been so fruitful for science) was opened to YaB only at the end of his life.

Under YaB's influence, I also began to become interested in astrophysics. M.Yu. Khlopov and I joined the group of my university friend V.S. Imshennik, who was working on the problem of supernova explosions. We were able to find a mechanism for neutrino ignition of thermonuclear reactions in stars. In self-consistent calculations, this mechanism could provide an explosion of a star with either the formation of a pulsar or the complete destruction of the star, with the release of a large amount of energy consistent with that observed. YaB was interested in our work and clarified many points in connection with it. It turned out that our mechanism for combustion and heating of material ahead of a front had been considered by YaB while he was still in his youth. In this way, it was remarkable that several of YaB's varied fields of interest came together in a single phenomenon: the physics of stars, the weak interaction with participation of the neutrino, combustion, and explosions.

****

My acquaintance with YaB has allowed me to learn of several episodes from his life, about which I'd like to write. In the pre-war years, and especially at the beginning of the Great Patriotic War,[80] he expended a lot of effort and energy on strengthening the defence of our country. In this area, his activities certainly weren't reduced to purely theoretical investigations, and frequently involved large risk to his life.

Once, when we were travelling together to Dubna and the driver of our

[80] The part of World War II beginning from the onset of Nazi aggression toward the USSR in June 1941. [Translator]

car turned out to be a middle-aged woman, YaB recalled how, at the beginning of the war, a woman driver had taken him and a colleague to a test site for experiments with high-explosive shells, with which the car was heavily loaded. At some point, the woman, exhausted by lack of sleep, fell asleep at the wheel, and the car veered off the road and turned over. Fortunately, the detonators, which were in his colleague's pockets, didn't react upon the impact, and there was no explosion. Another time, YaB and his colleague were miraculously saved during an explosion that took place when they were conducting research aimed at ensuring the stable burning of powder in reactive shells intended for the well known 'Katyusha' rockets.

And here's another episode in YaB's activities, about which I learned by chance. Looking over the just-published two-volume collection of Landau's works, YaB pronounced with respect: 'Dau has almost no incorrect papers.' Then, after a brief silence, he added: 'However, we did get stung on one of them. Landau proved that a tangential break of a supersonic flow was stable. Based on this, Khariton and I began to experiment with supersonic flows of hydrogen. We hoped to create on this basis flame-throwers capable of igniting Nazi tanks. However, it turned out that instabilities arose in supersonic flows in virtually the same way as in subsonic flows. At that time, we had many other problems to work on, and we didn't have time to spend on sorting out the theoretical side of this question. This was later worked out by S.I. Syrovatskii.' Listening to this story, I again was struck by YaB's phenomenal ability to apply the results of fundamental science to applied problems, and thought about the stupidity and incompetence of the bureaucrats who had forbidden work in multiple organizations, even for scientists of YaB's rank. (With the announcement of the official position banning multiple workplaces in scientific institutions, both Landau and Zeldovich were forced to leave the Institute of Theoretical and Experimental Physics, where they had worked part-time).

YaB never talked with me about what he called his 'special business.' For me, there is no doubt, however, that his role in the solution of the atomic problem includes far more than just his personal contribution. I believe that the breadth of his scientific interests was of invaluable importance, since it meant he knew people with various specialities, whom it was possible to attract to work on solving the huge number of specific problems that arose. In recent years, I've had occasion to meet some people inclined (after the event) to be sceptical about the value of this aspect of YaB's activity. I consider this to be deeply wrong and unjust. It is not possible to judge the past based on the moods and situations in the 1990s. In the first place, all this activity began during the war, in years of deadly peril. In the second place, in the post-war decade, many people working on the 'problem' (not only YaB, but also I.E. Tamm and A.D. Sakharov) sincerely believed that establishing a nuclear equilibrium could be the only way of preserving peace. They served this goal to the full measure of their talent and ability, devoting the best and most productive years of their lives. The situation changed at the beginning of the 1960s, when the danger of militarization and the opposition between global powers became clear. Here, I must again refer to words of Landau, spoken in the autumn of 1961. Dau talked about how he, perturbed by our new nuclear tests, which had resumed after a long moratorium, literally threw himself at YaB with the words: 'Is it your organization that has put the government up to these new tests?' 'No,' answered YaB, 'we didn't need this. We had people opposing it.' The recently published excerpts from Sakharov's memoirs shed some light on this story. As far as I know, it was at the beginning of the 1960s that YaB tried to move away from 'specialized' work. He himself let slip in his conversations with me that this was far from easy.

****

With the years, I developed a need for regular contact with YaB, calling him or stopping by to see him, in order to talk about scientific news, consult with him about my work, learn what he was currently working on. Sometimes, YaB wrote me letters (in general, he wrote easily — like he talked). Some of these contain funny drawings. These letters are very dear to me: the very personality of YaB is evident in them.

Once, on a Monday, returning home from lectures, I felt an irresistible desire to stop in to see YaB at the Institute of Physical Problems. There, I met S.I. Anisimov, who told me that YaB had left twenty minutes earlier. 'No problem, I'll stop by another time,' I said. There was no other time. Two days later, I learned about YaB's illness and sudden death.

He loved verses, knew and quoted many, and could himself compose them impromptu in both Russian and English. When I now think of YaB, I involuntarily recall the verses of D. Samoilov, written upon the death of Akhmatova:

> That's all. The eyes of genius are closed.
> And when the heavens dimmed and blurred,
> As if in a deserted abode,
> Our voices could be heard.

Afterword

The desire to write an afterword arose after I was able to become acquainted with the full text of A.D. Sakharov's memoirs, thanks to E.G. Bonner. It shows clearly what a huge role the mutual attraction between Sakharov and Zeldovich played in Sakharov's scientific activity, along with the support that Zeldovich gave him, both in the early stage of his 'speciality' and in later years, when he was caught up in the problems of elementary-particle physics, gravitation, and cosmology. AD spoke about this also in his valedictory address at YaB's funeral service. At the same time, in AD's memoirs, which were written during the most difficult years of exile, under conditions of almost total isolation, one encounters reproaches, and sometimes even open hostility, towards YaB. These parts of the text upset many people who knew both Sakharov and Zeldovich well. It seems to me that some of AD's reproaches stem from mutual misunderstanding (which under other circumstances could have been simply resolved by talking). This has prompted me to write an addition to my memoirs about YaB.

Of course, relations between people of the calibre of Zeldovich and Sakharov are often not simple. However, throughout my acquaintance with YaB, I was constantly aware of his delight with AD's talent, and an exclusively cautious, I would say anxious, attitude toward this talent.

This attitude was not at all disrupted by the Academy elections of 1953, when YaB's nomination was unsuccessful, and, according to AD's memoirs, Sakharov, having been elected, found himself in an awkward situation with regard to YaB. YaB's deep belief in Sakharov's talent and the fact that, in essence, his potential had not been realized in full, provoked in YaB a sort of anger against any circumstances distracting AD from his scientific activity. Once such distraction was AD's social activities and his struggle for human rights. I don't discount the possibility that YaB directed his anger toward those surrounding AD, considering them guilty of distracting Sakharov from his scientific work; this, in turn, could not help but call forth an especially painful reaction from AD.

I cannot say that YaB didn't sympathize with AD's goals and courage. Once, in the spring of 1970, in answer to a question about AD, YaB said, half joking and half serious: 'There is no doubt that AD is a great man, and great men sometimes make great 'mistakes'. For example, if somebody else had spoken out in the Academy against Lysenko, Khrushchev, no doubt, would have dissolved the Academy, but in the case of AD, Khrushchev had been removed. I think that AD should be awarded the Nobel Peace Prize: here, engaging in the activities in which he is engaging requires no less courage than that of Martin Luther King.'[81] YaB considered the human rights activities of Sakharov a 'mistake' because he didn't believe that they could be successful under the circumstances of that time, and could influence to some degree the situation in our country and in the world. He was also not spared anxiety about AD's personal fate and the loss to science if Sakharov's scientific potential were not fully developed. Once, in connection with this, YaB quoted to me the splendid verses of Pushkin: 'The desert sower of freedom, I went out early, before the star ...,' written, in the words of Pushkin, 'in imitation of the fable of the moderate democrat Jesus Christ,' and containing as a conclusion: 'But I lost time for nothing, happy thoughts and labour ...' In a very narrow circle, YaB also expressed annoyance at individual blunders which he perceived AD to have made. For example, he was genuinely

---

[81] It is curious that, in the end, YaB's prediction was correct, not only about the awarding of AD's Nobel Prize, but also with regard to the 'mistakes' of this great man. AD could easily have died in exile, but *perestroika* began, and the 'new thinking' emerged.

surprised that it was possible, when speaking in defence of Pablo Neruda to Pinochet — who had displaced a lawful, liberal president and shot thousands of people — to talk of the consolidation promised by his regime. In my opinion, this, and also YaB's concern for AD, can explain the rather sharp telephone conversation about this described by AD in his memoirs. I don't believe that YaB used this conversation over a tapped phone line to try to demonstrate his loyalty to the regime, since he displayed 'disloyalty' to a much greater degree, firmly refusing to sign any letters directed against AD. YaB understood that such letters from academicians would serve as a signal for Sakharov's general persecution, to which (as proved to be the case in reality) would be added the letters of workers, collective farmers, writers, and composers.

In difficult and stagnant times, YaB himself helped and literally saved many people. I know about some cases myself, and I am sure that there are many more which I don't know about because YaB never talked about them. He was a realist, and believed that, under the circumstances, the most effective method was 'quiet diplomacy,' that is, personal appeals to those on high, rather than public protests. In this regard, as YaB told me, he shared full mutual understanding with P.L. Kapitsa (several of whose letters have now been published). I won't try to judge whether or not YaB was right; however, he well understood the limits of his own abilities, and of the services and realities that existed independently of him. And his possibilities were very limited. Frequently, he was not even able to take on these researchers he would have liked to (if problems were posed by their personal data or by the absence of a *propiska*[82], for example). He went to great lengths to try to get these people settled in appropriate positions, in order to have the opportunity to continue working with them. (Many colleagues used him in this capacity, and he went to these lengths not only for his own students.) I believe that it was concern about the fate of the scientific fields that he had developed and about the fates of many people associated with him (rather than concern for his own position) that strongly limited YaB's freedom of activity. Sometimes, YaB's appeals 'higher up' were refused in rather unceremonious, and even humiliating ways. The following story is typical. In 1980, L.B. Okun' received an honourable invitation to give the concluding talk at the International Rochester Conference on high-energy physics in Madison

[82] In the Soviet Union, each citizen had a *propiska*, assigning them to some particular address. A citizen was allowed to live only at the address indicated on his *propiska*. Without a *propiska* in a particular region, it was not possible to obtain work in that region; at the same time, it was usually impossible to obtain a *propiska* in a region without employment there. This bureaucratic catch-22 was effectively used by the Soviet government to control where its citizens were living and working, and ultimately all aspects of their lives. In Russia in the early 1990s, the *propiska* system was officially replaced by a less restrictive system of registration, in which citizens are, in principle, free to live and work where they wish, but are required to register their addresses with the local government. Unfortunately, by the time of writing in 1999, this freedom is still at times only theoretical. [Translator]

(USA). After obtaining the required permission from the Committee on Atomic Energy, Okun' accepted the invitation and prepared his talk. In it, he emphasized the constant need for modern physics to search for scalar particles, including Higgs bosons. (Subsequently, this talk became famous, and in many ways determined the direction of future experimental studies, and even projects for new accelerators.) However, a highly placed bureaucrat in the Central Committee[83] forbade Okun's trip at the very last minute for reasons which were not 'clear'. After unsuccessful attempts by various people to try to change this decision, YaB called the secretary of the Central Committee, M.V. Zimyanin. In response, Zimyanin reprimanded A.P. Aleksandrov, the President of the USSR Academy of Sciences saying that he had let his academicians get so out of hand that they were shoving their noses into other people's business, and daring to telephone the secretary of the Central Committee. I should say that Okun's failure to appear at the conference and the disruption of the last talk became a scandal, and resulted in a substantial loss to our science.[84]

It stands to reason that our country underwent an even greater loss due to the exile of AD and his hunger strike. However, this was impossible to explain to the higher authorities of that time, who firmly maintained a completely different view of this matter. Therefore, when AD wrote to YaB and other academicians with a request to help send a person close to him abroad (Liza Alekseeva, the fiancée of E.G. Bonner's son), YaB said that he would do everything he could to try to ensure proper conditions for AD's scientific work, but it was not in his power to send somebody abroad. This response provoked a sharp reaction in AD. In turn, YaB, the only person to answer AD's appeal, suffered greatly when he learned of AD's reaction to his letter (and he learned of it from a foreign radio programme that quoted individual phrases from that letter). The entire affair gave YaB many bitter hours.

In conclusion, I would like to consider one other question touched upon in Sakharov's *Memoirs*. According to AD's thinking in the 1980s, the problem of creating thermonuclear weapons, on which YaB had worked, and to which AD had contributed in 1948, was 'stolen in full,' that is, based on information acquired by Soviet intelligence. AD's reading of foreign sources made his opinion about this even more firm. Due to a series of circumstances,

---

[83] Of the Communist Party of the Soviet Union. [Translator]

[84] In spite of the incomprehensible explanations of the Soviet delegates ('he didn't show up at the airport,' 'he couldn't manage to get a ticket,' and so forth), the 1,000 participants in the international conference understood perfectly well what had happened, and unanimously made an official protest. Understanding that they had made a mistake, high governmental officials themselves insisted on Okun's participation in another international conference a year later. However, the consequences of the scandal could not be wiped away. Together with the protest against Sakharov's exile, the Okun' scandal served as an additional argument against having the next Rochester Conference in the Soviet Union (though this had been planned earlier). In this way, a large number of Soviet physicists were deprived of the opportunity to participate in it.

material shedding additional light on this question came into my hands several months ago. In the August 1990 issue of *Priroda*, which was dedicated to Sakharov, I found in an article by Yu.A. Romanov a reference to the fact that the question of creating a hydrogen bomb was first raised in the USSR in 1946, in a special report presented to the government by I.I. Gurevich, Ya.B. Zeldovich, I.Ya. Pomeranchuk, and Yu.B. Khariton.

Interested by this, I stopped by to see Gurevich and asked him, if possible, to tell me about this report and to comment on AD's suggestion. Isai Isidorovich said that, in 1946, they had no indication that scientists in America were working on this question. It was simply that the deuteron and nuclear reactions between light nuclei were among his interests and those of I.Ya. Pomeranchuk, since they provide information about nuclear forces and represent the source of energy in stars. In their joint discussions, Zeldovich and Khariton noted that it was, in principle, possible to bring about thermonuclear fusion under terrestrial conditions by heating deuterium in a shock wave initiated by an atomic explosion. Their estimates showed that, under these conditions, it was possible to avoid the transfer of a large fraction of the energy released into electromagnetic radiation, and thereby to obtain an explosion of an unlimited quantity of the light element. This is how their joint proposal, which they made to Kurchatov, arose. 'It's possible that I can even show it to you,' said Gurevich, 'It's probably preserved in the archives of the Institute of Atomic Energy.'

Indeed, a few weeks later, I held in my hands a certified copy of this proposal, containing seven pages of typewritten text, with the formulas written in by Gurevich's hand and with the note '1946' made on the last page by Kurchatov. 'Here's your clear proof that we didn't know about any American projects,' said Gurevich, indicating the title page of the text. 'Imagine how deeply classified it would be, and behind how many seals it would have been preserved otherwise.[85] I agreed, though it remained unclear to me why it was not given any secret status at all, and was simply put in an archive. Gurevich explained this as follows: 'I think that, at that time, they simply waved us away. Stalin and Beriya urged on the creation of an atomic bomb. We didn't yet have an operational experimental reactor, and here the scientific "sages" show up with new projects that it might not even be possible to carry out. I didn't work on this any further, and how events developed after that, I don't know.'

---

[85] This report is published in full in *Uspekhi Fizicheskikh Nauk*, 1991, vol. 161, iss. 5, p. 170. Apparently, I.I. Gurevich was not aware that, as early as the middle of 1945, Fuchs had made reports about discussions of possibilities for creating a hydrogen bomb at Los Alamos. Therefore, he didn't know that a report he was involved in was preserved with the highest secrecy stamp in the archive of the Ministry of Middle Machinery (see I.I. Gurevich, Ya.B. Zeldovich, I.Ya. Pomeranchuk, and Yu.B. Khariton, *Soviet Physics-Uspekhi* **34 (5)**, 445–446, 1991; Goncharov, G.A. On the history of creation of the Soviet hydrogen bomb. *Physics-Uspekhi* **40 (8)**, 859–867 (1997); Adamskii, V.B. and Smirnov, Yu.N. Once again on the creation of the Soviet hydrogen bomb. *Physics-Uspekhi* **40 (8)**, 855–858 (1997)).

At the sport camp of the Space Research Institute on the Volga. 1973

Judging from Sakharov's *Memoirs*, in the middle of 1948, theoretical calculations in connection with this proposal were already being developed under the supervision of Zeldovich and A.S. Kompaneets in the Institute of Chemical Physics of the USSR Academy of Sciences. Tamm's group, which included AD, was created in order to verify them. Later experimental studies were conducted at the 'object.' They yielded extremely interesting scientific results; in particular, temperatures of about a billion degrees were achieved.[86] The fact that, in 1948, all work on nuclear fusion was carried out in the deepest secrecy (even the innocent report of AD on μ catalysis was classified) indicates that the government understood the importance of the problem by that time. It is possible that it was Khariton and Zeldovich who were able to convince the government of this (Landau once said with irony:

[86] See the contribution of L.P. Feoktistov in this collection.

'YuB and YaB are our Soviet saints. For the sake of the cause, they are prepared to argue with the big bosses, and to persist when the bosses don't understand.'). However, it is also possible that the government received intelligence data about similar projects in the USA. This is quite probable, given the inertia of the state machine, which was not inclined to especially trust its 'clever men.' (It is clear, for example, that no desperate letters from G.N. Flerov to Stalin about the need to create an atomic bomb could have been effective if there hadn't been intelligence data about similar work outside the country.)[87] The device based on the original idea expressed in the work done by YaB and his colleagues could not be realized, and YaB introduced substantial changes to its construction. However, his familiarity with calculations carried out at the Institute of Chemical Physics in connection with this project made it possible for AD to get up to speed, and during his study of the project, he had a new idea (or, as he called it in his *Memoirs*, his first idea) which, together with his second idea, led to the first successful test of a hydrogen bomb in 1953. AD and YaB worked together on the so-called third idea (radiative implosion), which led in 1955 to the virtually complete solution of the problem. In addition, a huge contribution was made by their younger colleagues (Yu.N. Babaev, Yu.A. Trutnev, L.P. Feoktistov, and others).[88]

Of course, today it is very difficult for people who grew up in the post-Stalin era, and especially outside the Soviet Union, to understand how such noble and extremely honest people as Tamm, Sakharov, Pomeranchuk, Zeldovich, and many of their colleagues could work so selflessly on the creation of a horrifying weapon, devoting all their knowledge and talent to this work, displaying extraordinary initiative, persistence and creativity, and not acknowledging what danger such weapons posed for the fate of the entire world in the hands of a totalitarian system. This was not simply the

[87] This calls to mind a not altogether decent comparison made by YaB: 'There is a sort of impotence when a person can't become aroused until he sees that somebody else is already acting.'

[88] In reality, A.D. Sakharov proved right by assuming that the original idea of a thermonuclear device was based on the 'tube.' It turned out that already in 1945 Soviet intelligence received reports of the discussions held in Los Alamos and, in particular, of the project of the 'Classical Super' proposed by E. Teller. I.I. Gurevich, undoubtedly, did not know about this, and their joint work was evidently initiated by I.V. Kurchatov himself.

As the government demanded the repetition of American samples in the first instance, YaB was forced to develop an analogue of the 'Super' However, in late 1953 he had the courage to insist on the closure of that project as a little-promising one. Sakharov did exactly the same by rejecting his own project (layer-cake or sloika) which, as he wrote in his memoirs, was imprudently announced by him and approved by the government.

Exactly after this, as a result of a brain storm organised by Ya.B. Zeldovich, A.D. Sakharov, and the team of their talented colleagues, an idea of radiation implosion was born (the third idea, in the terminology of A.D. Sakharov), which solved the problem of developing various systems of thermonuclear weapons.

interest of researchers in an unusual physics problem (although this was also very important). The motives driving these people are clearly laid out in Sakharov's *Memoirs*. Here there is nothing to add or remove.

This afterword does not in any way aim to somehow make excuses for YaB. He does not need this. After YaB's death, Sakharov wrote about the fact that all the disagreements between them seemed to him 'no more than foam carried by the flow of life.' I would like it if people's memories could preserve the creative collaboration between Sakharov and Zeldovich, which was so fruitful, and the fact that their signatures often stood side by side (as in the case of the manuscript of their joint paper 'On Reactions by μ Mesons in Hydrogen,' which is preserved among my documents).

Blessed with childhood

L.B. Okun'

On that frosty December day, the doors of the apartment were wide open. People came and went. Boris Yakovlevich Zeldovich sat at a large, round table and composed a list of awards his father had received during his life. In a large box lay a jumble of Hero of Socialist Labour gold stars, various decorations of merit, gold medals from various academies. This list was intended for an obituary in the *CERN Courier*. To go with the obituary, a photograph was chosen, a copy of which is now sitting in a frame on my bookshelf. It shows a man in a simple grey sweater sitting pensively and quietly, leaning on his elbows on a table.

It was rare to find him so quiet. He was usually active, and moved as quickly as mercury. I saw him for the first time in the Great Hall of the Polytechnical Museum in the early 1950s, at an evening dedicated to the fifth anniversary of the Moscow Mechanical Institute. He gave a talk about the theory of explosions. It was striking how merrily and energetically he could handle the formulas, blackboard, chalk, eraser rag, the audience. It was as though he was playing ping-pong, swiftly hitting away invisible balls.

In the middle of the 1950s, having returned to Moscow from the 'lair'[89] (his own term), he left his work on explosions and avidly threw his energy into elementary particle physics. At that time, he was often at ITEP and had long conversations with Chuk and with us, his students. (For those who don't know, I should explain that ITEP stands for the Institute of Theoretical and Experimental Physics in Cheremushki, and 'Chuk' was the nickname of Isaak Yakovlevich Pomeranchuk (1913–1966), the first student of L.D. Landau, and the founder and head of the Theoretical Physics Department at ITEP.) At one time, Yakov Borisovich was even on the ITEP

[89] The high-security establishment near Arzamas that was the centre for Soviet nuclear weapons research, also known as 'Ensk' or the 'object.' [Translator]

staff as a part-time researcher, but then decided to make his base at the Keldysh Institute of Applied Mathematics. It may be that two such outstanding and so dissimilar physicists as Chuk and YaB would feel too 'cramped' in a single research institute. In any case, they treated each other with great warmth and respect, in spite of the reprimands that were occasionally made by Chuk (described with such humour by YaB in his remembrances of him). Later, YaB often expressed regret that he had not remained at ITEP. But no matter where he worked, his connections with the physicists at ITEP — both scientific and personal — were never broken.

There were never any age or hierarchical barriers in contacts with YaB. He first came to my home at the beginning of 1958, when I, a young PhD scientist, lived with my wife and little daughter in a small room in a huge communal flat.[90] He wanted more detailed news about physics in the United States, where I had just been on a two-week trip and where, in particular, I had heard talks by R. Feynman and M. Gell-Mann about the theory of weak interactions. This theory revived the idea of conservation of the vector weak current, first put forward several years earlier by S.S. Gershtein and Ya.B. Zeldovich. When he went, he left a folder with reprints of his articles decorated with humorous inscriptions and drawings of little fishes.[91]

I think it was in that same year that he invited me to visit him on the evening of March 8, and I, unsuspecting, ended up at his birthday party. It was a merry evening. Funny, and sometimes not entirely decent, charades were enthusiastically performed (here, the efforts of A.S. Kompaneets were especially distinguished). On the wardrobe, stood a plaster bust of YaB coated with a bronze patina, given to him by one of his friends. It was a superb likeness. The sculptor had depicted with special care the bare hairy chest, decorated with three stars of a Hero of Socialist Labour. (The formal bronze bust was erected at the birthplace of Yakov Borisovich, Minsk, at the beginning of the 1980s.)[92]

Though people sometimes called him a 'hero three times over' behind his back, there was no trace of overbearing self-importance in him. He couldn't pass by a smooth, icy strip on a pavement without running up so that he could take a slide on it. After he came to ITEP, he immediately scolded I.Yu. Kobzarev and me for stooping, and proposed a special exercise, which he began to teach us immediately. 'Stand back to back with me,' he commanded, 'Move so that our spines are touching. Link arms bent at the elbows. Okay, I lean over and lift you in the air. Don't be afraid. Now you lift me. And now we'll lift each other in turns. But not too sharply, so that we don't drop each other.'

---

[90] A typical dwelling in the Soviet Union, consisting of a single apartment occupied by several families, with shared kitchen and bathroom facilities. [Translator]

[91] *Okun* in Russian means 'perch.' [Translator]

[92] It was traditional for a bust of a Double Hero to be erected at his birthplace. [Translator]

In general, he was very sensitive to others' ill health. Many years later, when I was in the cardiological unit, he brought me a book on deciphering cardiograms, and when I had finally recovered fully, he arrived with a frisbee, which we recklessly threw back and forth for a long time in the nearest empty lot. The spark of sporting ardour was sometimes kindled in him unexpectedly. I remember once when we were walking and discussing what happens to a black hole when its mass becomes zero as a result of radiation. At the metro bridge, he suddenly broke off our discussion and suggested, 'Well, who will be the first to run to the embankment along the stairway?' and we threw ourselves downward, bounding down the steps.

He especially loved to swim, and could swim for hours. I remember in May 1986 in Sukhumi, during breaks between sessions of the astrophysical symposium, he would swim far beyond the horizon. I still have several letters written by him over the years when he was in the Crimea. He thought them over while he was swimming. In these letters, he criticized himself for letting slip opportunities for scientific discoveries, reproached me for inactivity and scepticism, demanded that I persist in developing the quark model (this was still before quantum chromodynamics), and criticized me for my conservatism and excessive caution in the choice of articles for *Uspekhi Fizicheskikh Nauk*.

Uspekhi was an object of his constant attention. He often wrote articles himself; many of his reviews in *UFN* became classics. He persistently searched for young authors. He violently objected to the publication of articles that he considered to be poor. For example, he categorically objected to the appearance in *UFN* of an article whose essence was that the equations for the special theory of relativity could be written such that time would be the same in all reference frames but the velocities of light in the forward and reverse directions would be different. He, quite justifiably, did not consider this a 'success of the physical sciences'; when the article was accepted for publication, as a protest, he sent in his resignation. How he was convinced to return to the editorial board, I don't remember. I do remember very well how I once tried to convince him not to publish in one of his reviews a quatrain that purported to be the work of Khlebnikov. With a smile that I didn't understand at the time, he patiently listened to all my objections, but the verses (which turned out to be an acrostic composed by himself) remained in the text, and at the end of the review, he, not without a little venom, thanked me for discussions in connection with the work.

His last article, in defence of the general theory of relativity, written together with L.P. Grishchuk, was published in *UFN* after Yakov Borisovich's death (a day before his fatal heart attack, we had agreed to meet for a discussion of the article).

It was easy to discuss physics (and not only physics) with him, because he was never offended by criticism. It was always possible to tell him absolutely openly everything you thought about his latest idea or article. He was able to defend himself from attack very creatively and, by and large, successfully. However, if he was not able to defend his point of view, he honestly acknowledged his errors. I have met few people who could listen to critical comments with such enthusiasm.

Of course, he also didn't stand on ceremony in connection with his own evaluation of others' work. I remember once, in my office at the ITEP, we (several authors) told him, at his insistent request, about our work on neutrino oscillations. The work was purely phenomenological and, so to say, theoretically technical, while Yakov Borisovich kept waiting for some kind of physical punchline. When he was convinced that there was not going to be any punchline, he said, 'This reminds me of a tomcat who lured a female onto the roof and then took a long time depicting her at great length how he was castrated.' We didn't lure him onto the roof, but the comment was on target.

Discussions with him were unusually fruitful. In a magical way, he was able to compel his companion to transfer a vague half-thought from his subconscious to his conscious, and to think it through. If an answer to a question didn't satisfy him, he asked this question again a day later, a week later, a year later. He was organically incapable of forgetting it, of stopping halfway.

It was extremely interesting to discuss physics with him. An insatiable interest for new ideas and facts was always alive in him. He strove, and was able, to understand each new phenomenon in the simplest possible way, and to find new, very often completely unexpected, applications for it. It was precisely this talent that helped him to move several times into fields of physics that were new to him with surprising ease, and become a leader in these fields.

During one of our last walks (Vorobiev Road, Gorky Park, the Krymskii Bridge, lanes around Kropotkinskaya Street), he began to talk about the fact that he always strove to understand the essence of phenomena using simple arguments based on first principles, without using a complex mathematical apparatus. 'But maybe I shouldn't have aimed for this,' he said. 'Maybe it would have been better to try to use mathematical formalism as much as possible? After all, it's a very powerful tool.' However, more often than not, this question was merely rhetorical. I think that he understood perfectly well the uniqueness of the talent he possessed. Our discussion of this question remained unfinished, since we had arrived at the fence of the Institute of Forensic Psychiatry, and our conversation turned in another direction ...

In one interview, Einstein said about himself that he differed from most people in his preservation of the child-like ability to wonder at things that an ordinary adult would take in without wonderment. Yakov Borisovich possessed in full measure this rare gift. His ability to be amazed, openness to the discovery of new things, insatiable interest in life in general and science in particular, boyish excitement and mischievousness created a unique wondrous genius.

> 'Blessed with a kind of eternal childhood,
> With generosity and vigilance of stars,
> And the whole Earth was his inheritance,
> Which he shared with all lucky us.'

This was said by Akhmatova about Pasternak. The inheritance that Yakov Borisovich Zeldovich shared with all was the universe, the entire physical world.

'May I present the physicists of the twenty-first century ...'

D.A. Kirzhnits

I was acquainted with Yakov Borisovich for just under 40 years. However, we had neither any collaborative work, nor any kind of close personal relations. Therefore, it is better if I don't try to compete with the reminiscences of his friends, students, and colleagues. At the same time, we were linked by our many mutual scientific interests in widely varying areas of physics and cosmic physics (a fact noted by YaB himself at a seminar at the Lebedev Physical Institute on the day of my anniversary). Accordingly, there were many scientific contacts between us — at YaB's home, at the Lebedev Physical Institute, at various seminars and conferences. For this reason, I also have some memories. In these notes, I'll try, without pretensions to generalization, to write about several characteristic episodes of our relations, hoping to supplement or reinforce the image of YaB that has formed in the reader of this collection.

I learned about the existence of the physicist Zeldovich soon after the war, having received an invitation from the biology students of Moscow State University to tell them about the atom bomb. My own knowledge of this subject was restricted to my recollections of popular lectures by L.D. Landau and M.F. Shirokov and the little that I extracted from the bestseller of that time by H.D. Smythe, *Atomic Energy for Military Purposes*.[93] Therefore, I turned to senior undergraduate students that I knew and asked them if they knew of any other appropriate references. One of them remembered that he had seen an article about atomic energy in an old, pre-war, issue of *Uspekhi Fizicheskikh Nauk*. In this way, I stumbled upon the famous series of papers by Zeldovich and Khariton, written in 1939 and 1940, which lay at the root of atomic energetics.

My personal acquaintance with YaB was initiated several years later. I appeared in his flat on Vorobiev Road at the recommendation of my undergraduate thesis research supervisor, A.S. Kompaneets, who hoped to find me a place after my graduation from Moscow State University in the institute where YaB was working at that time. I was met by a not terribly young man (according to my criteria at that time), stocky, round-faced, wearing pre-war glasses, still with a very appreciable head of hair. Without any superfluous conversation, I was given five or six problems to solve on

[93] Princeton: Princeton University, 1945.

the spot simply, without much derivation (I remember the first — about the radius of formation of the statistical sum of a classical Coulomb gas). I wasn't able to deal with even one of them completely and clearly. Although I brought the correct answers the next day, it was clear that it was already impossible to make good this terrible first impression.

This episode is firmly etched in my memory not only because it marks the beginning of an acquaintance with one of the biggest and brightest people I have met in my life for the first time. In fact, I was sharply aware of a substantial gap in my professional education as a theoretical physicist, of a lack of intuitive physical thinking. Before this, I had not felt any insufficiencies, passing 'the minimum' with Landau, participating in N.N. Bogolyubov's seminar, and managing well with my undergraduate thesis work. It took years of factory engineering practice and work in the theoretical department of I.E. Tamm at the Lebedev Physical Institute for this gap to begin to close. But imagine my joy many years later when YaB said to my postgraduate student: 'Did DA really give you such a ... problem? After all, he's a good physicist' (the clip replaces the now forgotten indecent adjective). Therefore, I will always remember the goals for the future that I was set up by YaB, which were so important for all my professional activities.

In many ways YaB also helped me to realize these goals, especially in the first period of our contact, after my return to Moscow. It was typical for me to be more interested in the formal, theoretical side of things, and YaB 'swung me around' so that I was facing the physics of the question in hand. For example, at the end of the 1950s, after a talk by YaB at Landau's seminar, I got caught up in the problem of neutron fluidity (which still has not been clearly solved to this day): could the bulk of a system of neutrons keep itself from flying apart by its gravitational forces? I was fascinated by the unusual nature of the problem from the point of view of many-body theory: the mean distance between particles is large compared to the range of the forces, but small compared to the scattering length. YaB oriented me more closely toward the physics, in particular, toward the fact that the two-body problem differs from the many-body problem (for example, for point interactions, quantum effects are sufficient to avoid a collapse into the centre in the case of two particles, but not in the case of three or more particles; this is the so-called Tomas theorem). The physical depth and fruitfulness of this approach became clear ten years later, after the discovery of the Efimov effect.

Another example is the development of the Tomas–Fermi method, in which YaB became interested after listening to a talk by Kompaneets and myself at the same seminar series. For me, the most interesting aspect was the application of the non-trivial algebra of non-commutative variables (operators) in quantum mechanics. As far as applications to the physics of high pressures was concerned, I had worked in that area quite a lot, but only for illustrative purposes. Without overly pushing his own view, YaB firmly turned me toward the theory of equations of state, and forced me to understand its importance for geophysics, planetology, astrophysics, and the solution of a number of applied problems. In this way, he essentially determined one of the most important fields of my subsequent research activity, which proved to have its share of successes (two articles appeared in a list of the most cited papers on equations of state), and even included organizational activities (for many years I was the head of the periodically active All-Union Meeting on Equations of State). I refer to this with the single aim of underlining the role played by YaB in my life.

In connection with this last example, I remember that YaB refereed my review on the properties of matter at high pressures and temperatures. A typical referee report from a 'Captain' or even a 'Major' of science contains, along with a general evaluation, several concrete comments; as a rule, the referee avoids direct contact with the author. 'Field Marshall' Zeldovich invited me to his home, presented no fewer than a hundred comments, and worked for several hours on the clarity of the text. As we parted, he said, 'Well, we haven't achieved a Voltaire-like clarity, of course, but even so ...,' having in mind the citation closing the review presented in the *Micromegas* (a comparison of the small dukedoms in Europe with the empires of Turkey, Moscow, and China, as an example of contrasts intrinsic in nature).

Our contacts connected with astrophysics and cosmology were especially numerous. At the beginning, these were concerned with discussions of YaB's work from 1962, in which he draws the surprising conclusion that it is possible (albeit very slowly) to turn an arbitrarily small mass into a black hole. Many years later, this conclusion was refined by J. Beckenstein; according to the second law of thermodynamics, only a body with a mass greater than 10^{15} g (the mass of an average-sized mountain) can collapse to a black hole. I remember these discussions because, in this case, YaB didn't show his usual enthusiasm. He reacted to my delight with restraint, though, as I still believe now, there really is something admirable in this work — even the very fact that we have no information about the internal structure of a black hole means that it is perceived as a hot body. As Solzhenitsyn said in his Nobel-Prize lecture: 'Modern science knows that the suppression of information is the path to entropy, and to general disruption.'

Our astrophysical contacts include an exchange of 'courtesies' about priorities. At the end of the 1950s, I discovered that the matter of a 'white dwarf' star is not in a plasma, but in a crystalline state. Having become interested in nuclear reactions in the crystalline phase, I came across YaB's 1957 article about hydrogen burning in cold celestial bodies. Digging a bit further, I found the pre-war precursor to this article. Some time later, YaB solemnly presented me with a reference to earlier work for my article on white dwarfs.

We had especially detailed discussions about my activity with A.D. Linde on phase transitions in cosmology. As far as I know, it was YaB who first had the important idea of independent phase transitions in non-causally-connected regions (analogous to the independent crystallization in steel rapidly cooled in a mould, which leads to polycrystallization of the ingots). The consequences of this idea related to the domain structure of the vacuum and magnetic monopoles in the universe were successfully worked out by YaB and his co-authors. YaB suggested that I work on the most interesting consequence, connected with the formation of thread-like structures, but alas, I didn't heed his suggestion. And from this idea came the currently extremely topical problem of cosmic strings. We also discussed the problem of the cosmological constant in the Einstein equations, which, according to Linde, should change by many orders of magnitude during phase transitions. In spite of our efforts, YaB long doubted the possibility of distinguishing a cosmological constant from matter in a physically sensible way. Even when he agreed with our arguments, it was somehow reluctantly. This is all the more annoying because we were one step away from an inflationary scenario for the development of the universe, and who should have become the author of this scenario if not YaB, whose renaissance came about thanks to the cosmological constant?

In concluding these remarks, I'd like to touch on one less well known side of YaB's activities. While I was still working at a factory and coming up against the mathematical illiteracy of the average engineer, I thought about the need to reorganize the teaching of mathematics in high school and institutions of higher education with the aim of freeing it from a routine, rectilinear approach, and bringing it closer to the demands of real life. Therefore, I was especially interested in YaB's book *Higher Mathematics for Beginners*, which came into my hands by chance. At some point during a visit to YaB, we came to talk about this book. I asked whether the absence of traditional physical–mathematical approaches in it could be explained by the fact that YaB's higher education was in chemistry rather than physics. The answer 'I didn't study at an institution of higher education at all' completely amazed me. I then asked what moved YaB to take on the huge task of putting together the book. Instead of answering, he invited me into the dining room, where the young people were sitting — his children with their friends — and said, 'May I introduce you to the physicists of the twenty- first century, it was for their sake that I tried ...'

Not very long ago, when I first went to Minsk, the birthplace of Yakov Borisovich Zeldovich, I stood for a long time in front of his bust, depicting him as a Hero three times over, thinking about how generous nature is toward its chosen ones. And now that he is no longer among us, as well as the grief of the loss, I experience thankfulness that fate gave me the chance to have contact with such a powerful and bright talent.

Ten years of contact

A.A. Dolgov

It is December 1987. A clear, sunny day. A light frost. The weather couldn't do more to facilitate a good mood. I walk from the hotel in Protvino to a seminar at the Institute of High Energy Physics to give a talk on the relation between particle physics and cosmology. And suddenly — on the announcement board next to the auditorium, I see a small, handwritten note: 'Yakov Borisovich Zeldovich has died.' I re-read it several times with the sense that I've somehow ceased to be able to understand my native tongue. I remember that we had met a week earlier, had a heated discussion about problems in cosmology, talked about future work. YaB was, as usual, active, energetic, and, it seemed, full of strength.

I don't remember how I got through the seminar, which started with a minute's silence. I remember only that I was helped by the sacred respect for the seminar that was always taught by Yakov Borisovich.

One case was especially instructive for me. Several years ago, I was supposed to make a presentation at a seminar at the Sternberg Astronomical Institute, chaired by Zeldovich. One day before the seminar, I came down with a cold, called Yakov Borisovich at home in the evening, and asked him if it would be possible to move my talk to the next seminar. From the receiver came silence, followed by a polite inquiry as to whether or not I was on my death bed. When I answered in the negative, YaB said with a very icy tone that I never heard from him before or after, 'Then I don't understand why you are calling me.' He then digressed a bit and made a heated speech about how young people are losing their sense of respect for eternal values, and how there is a horrible decline in motivation, which has to be fought, otherwise society will simply perish. After this, of course, I couldn't even hint at the possibility of postponing my talk. Yakov Borisovich applied to himself the same, if not even higher, demands. At that time, I didn't yet know that that conversation would have something in common with his own fate ...

When I returned to Moscow from Protvino, I was told that on the morning of December 2, Yakov Borisovich had felt poorly and stopped in to see a doctor, who examined him and advised that he go to hospital for more extensive testing and treatment. 'I can't,' came the answer, 'I have a talk to give at a seminar today.' For any person (from student to experienced professor), a presentation is always a load, nervous pressure. I don't know how much this made an impact on Yakov Borisovich, but in the evening of that same day, his condition sharply deteriorated. The doctors said that there was nothing to be done.

I was acquainted with Zeldovich by Lev Borisovich Okun' in 1976. Yakov Borisovich had questions about the form of interaction of leptons in the then still relatively new unified theory of electroweak interactions. Our discussions led to my collaborative work with him and with M.I. Vysotskii on the cosmological lower limit[94] to the mass of a neutral stable lepton, which formed the beginning of our ten years of close contact.

[94] In Western scientific literature, this limit is called the Lee–Weinberg limit, from the names of the authors of this work, which appeared simultaneously with our own. This is quite understandable. However, I can't understand at all why the term 'Gershtein–Zeldovich limit' for the mass of the neutrino should never be used in the West, in spite of the fact that this result was obtained back in 1966, and all other cosmological limits on particle masses are essentially obtained on the basis of the arguments presented in this work.

Under the influence of Yakov Borisovich, I began to work in cosmology. My acquaintance with this fascinating area began with the writing of our joint review 'Cosmology and Particles,' in full correspondence with the stimulating credo of YaB: 'In order to study some new field of science, you have to write a review about it.' He could sense the importance of one or another direction of scientific research with surprising accuracy, took part in its development with enthusiasm, supervising, as it seems to me, with a lively, almost childlike, interest the new discoveries that arose before him. YaB never underestimated the importance of the symbiosis of cosmology and elementary-particle physics. Thanks to him, there arose a new field of science, based largely on the work and ideas of Zeldovich. Now, all major conferences on particle physics have sections on cosmology.

In remembering our collaborative work, I can't fail to mention how different his style was from the well-known stereotype of the master academician, a parasite on the work of his students.[95] Sometimes it even seemed to me that the situation was the opposite, and I had to argue against my inclusion as a co-author on papers for which my participation was limited to only insignificant discussions. In other cases, when I came to YaB with a new, sometimes very raw, idea, which fleshed out as a result of sometimes stormy discussions, he always categorically deleted his name from the author list if he felt that his role in the work was not fundamental. It seems to me that this is the result of the richness of physical ideas that he constantly generated over his entire life, and of the great generosity of his spirit.

The rare combination of the breadth and depth of his knowledge was striking. During a discussion of questions in elementary-particle physics, he could unconstrainedly invoke an analogy related to phenomena in the theory of combustion or statistical physics. In addition, he was always investigating something new, studying original papers and discussing questions of interest to him with those who he considered specialists in that area. Each such discussion was a non-trivial test for the other participants. YaB often had his own, non-standard view of a problem, having discovered it from a new angle, and it was not a simple thing to have a discussion with him 'on his own ground.' However, the result made all difficulties worthwhile. His ability to undertake encyclopaedic activity was described well by a prominent scientist, as far as I remember, Stephen Hawking, during his visit to Moscow: 'I thought that Zeldovich was a large group of Soviet physicists, like Bourbaki.'

One surprising quality of YaB that probably helped him work for so long and fruitfully was his always youthful, even boyish, spirit. Thanks to this, it was always easy to talk with him on any topic, in spite of the appreciable difference in our ages and positions. A number of episodes live

[95] By the way, I should confess that I've come across parasitic masters only in fictional literature and newspaper periodicals, and don't know whether I've simply been lucky in life or whether my experience reflects the general situation in real science.

in my memory. We are walking to his home along a rather steep staircase. YaB stops and suddenly jumps onto the fifth step. He looks at me — can I jump further? It's not pleasant to lose in the presence of a thirty-year age difference. I gather my strength and somehow jump onto the sixth step. YaB looks glum, is silent for a while, then solemnly turns to me: 'But my grandson can jump even higher!' The peace is restored.

Like a little boy, YaB hated losing. But one was not allowed to lose to him on purpose. I remember our tennis matches. In the warm-up, YaB might praise a good shot by his opponent, but as soon as we began to keep score, all efforts were applied to a single task — winning. When the game didn't go his way, and there was no hope of winning, YaB threw out a brief 'To hell with it!' and refused to keep score. It required time and a few successful shots for him to return to his usual good humour.

There is another episode that speaks of his ineradicable boyishness. We discussed the role of density fluctuations in the appearance of baryon asymmetry of the universe. One evening, it seemed to me that I had found a solution to one of the problems that arose at that time, and I called YaB without delay. As I was laying out my arguments, I realized that he wasn't listening to me. Indeed, he soon interrupted me with the words: 'You'd do better to turn on your television — I'm on TV now!' When he learned that I was at work, he was deeply and sincerely upset. By the way, I learned the next day, that he had already found a solution to this problem, and perhaps an even more successful one than mine.

On the other hand, YaB's ability to turn off his attention when a question didn't interest him was phenomenal. He could (I think unintentionally) fall asleep for a minute during a conversation, so that his companion either didn't notice immediately or didn't notice at all, then wake up when his companion fell silent, carrying on the conversation on a completely different topic. When I first came across this, I was very upset, but then thought that maybe there was something wrong with my reasoning. As a rule, that proved to be the case. Occasionally I could remove YaB's passive, and sometimes even active, resistance. The results of these battles often allowed me to understand much better the work in hand.

When YaB was interested by some problem, his ability to work seemed unbounded. Once, the doorbell to my apartment rang at eleven o'clock at night. I opened the door and saw a courier with a telegram: 'Call me without delay. Zeldovich.' I knew that YaB went to bed early, since he rises early. However, there was nothing to be done. I went to a phone box (at that time, I didn't have a telephone) and called. YaB was awake, and we spent nearly half the night talking about kinetic processes in the early universe. I confess that, at that time, I was more struck by the speed with which the telegram had been delivered. YaB had sent it by telephone, and it was delivered in less than an hour. Another case: I had caught a cold, and felt like taking it easy at home, all the more so given that I didn't have any presentation to make this time. When YaB heard about this, he came to visit me immediately, and my time of idleness came to an end. I decided not to let myself be ill after that.

When I finally got a telephone, contact became simpler, and I remember our long morning conversations. Fortunately, I also rose early, and this was easy for me. In addition, by this time, YaB would call not at six o'clock (as told by people who were acquainted with him earlier), but at seven. After the death of Yakov Borisovich, I very much missed our morning conversations, though by that time they were not quite so frequent. The point is not even the enormous scientific benefit I obtained from them, but the moral and emotional influence they exerted. (I'm afraid I can't express myself any more clearly.)

I remember the writing of our review about the baryon asymmetry of the universe in the journal *Priroda*. This was in 1981 (the article came out in 1982), and a reference to A.D. Sakharov was thrown out by the censor on the grounds that it was not appropriate to refer to his name in a positive light in a popular publication with a circulation of about 80,000, though the references to Sakharov in scientific papers were permitted. We refused to publish the article without this reference.[96] There followed a long battle. And only with the help of A.P. Aleksandrov (then president of the USSR Academy of Sciences, whom Zeldovich persistently pressed for help) was a compromise solution found: we take out the list of literature at the end of the article, and all references are given as footnotes. The article came out in this form.

Yakov Borisovich had a combination of discipline and a tendency to, so to speak, be naughty which I found amazing. This can be illustrated by two examples. As well as other physicists younger than YaB, I tried on multiple occasions to elucidate certain details in connection with his work on the atomic bomb. At times this was in a very free atmosphere, say, at dinner with a glass of cognac. But as soon as the conversation turned to these questions, it was as if someone threw a switch — YaB immediately became closed, and it was impossible to draw any information from him. Another example: YaB was talking about how he made a presentation in front of the electors some time in the early 1980s, in connection with the nomination of A.P. Aleksandrov to the Supreme Council. In his speech, he described how he made a slip of the tongue: 'Aleksandrov became a statesman, but remained a decent person.' However it was evident from the sly look on YaB's face that this 'slip of the tongue' was not entirely an accident.

I can't fail to remember Yakov Borisovich's love of poetry. He knew well and often cited a great variety of poets, but it seems to me that most of all he loved Pasternak. With his help, YaB won one of our arguments, when I insisted that, in our popular article for *Priroda*, formulae complicated the layout and made it unpleasant. YaB broke me with an excellent, though slightly distorted, quote from Pasternak:

> 'Although people need simplicity,
> Complexity is easier for them to understand.'

---

[96] In relation to this, I remember the words of F.G. Ranevskaya, who, when asked why she didn't participate in new films, answered, 'The money that you earn quickly disappears, but the shame remains forever.'

Of course, my conversations with YaB were not limited to physics. He talked a lot and very well about people whom he had encountered in his life. It's not my place to reproduce these talks, all the more so because many have already gone from my memory. A number of these tales (though far from all) can be found in YaB's published memoirs about his (alas) departed friends. I would like only to present one short conversation about P.L. Kapitsa, to whom YaB felt the deepest respect. At that time, Kapitsa was nearly 90 years old, and I asked how he could continue working at that age. YaB answered that, perhaps at scientific councils he falls asleep now and then, but in his laboratory, he skips along well enough; he fell silent, then added: 'Probably because he always lived with dignity.'

A flair for theory

B.L. Ioffe

Among all the qualities of Yakov Borisovich as a theoretical physicist, there was, in my opinion, one especially splendid one that distinguished him from other theoreticians — a flair for theory. I am referring to his incredible talent to feel the depth and perspective of a theoretical thought or idea, when this thought (or idea) is not yet well formed, raw, when it looks more absurd than reasonable, and nearly all others simply ignore it.

I'll present several examples.

In 1959 to 1961, the works of A. Salam, J. Word, and S. Glashow appeared, with the first attempts to unite the weak and electromagnetic interactions. At that time, they did not attract wide interest. However, they immediately caught YaB's attention. He arrived at the Institute of Theoretical and Experimental Physics (ITEP)[97] and said, 'What a splendid theory, why aren't you working on it?' We answered that the theory was not renormalizable, it would give too large a probability for the $\mu \rightarrow e\gamma$ decay, and so forth. This didn't stop YaB. In his opinion, the idea was so deep that it was necessary to work on it all the same, without worrying about such difficulties. And, indeed, he was right. In spite of the fact that, at that time (1961 and 1962) we didn't follow

[97] In the 1950s, over several years, YaB worked in the ITEP Theoretical Department in addition to his main position in Sarov, then was dismissed in accordance with a resolution from 'above.' A.Ya. Pomeranchuk (Chairman of the Theoretical Department) and A.I. Alikhanov (Director of ITEP) tried to fight this, but those on high were implacable: it was forbidden to hold more than one position. This same fate was shared by L.D. Landau — he was also dismissed from ITEP at that time. Our discussions with YaB were very useful for us. It seems to me that they were useful for YaB as well. To this day, I keep as a valued relic a reprint of an article of his from 1954 with the inscription 'To dear Boris Lazarevich, from a grateful student.' I like to think that this was not simply a joke.

his advice, it seems to me that his words (at least for me) weren't in vain — somewhat later, in 1963, I began to work on this theory with intermediate *W* bosons.

Another example is connected with the 1961 work of Goldstone, in which it was shown that a spontaneous breaking of symmetry leads to the appearance of massless particles — goldstones. The attitude toward this work in the ITEP was as follows: everybody agreed that the work was interesting, but nobody wanted to develop these ideas further. Perhaps the reason for this was that nearly everyone at the ITEP (especially I.Ya. Pomeranchuk) was keen on Regge theory. In numerous discussions, YaB emphasized the depth and perspective of the Goldstone's ideas and urged us to develop them. But, alas, his efforts were unsuccessful — we continued to be occupied with our own affairs. As is well known, the idea of spontaneous symmetry breaking and the appearance of goldstones now penetrates the whole of elementary- particle physics.

A third example of a similar kind concerns the cosmological constant in the theory of gravitation. Beginning in the 1970s, YaB said that in the existing field theories (including Grand Unification models), the cosmological constant calculated using perturbation theory, as a rule, diverges; even when it is possible in some theory to get it to converge, its value turns out to be many tens of orders of magnitude greater than the experimental limits. YaB's opinion was that the requirement for a theory to lead to a zero cosmological constant should lie at the foundation of our choice of acceptable theories. He returned to this idea many times, and for him it was a theme something like 'Carthage must be destroyed.' The problem of the cosmological constant remains unsolved till now, and the criterion proposed by YaB is now one of the main devices used in the selection of suitable theories attempting to unify all types of interactions, including gravity. By the way, there have also been attempts to solve this problem in other ways, in the framework of cosmology.

It seems to me that if it had not been for the turns of fate that left Yakov Borisovich little time to work on the physics of elementary particles, he could have accomplished much more in this field, and obtained much more satisfaction from his work.

Part V

Mechanics and Mathematics

Yakov Borisovich Zeldovich, as I knew him[98]

G.I. Barenblatt

As the end of the twentieth century approaches it is already possible to say with certainty who and what was unique in that century. YaB was unique in classical physics. His work spanned almost the whole field, and he understood it in its entirety. In this sense, he can be compared to Lord Rayleigh, who, after all, had 15 years of active work at the start of the twentieth century — just about 15 years remained in the century at YaB's death.

There were also substantial differences between them. At the end of his life, Rayleigh emphatically turned away from the new — quantum and relativistic physics, considering this a matter for the following generation. Until the very end, YaB raced through new areas of physics.

Moreover, the wave of glory that became universal toward the end of his life was primarily due to his work in recent decades, first and foremost his accomplishments in astrophysics and elementary-particle physics. To see this, it is sufficient merely to read YaB's obituary by A.D. Sakharov in *Nature*. He writes of their collaborative work on atomic and hydrogen weapons, talks about their lives in a certain city. However, when talking about YaB's scientific results and their discussions together, AD writes almost exclusively about astrophysics and particles, though it is obvious that it wasn't for such discussions that they were sent there. There won't be a second obituary in *Nature*, so for many, YaB will be remembered only as an astrophysicist and particle physicist, or, at best, people will ask, as M. Gell-Mann asked me in 1966 in Cambridge, 'Is it true that Zeldovich did something with flames and explosions?' I have no doubt, however, that YaB will always remain first and foremost in the grateful memory of mankind as a major classical physicist of the twentieth century.

I also had occasion to participate in nominating YaB for a very prestigious international prize. Now it is possible to talk about this, such prizes are not awarded to the dead. We had to decide how to phrase the nomination. After long discussions, we decided on the following: for a definitive contribution in a new area of physics — the physics of self-organizing systems in

---

[98] This paper was written in early 1988.

At the International Meeting on Combustion in Karpach, Poland, 1977. Professor A. Oppenheim presents Ya.B. Zeldovich with the N.P. Manson Medal.

active media. The matter was delicate. Very carefully, beating around the bush, I tried to find out what YaB's own opinion was: 'What do you think, what was your most important contribution to science? Is it correct to think that ...?' — 'Why do you want to know?' — 'Nothing special. Simply for intercalibration.' After some reflection, YaB agreed with the proposed formulation. Then I asked him directly, 'What about astrophysics, cosmology, "pancakes," particles, and so forth?' The answer was very definite: these results were more transient, and he shared his successes with others. To some extent, this point of view is reflected in the now famous autobiographical afterword to his collected works.[99]

And indeed: flames, ignition, detonation — these phenomena are very important in and of themselves. They have been an object of interest for all

[99] It so happened that, at a meeting of the Editorial Board for the collected works of YaB, I proposed to crown the second volume with such an author's afterword. This caught the chief editor, Yu.B. Khariton, off guard, and he reacted toward it with caution; however, the author, who was present, immediately agreed and rapidly wrote the afterword. In gratitude for the idea, I received from YaB four school notebooks with drafts. In general, I somehow came to understand early on that I shouldn't throw out even a small piece of paper that had been written by YaB's hand.

people, beginning from prehistoric times — to find evidence of this, it's enough to recall the legend of Prometheus. The creation of a theory for these processes was a matter of lasting importance. However, something much bigger happened because of YaB's magical encounter with these problems. The theoretical approaches to these phenomena developed in YaB's research (in a number of cases, with colleagues and students) in the 1930s and 1940s and, continuing until his death, became an important element of current natural science as a whole. One suitable comparison is the following. In their time, Rayleigh in England and L.I. Mandel'shtam and his school in the Soviet Union created a general approach to the study of oscillations. The useful concept of 'oscillation mutual aid' developed, when difficulties arising in investigations of one type of physical oscillation could be understood and eliminated via analyses of some other type of oscillation that was better understood for some reason. The same thing occurred with the physics of self-organizing, non-linear waves, with the definitive participation of YaB. Today, these concepts are successfully operating in dozens of fields, including some that are very distant from the combustion theory that gave birth to them.

****

For my part, I work in mechanics, and have done research primarily in hydrodynamics and fracture. However, over more than 30 years, I had the privilege of constant scientific and personal contact with YaB. Our collaborative research was connected first and foremost with intermediate asymptotics, beginning with a formulation of the main concepts in this area. We worked together on self-similar solutions of the second kind. In these, the power-law exponents of the self-similar variables are found by solving the eigenvalue problem, and not by dimensional analysis. At the end of his life, YaB saw that the ideas of scaling invariance, fractality, and incomplete self-similarity (self-similarity of the second kind) — developed, with his active participation, by different people in different places simultaneously and growing from different initial arguments — merged into a unified system of concepts. These concepts proved to be fundamental for a wide and continuously expanding circle of problems in the exact sciences. YaB was interested in this to the very end. In writing and orally, he expressed many splendid ideas that helped to determine the development of this class of problems. Together with YaB, we approached the stability of processes associated with self-excited waves in a new way, having discovered that their stability is not disrupted if the perturbed solution tends with time not to an unperturbed solution, but to a solution whose phase is shifted relative to the unperturbed solution. This changes the definition of the stability of invariant solutions, including self-similar solutions, whose spectra include zero eigenvalues. In the same year, in a completely different field of physics, a mathematically analogous idea was proposed by Goldstone. We also had other research, connected, in particular, with the stability of flames. YaB attached great importance to the book *The Mathematical Theory of Combustion and Explosions*, which we wrote together with V.B. Librovich and G.M. Makhviladze.

But this was not the only subject of our discussions and activities. YaB was interested in literally everything in classical physics. Having learned about a new burst of research in structural failure, he expressed the idea that the mutual strengthening of fractures in a defect cascade might be possible. This idea became especially important now, when multiple structural failure came to be the centre of attention. Here is another case: YaB learned about the suppression of turbulence by the addition of small amounts of polymers, and the explanation of this effect by the presence in the flows of aggregates consisting primarily of solvent molecules immobilized by the polymer molecules. He suggested an idea for verifying this explanation, which immediately worked. There were many such examples.

It is also appropriate for me to share my impressions of YaB as a person. He was completely and deeply indifferent towards material objects (furniture and so forth). In his study there are a desk and chair (on the desk, there is a child's inscription: 'Guga is a fool'[100]), they are functional, and that's good enough. I simply don't recall that the family ever undertook any serious renovation work. Money was of interest to YaB only in as much as he had a large family, many children, and then grandchildren. Each of them had to be helped somehow with something. He, like all other members of the family, dressed more than modestly.

I remember the following case. I happened to be at the entrance to the apartment of a fashionable Moscow dressmaker. She kindly let me have one of ten or twenty invitations she had to the opening of an international book fair at the VDNKh.[101] She agreed to do this after I convinced her that she couldn't use these invitations to obtain 'deficit books.'[102] By way of comparison: the department of scientific and technical literature of the Agency for Authors' Rights had one invitation for all its authors. I was told that it would be given to E.M. Lifshitz — a decision that was completely fair. They promised to scout around, and if they could get a second invitation, it would go to me. They didn't manage to get it. Waiting for the dressmaker to bring the invitation, I caught myself thinking that a single chandelier in her hallway cost more than YaB's entire set of household goods. With the exception, perhaps, of a plaster cast bust of YaB, which remained unbronzed. The

[100] I can testify that this assertion is deeply unfair. Guga (who wished to remain under her child's nickname) is now a doctor of mathematical and physical sciences, and an excellent physicist (here, I am referring to the opinion of many colleagues, first and foremost M.A. Leontovich). She graces any society in which she appears.

[101] The *Vystavka Dostizhenii Narodnogo Khozyaistva*, or 'Exhibition of National Economic Achievements,' a major exhibition in Moscow that was made up of various permanent and temporary displays. [Translator]

[102] In other words, books that were very difficult to find and buy in regular bookshops at that time. [Translator]

creator of this bust is now an eminent American sculptor. He came from Poland, but lived for a long time in Moscow. In accordance with the agreement in effect at that time, he went to Poland (this was at the end of the 1960s), but remained there only a short time: the times were not favourable for people such as him. Several YaB's friends, including myself, ordered this bust of YaB for one of his birthdays.

The bust was remarkable. YaB was naked, and the hairy chest was shaved in three places, where the stars of a Hero were attached. At that time, YaB's triple hero status was not made a show of, and he — and, according to him, a few other triple heros (I.V. Kurchatov, Yu.B. Khariton) — were asked not to appear in public wearing all three stars. (The only triple Hero of Socialist Labour whose status was discussed openly (one would like to think deservedly) was the chairman of an Uzbekistan collective farm, Khamrakul Tursunkulov.) Appearing rarely at gatherings with his three stars (more about one such case follows), and being reproached for this by those who looked after such things, YaB justified his action by the fact that he looked like Khamrakul.

On that festive day, the bust, covered by a sheet, was set up on a household stepladder. Vitya Goldanskii read into a tape recorder with the voice of Levitan[103] a 'decree' of the Presidium of the Supreme Soviet ordering the installment of a bust of Academician Ya.B. Zeldovich in the apartment of the honoured at such-and-such an address. My contribution was to pull away the cover: the bust nearly fell over. I remembered all this while I was being photographed with the bust of YaB which had been installed in accordance with a real decree in Minsk, where YaB was born. (YaB liked this photograph, and we sent it to many colleagues outside the Soviet Union. I don't remember if he and I were ever actually photographed together.)

In general, it was far from simple to give YaB presents for his birthday: a present had to be something useful, otherwise, he would just cart it off to his dacha or give it to somebody else. I remember once I gave him a blackboard (on which one could write in chalk): those who have been to YaB's apartment will remember it — it hangs to the left of the entrance to the living room. A.S. Kompaneets and I gave YaB a rotating armchair. The inscription (thought up by Kompaneets) delighted YaB: 'Genius is diligence plus agility.'[104] YaB had an inexhaustible supply of ideas when he gave

[103] Yurii Levitan was the most famous announcer on Moscow radio for some 50 years, from about 1930 through the 1970s. He read all official announcements on Moscow radio, including summaries from the *Sovinform* bureau during World War II and announcements concerning the launches of the first cosmonauts. Indeed, he had a very special voice (a low baritone) that was easily recognizable by virtually every citizen of the USSR. He was hated by Hitler to such a great extent that he was included as number two in the list of most wanted Soviet officials after Stalin. [Translator]

[104] In Russian, the word for agility is the same as that for rotatability. [Translator]

presents. For one of my birthdays, he gave me wine glasses and a compass. How do they fit together? The inscription read: 'Drink, but walk straight!' The inscriptions in books he gave me were very serious and deep. They support me in difficult times; these books will be preserved in our family, and then in the families of my children as the very dearest of relics.

****

YaB had courage that appeared in truly difficult situations. Several such situations are connected with myself; I'll talk about one of them.

My father, a volunteer in the Civil War and World War II, a Communist Party member since the war, was known in Moscow as a doctor and endocrinologist who was, according to many, excellent. After the XX Congress,[105] he expressed the opinion that Khrushchev didn't have the moral right to talk about Stalin without referring to himself, since his own arms were covered to the elbows in the blood of innocent victims. My father knew what he was talking about — in the 1930s, he had worked in the *Lechsanupr*[106] of the Kremlin. He shared this opinion with three of his friends from his first year at the Vitebsk Gymnasium;[107] I remember their names: Shur, Nemets, and Braude. Having freed themselves, as was said at that time, of their false feeling of comradeship, they denounced him to the right (or it would be better to say the wrong) authorities. To their denunciations were added (or it's possible that they came earlier, I was never able to establish the truth) those of two of my father's patients, who he had pulled back from the edge of the death — Dovgalevskaya (the niece of the former Soviet Ambassador in Paris) and old man Bolotin: they claimed he told them political jokes.[108]

In short, my father was arrested on April 8, 1957. YaB learned about the arrest from somebody other than me and quickly came to me. Three stars on his lapel. He criticized me for my silence and ordered me to get ready to go out.[109] Where were we going? To the Moscow city court: the affair would soon be discussed there. We arrived. YaB's bodyguard replaced the old woman with a pistol on duty at the entrance. We went up to the reception: I remember the extremely wide eyes of the secretary, bewitched by the three

---

[105] Of the Communist Party of the Soviet Union; this was the famous groundbreaking Congress in 1956 when Khrushchev publicly denounced Stalin. [Translator]

[106] *Lechebnoe-sanitarnoe Upravlenie*, a special closed medical establishment for the use of the top Soviet bureaucracy. [Translator]

[107] Schools of the highest level for the preparation of students for university studies. [Translator]

[108] These five were the witnesses for the prosecution at the trial. Since I knew my father's 'friends,' they tried to approach me. In accordance with old custom, I drew a circle around myself and warned them that I would spit on them if they came into it. When my father was released, all three were treated by him — my father felt he had no choice, he had taken the Hippocratic oath.

stars. YaB went through to the chairman: 'I demand that you allow me into the courtroom!' (the trial was supposed to be closed). 'What is it to you?' — 'I'll file a complaint against you if you give him an unfair sentence!' The chairman didn't let him into the courtroom, but the visit had made its impact.[110]

The first trial took place on June 22, 1957. In truth, one could sympathize with the dispensers of justice: they simply didn't know what verdict to choose. The point is that it was the day of the Plenary session of the Central Committee of the Communist Party, in which the decision of the Presidium of the Central Committee to remove Khrushchev was discussed. Later it became known that G.K. Zhukov supported Khrushchev — and, in fact, in the end, he wasn't removed in 1957. But that became known only in the evening — what about the court decisions during the day? Even so, the court knew its business — they sent my father off for a medical evaluation. YaB came to the trial, again with his stars, and sat for some time in the corridor together with our relatives. My father saw YaB (when he went to the courtroom under escort with his hands behind his back), and this gave him moral support.

Further, A.D. Sakharov took up the problem — again, at his own initiative. He wrote a letter to Khrushchev with a request to free my father. Khrushchev received this letter and told Suslov that he couldn't be objective in the matter, and that Mikhail Andreevich [Suslov] should decide the matter according to his own conscience. Once, in January 1958, AD was called from the city where he worked to Moscow for a talk with Suslov. The conversation was difficult for AD (I remember the Deputies' VIP waiting room at Kazan railroad station in Moscow, where AD called me that night before his departure on the return train, to tell me what had happened), but that spring, my father was released from the camp.

****

Several times, I had occasion to be with YaB on trips. Of these, I especially remember the first — to the alpine camp Belala-kaya in the Caucasus in the early spring of 1960. Four of us arrived together: YaB, his first wife Varvara Pavlovna Konstantinova, who died suddenly in 1976, his

[109] At that time, YaB constantly expressed himself in a very florid style; I couldn't find many of the words he used in either the Tolkovy Dictionary of V. Dahl' or the supplement to it, which was issued under the editorship of the known Russian Slavist Beaudoin de Courtenais. After his transition to work in the Academy of Sciences in 1964, this habit disappeared over the course of about a year. It seemed to disappear without a trace, but when he began an even purely technical conversation about various types of modern weapons (including the 'nuclear winter' problem), the habit reappeared as strong as before. It would be interesting to try to explain this phenomenon from the points of view of the theories of Pavlov and Freud.

[110] Some of my colleagues expressed their astonishment on seeing YaB's portrait with the stars in the study of my home at Berkeley. This explains why.

cousin Aleksander Grigor'evich Zeldovich, a physicist and specialist in low temperatures, who also, alas, has already passed away, and myself. We all lived in one room.

Varvara Pavlovna was a human miracle. Her photograph now hangs on the wall of my study, and her memory warms the spirit. YaB himself used to say: 'Varya is a person without faults.' She was older than YaB, and was at that time already in poor health. Nonetheless, she was the centre of our society, and all of us gladly subjected ourselves to her leadership. The camp authorities put us in a group of inexperienced skiers, and our excellent instructor (unfortunately, I don't remember his name) combined in his actions respect toward YaB and his companions and concern that we should not get hurt in our downhill ski runs. The camp was run on a self-catering basis, and we all took turns on duty, helping in the kitchen and bringing to the dining room food, tea, and stewed-fruit juice, which was given out two cups at a time. On the duty day for YaB and his team, everyone who had a camera brought it to the dining room — here was a unique opportunity to photograph YaB in the role of a waiter. This was not meant to be: A.G. Zeldovich and I took on a double load in the dining room, and YaB and Varvara Pavlovna concentrated their efforts in the kitchen. I remember the loud voice of YaB carrying from the kitchen making suggestions for improvements, and the admonishing voice of VP.

It was all splendid. In the evenings, there were even seminars with a small board that appeared out of nowhere. And then, suddenly, it all ended: we received news of the sudden death of I.V. Kurchatov. It was not possible for the shaken and saddened YaB to leave immediately, due to avalanches. I remember that we sat for a long time at the post office, waiting for a telephone connection with Moscow. Then, I told VP about my impressions of our conversation with the telegrapher and about some of her ideas. YaB listened and confirmed what I said. VP said, 'Come on! I saw her,' — at the camp, everybody ate together — 'A nice, pretty, lively girl. If it weren't for that, you two wouldn't have listened to her at all, much less have remembered the nonsense she came out with.'

The day before our departure, VP said, waking up: 'Oh God, now they're going to torture us with skis.' YaB had two cowboy shirts, one green and one red. They were called the shirt of the first kind and the shirt of the second kind, and were worn on our ski training trips on alternate days. VP and YaB argued about which shirt's turn it was today. Nobody was allowed to skip the lessons, or to skip their run down the hill. Once, I skipped my downhill run by hiding behind a tree; this incident long served as a subject of merriment in the Zeldovich home.

Our last trip together was in the summer of 1983, to a symposium on synergetics in Pushchino. We again lived together in one room, but this time, it was much bigger. There were a number of interesting people from outside the country at the symposium, among them I.R. Prigozhine and Haken; the discussions were moderated by B.B. Kadomtsev. YaB cursed about synergetics: as he used to say, he had never thought before that his whole life he'd been speaking in prose. On the contrary, it seemed to me that such a union was quite useful — an interdisciplinary, unified approach, the development of a common style of thinking. Consequently, although YaB did occasionally use the term 'synergetics,' a bit of irony always peeked through, at least in conversations in which I participated.

****

The atmosphere in the Zeldovich home was determined by the remarkable relationship between YaB and Varvara Pavlovna. Until 1964, YaB appeared in Moscow on flying visits; at first, he was even accompanied by the 'secretaries,' who influenced his routine. There were various secretaries; I remember two of them. One was an elegant, quiet young man. The second had very large teeth, and each time I saw him, I remembered the wolf in *Little Red Riding Hood*. Never, neither then nor afterwards, did our conversations touch upon YaB's activity in the other city.

At that time, YaB used to have very early meetings, usually at six-thirty or seven in the morning; further, his schedule was filled by the minute. In the autumn of 1957, I was giving a report in the living room. The telephone rang, and YaB said with some curiosity and surprise: 'It's for you!' That was how I learned of the birth of my older daughter Nadia.

Another time, parting at the entrance after wrapping up our business, YaB told me with a smile: 'I almost forgot. Your first lecture at Phystech is the day after tomorrow. The name of the course is 'Mathematical Theory of Combustion.' You can find out the details from Frank.' (D.A. Frank-Kamenetskii, then dean of the faculty of the Moscow Physical Technical Institute.) How? What? — I dashed back toward YaB. The 'secretary,' smiling kindly, didn't let me in (this was clearly agreed with YaB beforehand; the 'secretaries' didn't usually have that kind of power). Among the audience of this first course were A.G. Istratov and the late V.B. Librovich, who later became joint graduate students of YaB and myself. Subsequently, they did some splendid work on combustion theory; in particular, they resolved a fundamental paradox connected with Landau flame instability.

In 1964, YaB returned to Moscow, and his main place of work became the Institute of Applied Mathematics of the USSR Academy of Sciences. In my view this was a mistake — it would have been better to wait a little and come to an agreement with N.N. Semenov to return to the Institute of Chemical Physics. YaB was hesitant because of the clear shift in his interests toward astrophysics. However, the Institute of Chemical Physics was home territory for YaB, and they would have allowed him to do any kind of research. And at the IAM, YaB was an outsider, although, unquestionably, there were a few people there with whom he had good scientific and personal relations (among them, K.I. Babenko, O.V. Lokutsievskii, K.V. Brushlinskii, D.E. Okhotsimskii, and T.M. Eneev). He was aware of this. Therefore, without the slightest hesitation, he accepted P.L. Kapitsa's offer to move to the Institute of Physical Problems as the Head of the Theoretical Division after the death of I.M. Lifshitz.

****

Once, arriving at YaB's in the early morning, I saw someone sleeping on the sofa in the main room; we passed through it on tiptoe and worked in YaB's study. 'Who's that?' I asked Olya. 'Uncle Borya.' That was how I first saw B.P. Konstantinov. YaB dedicated an article to his memory, published in the second volume of YaB's collected works. BP, director of the Leningrad Phystech, hoping to do more for Phystech and the Academy of Sciences as a whole, accepted the position of vice-president and moved to Moscow. As a result, his influence decreased rather than increased. (I remember a conversation in the Zeldovich home about who had the greatest influence — the director of the only Ural Machine Company in the country, or the deputy minister of a certain industry. Further, now it was a difficult dependence, even in small things, an outwardly powerful person essentially without rights and weak. This affected his heart, which from childhood had not been healthy, as with all the Konstantinovs. Like M.D. Millionshchikov after him, he went to an early grave, and there was more than a little in common in their outwardly brilliant but essentially tragic fates. The maximum is not always the optimum.

YaB was a person who was *a priori* benevolent. He had quite a few friends, several of them close, though he always maintained distance. At the same time, he had enemies. Not antagonists in discussions or scientific views — in these situations, relations with YaB were not complicated — but enemies in terms of envy, anti-semitism, and so forth. YaB's attitude toward his enemies was not Christian all-forgivingness, nor the Old Testament view 'an eye for an eye and a tooth for a tooth,' nor somewhere between these two. He wanted to throw them out of his life, to not waste time and effort on them, when so many good things remained to be done. He addressed crude and inconsiderate criticism with disgust, when this became necessary. One such response, to criticism of YaB's book on higher mathematics for beginners, is especially memorable. YaB wrote this book when he saw how unnaturally, in his view, they taught his children mathematics in the senior classes at school. He wanted to show in this book that in fact, mathematics is a discipline of the natural sciences. YaB oriented the book, as he himself wrote, not at a resistant reader, but at a reader who himself was straining to go forward, who didn't need to be 'nudged,' and with whom it was possible to walk side by side and have a friendly conversation. It is characteristic that the enthusiastic acclaim of the American mathematician and teacher of mathematics Lipman Bers named YaB's book in the foreword to his own course in analysis as one of two that had most influenced his pedagogical views.

Recently, I was reading about Fellini. In an Italian newspaper after a new film, there was a critical article: 'Fellini as an artist is dead.' Stupid, rude, and the main thing — scandalously unjust. What did Fellini do, how did he respond? He didn't — without asking anyone, he made a new film. In any case, he didn't go to the Central Committee of the Christian Democratic Party with a complaint. But YaB was not in Italy (he travelled there later for several days, where he had an audience with the Pope; I've preserved an excellent photograph: the holy father holds the first volume of YaB's collected works under one arm, and YaB is clearly excited). They[111] didn't allow the book to be reissued — that was a period of deep stagnation. With difficulty they were convinced to allow him to respond to the critical article in the same issue of the journal. Who won? No one. Who lost? — Our school children, who couldn't obtain YaB's book. And this is YaB, with his glory and stars ...

****

I remember well the day I first saw YaB with my own eyes. It was the spring of 1952, and I was at that time a postgraduate student with A.N. Kolmogorov. N.N. Semenov and A.S. Kompaneets came to AN. They requested AN's help — there was going to be a public debate with N.S. Akulov about chain reactions. Everyone clearly remembered the 1948 session of the VASKHNIL;[112] those pretending to the role of Lysenko, who needed scientific authorities as victims, appeared in many areas.[113]

Kolmogorov was a man of few words. He dryly promised to come and, if it seemed useful to him, to make a speech. Therefore, I went to the debate. I remember the presentation by Lebedev from the Chemical Department; he showed an article by V.I. Gol'danskii on the epidiascope.[114] Everybody noticed numbers written on the pages with a red pencil. What was that? 'And here,' solemnly proclaimed the speaker, 'I show how many times Gol'danskii, a beloved and close student of N.N. Semenov (the implication here being that Goldanskii was his son-in-law as well!) cited Hinshelwood.' Four years later, Sir Cyril Hinshelwood would share the Nobel Prize for chemistry with N.N. Semenov. But at that time, the war against cosmopolitanism and everything foreign was at its height ...

---

[111] To Russians, 'they' (*oni*) always means the authorities. [Translator]

[112] All Union Academy of Agricultural Sciences. [Translator]

[113] Still earlier, such a pretender appeared in mathematics. He gave a lecture to the Moscow Mathematical Society on Weissmanism–Morganism in mathematics, beginning with the words: 'When one well-known Moscow mathematician became acquainted with the outline of my talk, he told me I was a fool ...' L.A. Lyusternik, a splendid mathematician and truly good poet, bit his lip and came out with this epigram:

> They told me that I was a fool,
> And I proved consistently,
> As I talked long and impudently
> That this is, indeed, quite true!

The author of this talk is now rather well known for his left-wing presentations. Back in the time of I.G. Petrovskii, he wanted to transfer to the Philosophy Department of Moscow University. I.G. refused his application.

[114] A now old-fashioned device for presenting texts on a screen. [Translator]

YaB appeared on the rostrum, a star on his chest, his tone very serious. 'I don't see a discussion of two directions,' he said. 'On the one hand, indeed, there is a scientific direction, brilliant theoretical investigations, experiments, new theoretical developments. On the other hand, there is only abstract reasoning of doubtful validity about kinetic equations.' Immediately after YaB was A.N. Kolmogorov. One had to know Andrei Nikolaevich in order to appreciate his devastating irony. He began by saying that he did not agree with the previous speaker. On the contrary, he saw here a well defined new direction, not only in physical chemistry, but also in mathematics: the solutions for sufficiently clear equations proposed by N.S. Akulov led to results that clearly did not fit into the traditional understanding of a modern mathematician. The representatives of the new direction could turn to the Chair of Mathematics in the Physics Department, and he hoped and was simply sure, that there, they would be able to explain the existing approaches in usual mathematics. Maybe they could even simply show how to solve these equations ... and so forth.

AN didn't take into account the fact that his speech, which was being recorded (during recording, nuances disappear), could be made to sound, by the small omission of a tone, like it was praising the 'new direction.' He was upset when he learned from me that it had turned out that way; the recording of his speech appeared in the wall-newspaper at the Physics Department. At the debate itself, however, Kolmogorov's position was clear, and he offered Semenov appreciable support.

****

I remember well the day when YaB and I sat for a long while on the balcony of his home for the last time: June 10, 1987. He gave me a reprint, wrote a date on it, and, something he had never done before, indicated a place: 'balcony, No. 6.' Contrary to our usual practice, we had something to drink. In our conversation, we were careful to avoid deviation from things that were truly important for us both. What he said, I will carry with me forever. Later, we saw each other many times, but always on the run: at his home, at work. We talked on the telephone. YaB came to my birthday party with his wife Inna Yur'evna. Together we made speeches at the funeral service of V.B. Librovich — it is an unnatural thing to have to bury one's own student. On November 26, there was a seminar at the Ishlinskii Institute of Problems in Mechanics. I gave a talk about micromechanics — a new area in the mechanics of continuous media. My students also made presentations, and the seminar lasted the entire day. YaB was supposed to have come, that goes without saying. In the morning, he had called: 'You won't be offended if Inna and I don't come?' — 'Come now, Yakov Borisovich, I'm ready to repeat the talk for you personally whenever you wish!' A certain anxiety crept over me, but I drove it away. In the evening, YaB called, asked how it had gone, and was interested in details. He called once more ... I never heard from him or saw him alive again.

Yakov Borisovich died on the wing. He didn't want to be old, to be a burden, to gradually lose his strength and class. In the memory of those who knew him, he remains always on the move, running and striving for high peaks visible only to a few. I re-read the end of his autobiographical afterword in the collected works: 'In the middle of the 1980s, the most difficult and fundamental questions in science are woven together in a tight knot. I have no desire stronger than to live long enough to see the answer and understand it.' It is bitter to read this — his wish didn't come true. But if we think about it, it was all as he wanted. The shortest path from peak to peak is along a straight line, as Zarathustra said. For this, one must have long legs, and YaB had them. From one peak reached, new ones open up, and there's no end. He didn't want to wait until further motion would cease to interest him. Now YaB is gone — into eternity. The little weaknesses will soon disappear in our memory. His influence, which has made us into different people, will always remain, together with a constant feeling of thanks to fate for allowing us to get close to a great man and giving us friendship with him.

Three encounters[115]

G.S. Golitsyn

This is the name of one of the I.S. Turgenev's splendid stories about love. I had three meaningful encounters with Yakov Borisovich Zeldovich, which in some way supplement what has already been written in the book of reminiscences about him published in 1993 (in Russian).

I first heard of YaB from my supervisor Kirill Petrovich Stanyukovich, who arrived at the Physics Department of Moscow State University in the autumn of 1955 at the invitation of Academician Mikhail Aleksandrovich Leontovich. Stanyuk — as he was called by his friends, but certainly not by us forth-year students — began to give a wonderfully brilliant course on hydro- and gas-dynamics. Having come to listen to him initially mostly out of curiosity, I remained to do a preliminary research project with him, then my undergraduate thesis. For six years, I regularly went to his home, as he loved receiving guests. Once, he merrily told me about a birthday celebration for YaB, as he called him, who was then still a Corresponding Member of the Academy of Sciences, but who had recently become a triple Hero of the Socialist Labour. A group of friends had ordered a bust of YaB from Sarra Lebedeva, a major Soviet sculptor of that time (this bust is referred to several times in other contributions in this book). During the presentation, a 'Decree of the Supreme Soviet of the USSR' was read, ordering the installation of the bust in the *flat* of the Hero (in real decrees of this sort, this would read the

[115] This article was not included in the original Russian version of this book, published in 1993. [Translator]

birthplace of the Hero; in the other contributions, no one noted this). There were some very amusing details, but I can't remember them.

After graduating from the Physics Department, at the recommendation of M.A. Leontovich, who had followed the course of my thesis work on magnetic hydrodynamics, I ended up in the recently founded Institute of Atmospheric Physics of the USSR Academy of Sciences. However, Kirill Petrovich continued to invite me to visit him frequently for several more years. At the beginning of June 1958, he told me that YaB had asked that I telephone him at home the following morning at 5.30. I literally couldn't believe that Zeldovich himself had asked that I call him — I had already heard from others that he often telephoned very early. Kirill Petrovich explained to me that YaB had told him that he had worked on the structure of a magnetohydrodynamic shockwave with finite conductivity. At this, KP told him that an article on this topic by his research student Goga Golitsyn should be coming out soon in the *Journal of Experimental and Theoretical Physics*. Then, YaB decided he wanted to see me, in order to find out whether the work was complete.

That evening, I went to bed early and set my alarm clock for 5.15. There were two payphones several minutes' walk from the building where I lived at that time. I had checked the previous evening that both worked. At 5.30 precisely, I dialed the number that had been given to me. In a cheerful and businesslike voice, YaB proposed that I come to see him at ten, and explained how to get there. Our conversation at his home lasted 15 minutes. For the first few minutes, he looked over my typewritten paper. Then, he said that he had arrived at the same result in a somewhat better way, but he was glad to see that everything had been done. He then spent ten minutes asking me about what I had already done in magneto-hydrodynamics. He spoke approvingly of the isomagnetic jump that I had introduced for poorly conducting media in a magnetic field (by analogy with the isothermal jump described by Landau and Lifshitz in *Mechanics of Continuous Media*). In this case, weak shocks do not have a shock front, and all of their parameters vary smoothly on the scale determining the magnetic viscosity and sound speed, while strong shocks have a discontinuity in their thermodynamic quantities and velocity. The magnetic field remains continuous and propagates ahead of the wave, leading the front on the same scale as that for weak waves. Having briefly questioned me about my work at the institute with A.M. Obukhov (then also a Corresponding Member), YaB let me go.

It was only 20 years later, when I was already a Corresponding Member, that Kirill Petrovich told me that he had worked hard to convince YaB not to take me to work at the 'object.' I naively commented that I had already received my assignment and was working in an institute of the Academy of Sciences. KP informed me that if YaB had only said the word, I would have been at the 'object' the next day, in 'Kontora-400,' a place which was mythological to me at that time. In 1997, Gurii Ivanovich Marchuk told me how, in 1951 as a brand new Candidate of Science, he was visited at his home by two people in a car, who told him he had a half hour to get his things together, he shouldn't worry about his family, everything would be fine. They didn't answer his questions during the subsequent two-hour car trip. It was only on the following day that he learned that the director at the 'object' was Dmitrii Ivanovich Blokhintsev, and that he would be working in the department of Evgraf Sergeevich Kuznetsov; both these names were well known to him. The 'object' then came to be called the Physics Energy Institute; three years later, the world's first reactor providing five megawatts of electrical energy was put into operation there, and the city came to be known as Obninsk of the Kaluga Region.

Our next contact occurred nearly 13 years later at the Sternberg Astronomical Institute in January 1971. There, I was supposed to defend my doctoral dissertation on the dynamics of planetary atmospheres. I had been able to think of a way to approach this problem, which was already of practical interest in connection with the landing of automated stations on the surfaces of Mars and Venus, from the point of view of similarity theory. My defence, initially scheduled for December 1970, was moved to January 1971 (I now know that I was 'displaced' by I.D. Novikov with his doctoral dissertation). However, in January, my opponent Andrei Sergeevich Monin had to leave for a maritime expedition. I had to urgently find a new opponent. My director, Aleksander Mikhailovich Obukhov, who was also the head of the theoretical department of our institute, probably knew why I had been 'moved,' and immediately said that he would ask Zeldovich.

Thus, I telephoned him at home for the second time. This time, a meeting was set at the SAI (Sternberg Astronomical Institute) during the afternoon. YaB, not even having glanced at the 200-page dissertation, immediately took up the summary, read the first pages attentively, and skimmed more quickly through the middle and end. This took him about ten minutes. For another five or seven minutes, I answered his questions, and was asked what I myself considered to be my most important result. He then asked me for clean paper and immediately sat and wrote an evaluation. He filled four pages with his rather sweeping handwriting. When he finished, he said that he would have the evaluation typed, I had nothing more to worry about, and we parted. The entire procedure of becoming acquainted with a dissertation and writing the evaluation had taken him just under forty minutes! Since then, whenever anyone tells me that they don't have the time to get familiar with a dissertation or to write an evaluation, I always present the example of YaB.

I remember from his presentation at the defence how YaB characterised similarity theory, being himself a master in its use. His words were roughly as follows: 'Some say that similarity theory bears only a similarity to a theory. All things considered, this is, of course, not so. At the same time, however, the ability to use it correctly is more an art than a science, and results obtained using similarity theory seem to have arisen from virtually nothing.'

Our third encounter was some time in the middle of the 1980s, again at the SAI. At that time, YaB's students Sasha Ruzmaikin and Anvar Shukurov occasionally, though fairly regularly, got in contact with me about questions connected with turbulence, convection, and so forth. For me, the 1980s were years of convection in rotating fluids (with both coworkers and colleagues from other institutes, I also worked at the time on theories of climate, and especially of the nuclear winter). Again using similarity theory (together with other methods), an estimate of velocities during convection taking into account rotation had been obtained. I verified this estimate using simple experiments in my kitchen at home, without any special equipment except for a mercury thermometer. Later, Boris Bubnov, a colleague at our institute, and I developed a whole programme of experimental and theoretical work; at the end of the 1980s, similar research was initiated in several universities in the USA, then in Australia, Canada, and Germany.

Due to my meetings with Ruzmaikin and Shukurov, I was called to give a seminar on various obvious applications of this work. As always, YaB listened attentively, and asked questions. At the end of my talk, he got up and noted the use of experiments by theorists. He especially liked my use of three rotation speeds — 33, 45, and 78 revolutions per minute.[116]

Of course, I saw YaB many times at seminars and in meetings at the Academy, but this was from a distance, though here also, I made two observations of behaviour that was instructive about and characteristic of him. In the spring semester of 1959, L.D. Landau began to give a course on quantum electrodynamics in the Physics Department of Moscow State University. The large physics auditorium was overflowing, and one had to arrive long before the beginning of the lecture in order to get a place. YaB always sat in the front row.

The second example was associated with A.D. Sakharov. This was at the end of the 1970s. AD gave a talk at YaB's seminar. When the talk was over, Vitalii Lazarevich Ginzburg began to ask pointed questions, clearly shot through with irony. YaB, characteristically firm ('Vitya, don't!'), made Ginzburg back off, and the discussion proceeded normally and constructively.

My three encounters described above show very characteristic features of YaB as a scientist and a person. It seems to me that they augment the substantial information already supplied by other contributions.

---

[116] Record players are now a thing of the distant past. If I had decided to work on convection with rotation fifteen years later (I carried out my measurements over several weekends in January 1980), I would have begun by writing a proposal to the Russian Foundation for Basic Research or INTAS, then waited a year for their answer. After a positive beginning I would have had to find additional money for the development, construction, and manufacture of a rotating device, searched for an undergraduate or postgraduate student to carry out the measurements, and so on and so forth. It would have been years until I could have begun taking actual measurements.

YaB and mathematics

V.I. Arnol'd

Usually, Yakov Borisovich telephoned me at seven in the morning. 'Doesn't it seem to you ...' he would say, and there would then follow some sort of paradox. The last such call was two weeks before his death. He talked about the strange, 'chaotic' behaviour of solutions for the Riccati equations with periodic coefficients.[117] This work (together with a coauthor whose name I have forgotten) remained unwritten, and I will write about it here in a little detail.

The Riccati equation is a differential equation whose right-hand side is a second-order polynomial in the dependent variable. It is characteristic of the solutions of this equation to approach infinity over a finite time. In a traditional approach in the framework of the usual theory of differential equations, there could not be any chaos here. The chaos about which Yakov Borisovich was speaking arises if we curl the axis of the dependent variable into a circle, adding one infinitely distant point.

Periodicity of the coefficients makes it possible to also bend the time axis (the axis of the independent variable) into a circle. This leads to a dynamical system on a closed manifold — the surface of a torus. The properties of this system were studied by H. Poincaré, who discovered that they strongly depend on whether the 'rotation number' — the mean deviation of trajectories in time — is irrational or rational. If the rotation number is rational ('resonance'), some (and for the Riccati equation, usually all) trajectories are closed. In the irrational ('general') case, the trajectories densely cover the torus and return infinitely many times to the neighbourhood of an initial point, never exactly repeating their path. The time average of any function along a trajectory coincides with its average on the surface of the torus ('ergodicity', i.e., the first degree of 'chaos').

In this way, we were speaking of the rediscovery of an important area in the modern mathematics of the theory of dynamical systems. Applications of Poincaré theory to the Riccati equation should have been in textbooks, but as far as I remember, none of the mathematicians noticed them. The psychological difficulty here — the variation of the topology of phase space (a transition from an affine to a projected line) — is akin to the description of the Schwarzschild solution using the topology of a black hole. The removal of such difficulties is an essentially mathematical activity, but most mathematicians are slower to move away from stereotypical thinking in the framework of a precisely posed problem, and are not eager to radically change their point of view.

Yakov Borisovich, on the contrary, was always prepared to change his point of view. I remember how he first called me to his home on Vorobiev Road in the early 1970s, and I told him about the then recent achievements

---

[117] In modern terms, this equation is SL(2) connection over a circle.

in the theory of dynamical systems (unpredictability, chaoticity, turbulence, strange attractors, invariant tori, and so forth). For some time, Yakov Borisovich tried to resist, holding on to the old dogmas. Fortunately, I wasn't intimidated by his authoritative tone, nor by his references to Landau, and (shyly) said: 'But, Yakov Borisovich, one can look at this from a different point of view.'

'Yes?' said Yakov Borisovich and immediately did a headstand. For several minutes, he looked at the chalk-written board from bottom to top, then righted himself and began to discuss on which physical problems the new mathematical theories should be tried without delay.

Being first and foremost a physicist,[118] Yakov Borisovich had his own view of mathematics, which differed sharply from the view of most mathematicians of his generation (educated on axiomatic–deductive set theoretical concepts, rising to Hilbert and Bourbaki), and from the view of most physicists — namely that the only useful aspect of mathematics were analytical techniques, a sort of continuation of the mastering of arithmetic calculations. Yakov Borisovich's point of view was closer to the position of the younger generation of mathematicians and theoretical physicists (L.D. Faddeev, A.M. Polyakov, S.P. Novikov), for whom a qualitative, geometric, conceptual mathematics merges with theoretical physics. Mathematical concepts and ideas, and not only computations, were his element.

However, also in a technical sense, some of Yakov Borisovich's achievements anticipated subsequent mathematical investigations, sometimes by tens of years. This is especially true of the theory of singularities, bifurcation and catastrophe — the area of mathematics that describes the appearance of discrete structures and all possible jumps and discontinuities from regular, smooth variations.

For example, YaB's 1941 work on reactions in a flow essentially constructs a theory of bifurcation of an equilibrium curve in a product of phase space on a parameter axis — the theory of the birth and death of new 'islands' of this curve. In the modern mathematical theory of equations depending on a small parameter, these phenomena were studied only at the end of the 1970s (in the work of French mathematicians on so-called 'non-standard analysis' and 'ducks theory'). When these works appeared, YaB immediately recognized in them a small generalization of his old theory. One splendid property of this theory was that, although a specific system specified by explicit formulae had been investigated, the qualitative character of the results didn't depend on the details of these formulae, and remained the same for a wide class of 'general position' systems. The strict mathematical proof that systems that behave otherwise are exceptions was

[118] 'I'll have to challenge you to a duel,' YaB told me when I quoted Newton to him: 'Mathematicians, who all discover, investigate, and prove, must be satisfied with the role of dry calculators and unskilled workers; the other [the physicist — V.A.], who cannot prove anything, grasp everything on the fly and has pretences to everything, carries all the glory of both his ancestors and his descendants.'

obtained by mathematicians only recently. However, the character of the phenomenon was discovered by YaB fifty years ago, and its universality was, of course, clear to him.

The same universality and independence of specific details was characteristic of YaB's research explaining the large-scale structure of the universe as the effect of small, smooth inhomogeneities in the initial velocity field of a dust-like medium.

The appearance of singularities in caustics in this problem was first detected by E.M. Lifshitz, I.M. Khalatnikov, and V.V. Sudakov. Yakov Borisovich's 'pancake theory,' in essence, is equivalent to a theory for the simplest, so-called Lagrange singularities in symplectic geometry — singularities of projections of Lagrangian manifolds (on which the Poincaré invariant vanishes) from a phase space onto a configurational space.

This same theory provides a description of typical singularities of a caustic and their restructuring during variations of a parameter in optics. Its mathematical difficulties are so great that many questions remain unresolved to this day, and results achieved (in recent years) have been obtained only after a series of laser-optics experiments and computer simulations. This is all the more to the credit of YaB, who immediately sensed the importance of his 'hydrodynamics of the universe' as a general mathematical theory.

The transition in YaB's work from locally analytical studies to analyses of globally topological and statistical percolation properties arising in structures also fails to call forth delight among mathematicians. In these studies, it is more accurate to say that physics becomes a servant of mathematics, rather than the opposite.

Like all mathematicians, YaB loved to distinguish a precisely formulated mathematical question in a physical problem. He believed that it was worth precisely formulating the problem mathematically — and that the mathematicians, 'who, like flies, can walk on the ceiling,' would find a solution! He was especially annoyed at modern mathematics' inability to solve the questions of a frozen magnetic field of minimal energy and a rapid magnetic dynamo.

I remember a discussion of the first of these questions during a talk I gave about the Hopf asymptotic invariant[119] at YaB's seminar at the Institute of Applied Mathematics (this was probably in 1973): YaB and A.D. Sakharov interrupted me, waving their hands, and explained that the frozenness of the force lines would prevent the decrease of the energy of the frozen field to zero.

YaB's mathematical problem is posed as follows: among all vector fields with zero divergence on a three-dimensional manifold obtained from a given vector field via diffeomorphisms conserving volume elements, what is the

[119] At that time, the connection of this subject with anomalies in quantum field theory and the multivalued action of Polyakov was, of course, unknown; this connection was shown by S.P. Novikov only ten years later.

field with the minimum integral of the square? (This minimizing field can have a singularity.)

This problem, which remains unsolved today, models the evolution of the magnetic field of a star, neglecting magnetic viscosity (Ohmic dissipation or reconnection of force lines). It is assumed that the flow of energy being not mimimum, the field would give rise to a Lorentz force, which would move the medium; as a consequence, the excess energy would be dissipated by hydrodynamical viscosity until the medium becomes stationary and the field energy would be minimized.

YaB and AD maintained that, for example, the energy of an axially symmetrical field in a sphere (the field being frozen) can be made arbitrarily small by means of an appropriate diffeomorphism (it seems that, even today, this has not been rigorously proven).[120] The question of the topology of the minimizing field in the general case is also unresolved, as far as I know (even in the simplest two-dimensional model, which requires minimization of the integral of the square in a disc of the gradient of a smooth function with more than one maximum and equal to zero at the boundary, by means of conservation of area elements for the transformation of the disc into itself).

The other problem precisely posed by YaB of a rapid, steady-state, kinematic dynamo can be formulated as follows: does there exist a divergence-less, velocity vector field **v** that is steady-state in time and periodic in space for which the induction equation

$$\partial B/\partial t + \{v, B\} = \varepsilon \Delta B, \qquad \operatorname{div} B = 0,$$

(where $\{v, B\} = \operatorname{rot}(v \times B)$ is the Poisson bracket), has a solution that grows in time and is periodic in space $B = \varepsilon^{\lambda t} B_0(x,y,z)$ with increment $\operatorname{Re}\lambda > 0$ that does not tend to zero as the magnetic viscosity ε decreases to zero?

The stretching of the magnetic lines by a flow with an exponentially extended particle distribution leads to the exponential growth of the field (YaB explained this effect in this way: the vicinity of a closed magnetic line is stretched by a factor of two and fits in its place, similar to the case of a simple rubber band).

However, the growing field has ripples, and a viscous term can suppress the initial growth. The question of which of these effects — the extension of the particles (giving rise to randomness in the magnetic field) or the small amount of diffusion — will win out remains open.

Numerical experiments with the 'ABC field' of Beltrami

$$v = (A\cos y + B\sin z)\partial/\partial x + \text{(cyclic permutations)}$$

[120] By the time of the present new edition of the paper the proof has already been published by M.H. Freedman (see the discussion in the book *Topological Methods in Hydrodynamics* by V. Arnold and B. Khesin, Springer 1998, pp. 119–193).

indicate, for example, that when $A = B = C = 1$, the dynamo effect can operate ($\mathrm{Re}\,\lambda > 0$) when the magnetic Reynolds number is $1/\varepsilon$ in the intervals from 10 to 20 and from 30 to 100 (Frisch, Galloway).

For Reynolds numbers from 50 to 100, the increment changes only slightly, and is close to the (empirical) index for the extension of the particle distribution by the flow. For some reason, YaB was inclined toward the hypothesis that upon further increase in the Reynolds number (to 400?), the dynamo effect would cease operating (the mode would begin to decay), and further would possibly appear and disappear again.

No theory for these phenomena has been developed, and numerical experiments require the calculation of eigennumbers of matrices whose order is much greater than a million, making the problem unsolvable in practical terms so far.

YaB considered the existence of a rapidly kinematic magnetic dynamo that is periodic in time to be reliably established (v is a divergenceless velocity field that is periodic in time and space and B is also periodic in space). As far as I know, this theorem of YaB's has not yet been fully digested by modern mathematics.

In the last ten years of YaB's life, I had the pleasure of working with him quite a lot. Most often I took on the role of a listener or reader (YaB's letters usually ran to eight pages, and I received a letter every week).

'You can throw out this letter without reading it. The reason is that for me, writing to you has become a psychotherapeutic act, and a means of checking myself, clearing up something to the end. I write, and I see your sceptical look ('the eyes of Major Pronin'), and my hand refrains from writing something doubtful ... Do many psychos write to you? They write to me very often.

And so, what we know about singularities is true, but only locally. At the same time, there are some global properties of the system, that ...'

'It seems that Dubrovskii wrote to Masha Troekurovaya:[121] 'The sweet habit of turning to you every day, not waiting for an answer to my letter, has become for me a law' (in the period when they communicated via a tree hollow).

And so, as far as I know, the comment ...'

One of our last joint activities was a commentary on the 'works' of YaB. 'I am writing,' said YaB to me on the phone, 'a necrological composition. It's sad, of course, but in my view necessary. As Oscar Wilde said, "everyone has students, but the biography is always written by Judas." Please write about mathematics.'

Re-reading *Higher Mathematics for Beginners* at that time, I saw how much there was in the first edition of YaB's textbook of what the mathematicians of my generation are trying to introduce into the emasculated, dead teaching of our science (facing enormous resistance as they do so).

---

[121] The main characters of Pushkin's novel *Dubrovskii*. [Translator]

Ya.B. Zeldovich and his wife A.Ya. Vasilyeva and Bulgarian colleague Luchezar Filipov with his father and wife.

The book begins with a shocking definition of a derivative as the ratio of the increments 'assuming that they are sufficiently small.' This definition, blasphemous from the point of view of orthodox mathematics, is, of course, 'physically' completely justified, since the increment of a physical quantity by less than, say 10^{-100} is purely a fiction — the structure of space and time on such scales can be very far from a mathematical continuum.

However, this simple reasoning destroys such a substantial part of modern mathematical investigations that is it dangerous even to make reference to it here. The censors of the Russian mathematical books at that time, the topologist L.S. Pontryagin and the mechanics researcher L.I. Sedov, let loose a flood of accusations toward YaB, which YaB (with his somewhat boyish ambition) suffered more painfully than they deserved. I believe that the fight with these censors, whose extreme power still needs to be explained, its reasons being nonscientific, and with their incompetent allies, took a great toll on YaB. This fight for the republication of a clearly much-needed book — which YaB conducted, like always, with his full effort — and the associated difficulties, which affected him very emotionally, shortened his life.

This fight ended in complete victory for YaB. In the exposition of his analysis for school pupils (1980), L.S. Pontryagin wrote: 'Some physicists believe that a so-called rigorous definition of derivatives and integrals is not necessary for a good understanding of differential and integral calculus. I share this view.'

The return of mathematics teaching from scholastic, formally-worded computational exercises (be it $\partial^k/\partial x^k$ — the language of Leibnitz, $\varepsilon - \delta$ —

the language of set theory, Ext-Tor — the language of homological algebra, or IF-GOTO — the language of programming) to a substantial mathematics of the ideas and concepts of Newton, Riemann, and Poincaré was an absolutely necessary step. YaB was the first to find the courage to say this openly and help achieve this step in a timely fashion.

YaB's time was scheduled to the minute. Plutarch writes that Phemistocles gave the same meeting time to all of his clients, so that each of them, seeing the others and waiting for his own turn, would be impressed with a sense of the importance of his patron. Yakov Borisovich, on the contrary, assigned each his own time, but therefore couldn't extend the conversation for even one minute. Having become used to the humiliating attitude[122] that was usual among mathematicians, especially toward young colleagues, I was pleasantly surprised at YaB's correctness and peculiar delicacy, which clearly contradicted his naturally heated temperament. 'Zin, you're coming up against my crudeness'[123] was his expression of extreme anger. Although YaB called himself a student of Landau, he didn't follow him in everything.

Preparing commentaries for some of YaB's works, I looked in the Science Citation Index and found, I think, about seven thousand references to his papers in one year (a number second only to Landau). If we take into account that the Index ascribes all joint works to the author whose name comes first in the alphabet, and that Zeldovich's name in the Latin alphabet begins with Z, the true number of references rises substantially. I don't know how many references there are to his works now,[124] but it is clear that the YaB's influence in both physics and mathematics remains absolutely exceptional.

[122] I ascribed this attitude to the influence on mathematicians of L.D. Landau until I learned that an Academician who didn't have any connection with Landau — the chairman of the mathematicians' scientific council — muddled the name of the opponent during a thesis defence, then justified his mistake with the words: 'Well, it does not matter, he's not a big fish.'

[123] An excerpt from the famous song by V. Vysotskii (1938–1980) entitled 'A chat in front of the TV.' [Translator]

[124] The citation number quoted in <http://www.scientific.ru/journal/news/n030701.html > is 10497 (November 2001)

My reminiscences about Ya.B. Zeldovich[125]

V.E. Zakharov

In 1956, when I graduated from high school, many of my fellow students and I were mad on physics (and, to some degree, mathematics). In my circle of friends, the famous Soviet physicists were idols. Yakov Borisovich Zeldovich was a particularly legendary personality. Just fancy — a Corresponding Member, a Hero[126] several times over and a Laureate at the age of 32; a laboratory assistant who had not graduated from a university and had leap-frogged straight into a doctorate. It was he, not Sakharov (who was much less well known at that time) who was described as 'the father of our bomb.' Still, nobody had access to any real information, and people only talked to their own kind, in low voices and with caution. Stalin had only just died, and it was not safe to discuss topics like this with just anybody. Alas, on one occasion I found this out to my cost. But that is a different story altogether.

In 1958, I became a student of R.Z. Sagdeev, then a promising Candidate of Sciences. The first thing he told me to do was to read a monograph on shock waves by Ya.B. Zeldovich. I did so rather quickly and with pleasure. The book proved to be clearly written and fascinating. It instilled in me a love of hydrodynamics and affected my scientific taste in many ways. So I believe it is a privilege to consider Yakov Borisovich as one of my scientific tutors.

My life went on its way. In the summer of 1960, I found myself at the Institute of Atomic Energy, in Budker's department, working as a fifth grade laboratory assistant. It was not a high-ranking position, but Zeldovich's example inspired me. Besides, the salary I was paid was not too bad (better, I think, than what YaB was paid in his time); on top of that, I was exempted from army service. A year later, following Sagdeev, I moved to the Novosibirsk Academgorodok, where I entered the fourth year of the physics faculty as a full-time student. The position of laboratory assistant at the Nuclear Physics Institute (now known as the Budker Nuclear Physics Institute) was kept open for me until I graduated from the university. This was possible only at that institute and in those days. This text is not about Budker or Sagdeev, but I must include a word of gratitude for what they did for me.

Thus, I was completely immersed in the world of professional physicists for a long time. In this world Zeldovich was not merely respected — he was adored. Among ourselves, we called him YaB or Zeld. He was loved not only for his talent, erudition, and incredible energy, but also for his constant kindness and, especially, his goodwill towards the young. As with all famous people, many stories were told about him. They said that when he had been a lab assistant, the head of another laboratory had swapped him for

[125] This article was not included in the original Russian version of this book, published in 1993. [Translator]

[126] Hero of Socialist Labour.

a vacuum pump, and that decided his fate. (Having had experience of the life of a lab assistant myself, I can well believe this.) It was also said that once he put on all three of his stars[127] to impress the actresses at Mosfilm,[128] but was not allowed to enter the studio — the stars were assumed to be fake. In fact, he was very successful with women without the stars too, and that caused admiring envy in all of us. He was also loved for his wit; for his brilliant speech — juicy, sometimes rather salty, yet sparkling with erudition. I apologise to the reader, but here is a typical episode.

A discussion is underway at the blackboard. Somebody makes an assertion and enthusiastically tries to prove his point. At last, the man exclaims, 'As YaB says — you can cut off my member if I'm wrong!'

Another story. Of one talented mathematician, who was flirting with party bosses, YaB said, 'How long is he going to be spraying himself with pig's perfume?'[129]

I saw Yakov Borisovich in person for the first time in the winter of 1961–62 in Academgorodok. Winters are long there, and it is difficult to be more exact about the time. Roald Sagdeev was defending his doctoral dissertation, and YaB, who had been a full academician for three years already, was his official opponent. I have completely forgotten the details of the defence, but the banquet I remember perfectly well, thanks to YaB.

He came into the middle of the hall — a middle-aged man, short, powerfully built, in round glasses, buoyant and energetic — and he proposed the following toast: 'There were two thieves, one young and one old. They arranged a contest: each of them had to climb a tree and steal eggs from the crow's nest without the hen crow, who was sitting on the eggs, noticing anything. The young thief climbed up just as he was — in his jacket and boots. The hen crow noticed him and made a noise. The old thief said: "What a fellow! Now watch how it should be done!" He took off his boots and jacket, climbed the tree and stole the eggs. Down he came — to find that the young thief had made off with his jacket and boots. So, let us drink to the young generation of scientists!'

I was not introduced to Zeldovich during that visit of his. I was only a beginner, a student, 'nobody, nothing, and nameless.' We became acquainted during his next visit to Academgorodok, towards the end of the sixties. YaB said to me then, 'I've heard a lot about you. Would you like to get involved in working on these problems ...?' and a series of suggestions ensued. In response, I began to tell him about my work, about solitons, about wave collapses. He was interested, listened very carefully, asked profound questions. He understood that I had enough problems of my own to be getting on with and regarded this fact with utmost respect. However, on a

127 As worn by a Hero of Socialist Labour.

128 Moscow Film Studio.

129 From Saltykov-Shchedrin's *The History of a Town*.

number of occasions after this he formulated important unresolved problems for me. I want to mention two of them.

Once, Sagdeev said, 'YaB wants us to develop the non-linear theory of Jeans instability of the universe. Then it will be clear how to explain the distribution of galaxies by their masses.' As far as I understood, YaB considered this to be one of the major problems. Subsequently, he made considerable progress in solving this problem, by explaining the formation of flat galaxies by the occurrence of caustics due to the crossing trajectories of non-interacting cosmic dust. From a qualitative point of view, this explanation is faultless. Still, I think that we have a long way to go before a quantitative theory is developed to describe the observed range of galaxies in a satisfactory way. In fact, this problem belongs to the theory of wave collapses (from the physicist's point of view) or the theory of catastrophies (if one approaches it from the mathematician's point of view).

The other problem was put to me by YaB personally. 'It is rumoured,' he said, 'that Navier–Stokes equations have no global solutions but describe the formation of features. I think that this is harmful nonsense, but it needs sorting out.' YaB meant the works of Olga Aleksandrovna Ladyzhenskaya, a close friend of Anna Akhmatova,[130] nowadays a famous mathematician and academician, and still a beauty at eighty. It was she who first aroused doubts about the absolute validity of Navier–Stokes equations, which had been in use for 150 years. These doubts are so serious that four years ago a private US foundation declared that one million dollars would be paid to the person who cleared up the doubts. What can I say — I am still working on this problem. It is a problem of the theory of collapses, too. A solution has not yet been found, and it could be that YaB was right!

The late 1960s was an interesting time in many respects. Our dissident movement was blossoming then. Now, when everybody wants their name to be added to the list of dissidents retrospectively, very few people remember that the dissident movement began with scientists' letters. The first was Kapitsa's letter, which was very reserved, but unprecedentedly impudent in those days. Then there was a letter written by mathematicians in defence of Yesenin-Volpin, who was put into a mental hospital, and our Novosibirsk Academgorodok letter in defence of Ginzburg, Galanskov, and Dobrovolsky. After that, like an artillery volley, came Sakharov's memorandum. I was most active in the dissident movement and in 1968, after the occupation of Czechoslovakia, was sure that they would put me into prison. As I found out much later, I was saved then by Budker and the Committee of Concerned Scientists.

YaB did not take a direct part in the human rights movement, but it was clear to everybody whose side he was on. YaB stopped his classified works then and declined the high posts he was offered. 'Zeldovich is a wimp,' a well-known member of the scientific community told me then. Instead,

---

[130] Russian poet (1889–1966).

Zeldovich began to give lectures on cosmology at Moscow University, which became famous.

In the late 1960s, there was another significant event. After many decades of being separated, physics and mathematics again embraced each other. Prominent mathematicians, including S. Novikov and Ya. Sinai, started to deal with physical problems. The Department of Mathematics was opened at the Landau Institute of Theoretical Physics. YaB himself became head of a department at the Institute of Applied Mathematics. He began to establish his scientific school there in the field of astrophysics and cosmology. The advances made by this school brought him new glory.

At the same time, YaB wrote a new, completely ground-breaking textbook: *Higher Mathematics for Beginners*. As always, being an extremely talented person, he was ahead of his time. At that time, similar textbooks, but of a much poorer quality, were appearing in the USA. YaB's textbook had no rigorous theorems. The major mathematical facts were presented using examples and a common-sense style. This was a remarkable book, which was a fine addition to standard mathematical textbooks.

Unfortunately, many mathematicians gave Zeldovich's book a hostile reception. The management of the Steklov Institute of Mathematics was especially irritated. These people — Academicians I. Vinogradov, L. Pontryagin and others — were known not only for their inveterate political conservatism, but also for their open, aggressive anti-Semitism. Undoubtedly, they also had great scientific ability. It is for philosophers and psychologists to explain how such things can be combined. Perhaps they will warn us if similar things could happen in the future. But in real life there was a sharp debate between YaB and Academician L. Sedov, which, I think, caused Zeldovich much chagrin.

Due to all these conflicts, and also to the general worship of the physicists, which was like a national cult in those days, YaB became a public figure for some time. I do not think that he wanted this, or that it did him good. In humanitarian circles, people were already becoming irritated by the dominating role of physicists. Slutsky's verses about physicists being honoured and lyric poets being out of fashion became a touchstone for many. Little-educated people, believing themselves to be humanitarians, but having no knowledge even of the English language, became infected with a banal fear of science. I recall how, at a party at Fazil Iskander's, the wife of literary critic B. Sarnov — her name was Slava — angrily condemned all of us physicists, especially YaB, whom she believed to be our apostle. Thus, YaB came under attack from both sides simultaneously.

By the end of the 1960s, my scientific interests began to deviate strongly from the orthodoxy accepted at the Institute of Nuclear Physics. Solitons and wave collapses pulled me strongly towards mathematics. This did not go unnoticed by Budker, who offered me the choice: deal seriously with lasers on free electrons or leave. I chose the latter. I already had quite a large group of students, but an amicable divorce took place — the group left for Nesterikhin at the Institute of Automation, and I went to Chernogolovka, where I became head of a sector at the Landau Institute of Theoretical Physics. It was a cause of pride for Budker — what great people we produce! So, my move to mathematics continued and went so far that in 1974 I was put forward as a Corresponding Member of the Academy of Sciences — not in physics, but in mathematics!

I turned to YaB, and he immediately offered to give me an additional recommendation. But he told me frankly, 'I do not know if it will do you good or harm.' The result of the voting was quite good, and I got through to the second round. However, it was a lost cause, and it was not until 1984 that I became a Corresponding Member — in physics — again with great support from YaB. I keep his first recommendation, which I watched him write by hand on a window sill at the Physical Faculty of Moscow State University, as a most dear memento.

In the mid-1970s, it became clear that the scientific direction we had been developing for all those years — the physics of non-linear phenomena — was an independent field of science with its own problems and methods in the established international community. International congresses on non-linear sciences were starting to be held. As many of us, YaB included, were not allowed to go abroad, we organised these congresses in the USSR. Four international conferences were held in Kiev, in 1979, 1983, 1987, and 1989. All of them were rather successful, attracting many well-known scientists from all over the world. The conference proceedings were subsequently published.

Yakov Borisovich's new students, dealing in astrophysics, magneto-hydrodynamics, and cosmology, took part in those congresses. I was and continue to be friends with many of them, primarily with Rashid Sunyaev. YaB himself was present at two congresses — in 1983 and 1987 — but I remember especially the congress in 1983. It was held in October, at a difficult and disturbing time. A Korean airliner had recently been shot down, and the Cold War was at its height. One of the participants in the congress, the famous American scientist Norman Zabusky, was expelled from the country for visiting a domestic seminar of refusenik Jews. The KGB was omnipresent, but we pointedly ignored the presence of our guards. Against this background, our activity was like a feast during a famine: lectures were given, seminars were delivered continuously, and parties in the evenings were just as frequent.

YaB simply reigned over it all. His lecture on the formation of galaxies from caustics of cosmic dust was brilliant and attracted a huge audience. His wit during evening 'informal sessions' was inexhaustible. His second wife, a cheerful and sociable woman, was a perfect complement to him.

The next congress, held in April 1987, was somehow less bright. YaB's wife had died by that time, and he was more silent and more subdued than usual. Probably, his illness had already taken effect. On December 2, 1987 we learned that Yakov Borisovich was no more. His funeral brought together a multitude of people, and not only from Moscow.

By modern standards, the life he lived — 73 years — was not so long.

But what a life it was — bright, full, and completely honourable! YaB was the leader of a great science in a country which was only great in so far as it was home to such a science. It is surely a sin, but sometimes I catch myself thinking such thoughts. It is a good thing that YaB did not live to see how the conscious or unconscious desires of those who fear science have turned to reality. How rapidly and, probably, irreversibly, the Russian science he loved so much has disintegrated.

The 'humanizing' of mathematics

A.D. Myshkis

My first meeting with Ya.B. Zeldovich was fleeting, and, although it would completely change the course of my life, it has now nearly vanished from my memory. In 1947, when I worked in the Department of Higher Mathematics of the Zhukovskii Air Force Engineering Academy, I was recommended to YaB (I think by I.G. Petrovskii, then my scientific boss) as a possible collaborator. My meeting with YaB was very short; as far as I remember, he said only that I would work with applications of mathematics outside Moscow. The result of this conversation was that, at the end of 1947, when I was already working in Riga at the newly founded Higher Military School, I — an ordinary captain — was summoned by the Air Force Commander-in-Chief Marshall K.A. Vershinin; I remember the night-time commotion this caused. However, Vershinin briefly asked me about the conditions at the school and sent me back. Later, someone told me that a long telegram was sent by the school asking that I be allowed to stay there, and that in the end, of the two candidates requested by YaB, the Commander-in-Chief confirmed only one: this was E.I. Zababakhin, the future Academician, with whom I had studied at the Air Force Academy in the same section.

More than ten years went by. My constant work with both 'pure' and applied mathematics, teaching various topics in mathematics to both mathematicians and students in applied sciences, gradually led me to thoughts of a relativity of the concept of rigorousness, and of legitimacy of the simultaneous coexistence of various levels of rigorousness, with all the consequences that resulted from this.[131] I set myself the goal of writing a major mathematics course for engineers at a level of rigorousness that was inherent in them, getting rid of ineffective material and developing applied mathematical concepts as much as possible without detriment to the main ideas. Hardly any of my mathematician friends shared in this idea, and some even sharply objected, but, in early 1959, I.G. Petrovskii, without expressing his

[131] See I.I. Blekhman, A.D. Myshkis, and Ya.G. Panovko, *Mekhanika i prikladnaya matematika. Logika i osobennosti prilozhenii matematiki* (*Mechanics and Applied Mathematics. The Logic and Peculiarities of Mathematical Applications*). Moscow, Nauka, 1983.

own point of view on the subject, told me that YaB, based on similar reasoning, had written an introductory course in higher mathematics at a physical level of rigorousness. Petrovskii also told YaB about my existence.

In my eyes, the name of YaB was always surrounded by a sort of aura. All the same, I wrote to him from Khar'kov, where I lived then, and described my views, the beginning of the course, and the possibility of collaboration. I soon received an answer, written by YaB on February 2, 1959;[132] I'll present an excerpt from it: 'Dear Comrade Myshkis! ... I have indeed written a course in higher mathematics. I should say that, in the process of writing it, it turned out much larger than I had expected.' (There followed a brief description of the contents, including Part VIII, 'Mathematical Supplement,' which was not included in that course. — A.M.). 'Now this has all been written by myself, Parts I–III are already edited, and the remaining parts printed for the first time. These are mathematics + the physics chapters, as they must be presented if the mathematics is known. At first, I supposed this would all be for school level. However, the volume (750–1,000 pages) has now gone beyond all frameworks ... I would be very glad for some type of co-operation, and can only regret that I didn't receive your offer earlier ... In spite of our similar views, I think that our different biographies — you are a mathematician and I am a physicist — will unavoidably lead to some differences between your book and mine, which justifies their separate existence ... Most of all, I would like to fundamentally rework my own — having just written it, I can see how it could have been done. However, given my time constraints, this is quite impossible for me. Come to visit, I'd like very much if you could read and criticize my book.' (Signature, home telephone and address. — A.M.).

Thus began my long period of contact with YaB, which was nearly completely determined by the work situation, and at times became extremely strong and at times temporarily ceased. He was one of the most colourful people (if not the most colourful) that I have known. My constant awareness of this always restrained me somewhat in my relations with YaB, interfered with my asking questions about things that were of interest to me, but was not directly related to our affairs, etc. I'm afraid that this may be reflected in these notes as well.

I first visited YaB on May 21 and 23 of that same year, and became acquainted with the manuscript of his eminent book *Higher Mathematics for Beginners* (hereafter, HMB). I think that this unique book, in spite of its five Russian editions and numerous translations, has not been appreciated as much as it should have been. Certain mathematically inclined readers acquainted with modern presentation of mathematical disciplines concentrated on minor inaccuracies and evident lack of mathematical rigour, while 'failing to notice the elephant.' At the basis of such criticism usually lies the widespread naive belief in the objective existence of a certain 'absolute'

---

[132] All dates have been established using my own diary, postmarks on letters, etc.

rigour, which must always be present in mathematical reasoning. This doesn't allow for the fact that all types of rigour are relative, depending on the field and goals of the reasoning: something that is not rigorous for a mathematician could be quite rigorous for a physicist or engineer, especially in the early stages of their education. And, most importantly, it does not take into account the fact that the reader of HMB does not only become acquainted with the meaning of the main concepts of mathematics (and such acquaintance is more important for an applied researcher than formal proofs), but also studies these concepts as applied to the solution of *real* problems, not just problems selected from a workbook. In essence, YaB wrote an introductory course in mathematical modelling — the first in the world. The reader learns not only how to solve equations, but also, more importantly, how to work them out and draw conclusions from their solutions, refine the field of applicability of certain assertions, perform dimensional and order-of-magnitude analyses, compare approximation formulae and asymptotic expressions, and so forth. In connection with all these, the author expresses a large number of considerations, making the book useful not only for high school pupils and university students, but, I believe, also for those beginning research. The lively style of the writing is also noteworthy, and conveys the character of a conversation with the reader and YaB's general nature, without becoming annoying.[133]

On July 31, 1959, YaB sent me a second letter: 'Dear Anatolii Danilovich! ... I now have in my hands in Moscow a finished manuscript of the book ... On the whole, of course, I would rather not change the book appreciably. I am very interested in somehow using concepts that are not included in the book (complex variables, summing of series, numerical ∫, probability theory). Maybe here some contact with your book is appropriate? However, I don't yet have it ...'

On October 12, 1959, YaB sent me a complete manuscript of HMB, already edited at Fizmatlit [the publisher], with the note, 'I hope to print bits not included in the book (probability theory, numerical methods, and others) by the end of 1959 and to send you a copy ...' I immediately began to read the manuscript. There soon arrived another letter from YaB (sent on October 21, 1959: 'Dear Anatolii Dmitrievich! ... Write to me about your general impressions, whether you have many comments ... The following chapters are now being typed. I know that I should have written about Fourier series, the Fourier integral — and I just haven't been able to get to it.'

I've written about my general impressions of HMB above, and I had rather a lot of specific comments; by the way, they in no way were aimed at

[133] It is curious that HMB was one of two books cited by L. Bers (the president of the American Mathematical Society and head of the Mathematical Division of the National Academy of Sciences of the USA) as 'very non-standard' works that had strongly influenced his interesting course in mathematical analysis, which was published in Russian in 1975. Due to the small number of copies printed (33,000), this useful course is little known among Russian teachers.

introducing a mathematical gloss alien to YaB's style. At the end of October, I sent YaB the manuscript of HMB with comments and thoughts about further contacts, and on November 12, 1959, he wrote to me: 'Dear Anatolii Dmitrievich! (I am horribly confused: is it Dmitrievich or Danilovich? I'm very ashamed, but I ask you to clarify.) Thank you for your review and comments. I will use them soon, when I correct the manuscript after Norkin [the editor of HMB — A.M.] and the publisher ... We should already have a serious conversation about the next book. Now, I can only send you an outline. This indicates what is already written and what is not yet done. Over the next 2 or 3 months, the handwritten text will be typed, and I'll send you a copy. Then we should meet and think about the final volume, what parts you will be writing and from what viewpoint. I've worked out the question of the eigenfunctions of an equation with non-Hermitean boundary conditions. Are you interested in this?' To the letter was added the outline, consisting of the titles of the six chapters he had already written, divided into sections, and three sections which needed a writer.

Soon I discussed all these questions with YaB at his home. Incidentally, I asked him for whom he wrote HMB. I expected that YaB would indicate one group of students or another, but he answered with certainty: 'For my son.' As is known, YaB's son Boris is now a well-known physicist. Discussing with YaB his own work on eigenfunctions, I pointed out that, from the position of pure mathematics, an argument in which the coefficients of some expansion are derived assuming the possibility of such an expansion cannot itself be considered the foundation for this possibility. YaB seemed to like this, and, with his characteristic impetuosity, offered me a job with him, though he added some time later: 'Though I'm not giving out application forms now ...'

The reprinting of materials for the continuation of HMB was delayed. In August 1960, YaB sent me a copy of the freshly published HMB with the inscription: 'To the esteemed A.D. Myshkis, from the author, a like-minded person. — I'll be waiting for specific comments! Call me!' It was not until August 21, 1961 that he sent me a letter: 'Dear Aleksandr Dmitrievich! The last six months has been a very difficult period for me, I was in Moscow only for a few days, and this will continue until November. Semendyaev can give you the manuscript ... Please forgive me!'

In the middle of October 1961, I began to edit the material written by YaB for HMB but not included in it, which was now to comprise a substantial part of a new book (its first working title was *Introduction to Mathematical Physics*, but gradually it changed to *Elements of Applied Mathmatics* (henceforth, EAM)). I also began to write small and large additions. The work required more time than we originally assumed (we even had to extend the deadline for submission of the manuscript that had been indicated in the contract), and continued with some interruptions until the middle of November 1963, and even later.

YaB actively participated in the preparation of the materials. He re-read his own and my texts again and again, writing comments on individual pages and in entire notebooks (including his general notebook No. 43 — apparently, YaB wrote and kept some of his drafts in a well-defined order). His comments were impetuous and temperamental: 'A muddle!,' 'Maybe we should give a generalization of the scalar product to *n*-dimensional space?,' 'The vector product in connection with rotation and three dimensions. NB: chat a bit about parity' (he was referring to the law of conservation of parity. — A.M.) and so forth. In all cases, where possible, YaB tried to push away from physics and not from mathematics, and to move from the specific to the general. Some fragments of a physical character he quickly wrote himself; naturally, I tried as far as possible to preserve his text and satisfy his desires, even when it seemed to me that some fragment could be presented more smoothly in some other way. (For example, this is precisely why the integral Fourier transform precedes Fourier series in the book, while, in mathematical books, they are always presented in the opposite order.) The prepared texts were reprinted again, since upon new readings, YaB came up with new comments and additions. The preserved drafts show that this sort of reworking of the text from beginning to end occurred at least three times. Usually, YaB read the text and wrote comments at some time that was convenient for him, after which we would meet and discuss them in detail; then, I would use these comments and my own notes to write a new text or to edit YaB's text, and so forth. To arrange new meetings, YaB told me when he would be in Moscow. I spent several days at his home from morning until late evening, and toward the end of all the work, in order to maintain more direct contact, YaB simply had me stay in his apartment, where I lived from November 1 to 4, 1963. We were occupied for the entire day, taking breaks to eat and exercise, during which we threw each other a medicine ball in the stairwell, one standing above and the other below.

The work was accompanied by correspondence. For example, in a letter dated March 29, 1962, YaB continued a discussion about the conditions of applicability of the equality

$$\lim_{x_0 \to \infty} \int_0^\infty e^{ix} f\left(\frac{x}{x_0}\right) dx = if(0),$$

proposing to conclude the duel 'with the cutting off of any part of the body' of the loser. In place of a signature, the letter ends with a clearly recognizable drawing of YaB with his son throwing me a glove; to confirm the identification, beside me are drawn a mouse and a cat.[134] In a letter dated March 3, 1963, he informed me: 'Very soon, the second edition of *HM for babies* will come out' and made specific suggestions about EAM, such as 'A tedious solution [of a linear differential equation — A.M.], need to exchange

[134] A play on the author's name; *mysh* means mouse and *kisa* is an informal term for cat in spoken Russian. [Translator]

variations of an arbitrary constant to a Green function' and so forth. These suggestions were developed in a letter dated March 31, 1963, and on April 6, 1963 I received a copy of the new HMB with the inscription: 'To dear Anatolii Dmitrievich, a companion-in-arms in the battle for humanizing mathematics.' The distribution of the book rose from 75,000 to 150,000, testifying to its success. Further comments from YaB about EAM (in particular, about the 'philosophy' of eigenvalues), are contained in a letter I received on June 15, 1963. It ends with the words: 'It's time to rid ourselves of the book ... Forgive my carelessness. I've had to deal with a mass of scientific unpleasantness.' In the last letter of this cycle (July 24, 1963), YaB wrote about the need to officially extend the deadline for submitting the manuscript from September 1, 1963 to January 1, 1964, which I did.

After the middle of November 1963, when the entire text had been agreed upon, I thought that the manuscript was ready. However in a letter sent on December 16, 1963, I read: 'The day before yesterday, I submitted it, having done major revisions on the Fourier transform ... It would be good to add more to the variational section ... We don't have any exercises?! ... Everyone remembers your merry confinement in the kitchen. We'll tune the piano ...' (About the last two phrases: YaB's large family lived in a double apartment, and they had me stay in the unused kitchen; when I became tired, I sometimes went to the dining room and tried to play on their poorly tuned piano.) I had to incorporate these comments during the checking of the proofs printed at Fizmatlit. And I had to check everything over once again, when, in April 1964, a very detailed review of the manuscript by K.A. Semendyaev arrived. I performed all this work without YaB, and only finished in August.

During my 'vigils' at YaB's, I to some extent got close to his family, and with pleasure took part in lively general conversations at the dinner table. Once, I came across a rather unusual business — a family physical seminar. (YaB's entire family, apart from his young grandchildren, consisted of physicists — he himself, his wife, both daughters and their husbands, and his son, who at that time was a senior university student.) I was present at a talk by BYa at this seminar; YaB participated in the discussion very actively during the whole talk. I remember that I was struck by his ability to draw correct conclusions from formulae that were senseless from the point of view of pure mathematics ...

On November 10, 1964, YaB wrote to me: '1. Our book has essentially got stuck: the publishers can't find an editor ... Maybe you have colleagues who could do this, or some other idea? ... 2. The third edition of *Higher Mathematics for Beginners* is being prepared. I made use of the preface to advertise our book. I put in the preface an excellent citation from Courant — about the relation of dogmatic axioms on the one hand, and our method of proceeding "from the general to the specific," on the other. [This is an obvious mistake, he should have said "from the specific to the general," as does Courant. — A.M.] This is from the popular journal *Scientific American* from October 1964, which is entirely devoted to mathematics. It's curious — take a look. 3. In the third edition, I added a chapter on δ function at the end, in the form of an appendix. This partially overlaps with our chapter, but I don't think this is a disaster. I hope you don't object? I rewrote it, with a different spiel... 4. The publisher wants to send you the chapter on δ function for HMB to check and edit. I'd be grateful for any comments — send them both to the editors and to me. If you feel certain that this addition to HMB is inappropriate for various reasons, including our joint book — let's discuss this together first, without involving the publisher ...' Of course, I immediately wrote that I agreed, and on December 21, 1964, YaB responded: '... The part of your letter where you agreed about the overlapping δ function material in HMB and Elements took a weight off my chest. Thank you ...'

At the end of April 1965, I began to receive the manuscript of EAM from Fizmatlit for final checking, and I received the proofs in August of that year. YaB also received a copy of the typeset manuscript, and here, too, he introduced some refinements. It was only in January of 1966 that EAM saw the light of day, and on January 30, 1966, YaB wrote to me: '... Let me congratulate you on the publication of our book. It has already begun to be sold. Several "management" questions. We received ten authors' copies + they'll buy me 20 copies in the store. How many should I send you? Or will you come and collect them — should I keep them at home? Will 30 be enough? Who would you like to give them to in Moscow? (My rough list: Petrovskii, Kolmogorov, Keldysh, Tikhonov, Samarskii, Semendyaev, Yaglom Isaak, Markushevich.) With what inscriptions? Should I fake your signature or wait for you? (I'm not planning on going anywhere until May.) They will transfer to us 50% of the remaining 50% of the fee. But it seems to me that you have worked more, and have spent a lot on reprinting and setup, and it's fair for me to return some of the sum to you. How much? [This was not necessary. — A.M.] They've asked Akiva Yaglom to write the review of our book and HMB, but he hasn't calved yet (though he has begun to moo). Do you have any idea who else could do this and would like to? In Baku, they have suggested publishing the book with minimum revisions in both Azerbaijan and Russian. Would you agree, say to divide the material into mathematics and physics? Here, the point is not my general greediness — inspiration is not for sale — but why not sell the manuscript with a minimum amount of extra labour?...' [This publication never came about. — A.M.]

The question of the third edition of EAM was raised at the beginning of September 1970, and after rather intense work, including three days working directly with YaB, I gave the text of revisions and additions to Fizmatlit in the middle of October. There were many more of these than for the second edition. On the advice of A.N. Tikhonov, a chapter dedicated to computer applications was added. However, there were also many additions to the old chapters, some being written completely by YaB, as always, in a very non-standard form for mathematics courses. In July and August of the following year, the typeset proofs started to be sent to me and YaB, and he again introduced various improvements. The book came out in March 1972, and at the end of the newly written preface it said: 'The authors are currently concluding work on the book *A Medium of Non-interacting Particles*, which directly touches upon material in this book, and comprises the first part of a course in mathematical physics.'

YaB already had clear ideas about the possible content of this new book at the end of 1968. He wrote to me: '... First of all, let me wish you a Happy New Year 1969. Let this year see a mathematical physics manuscript. Specifically, I propose to you as a first chapter, for a prelude to the main song, the title "An Ensemble of Non-interacting Particles." I don't believe there has been such a thing in any textbook yet! To start with, not even dynamics ($\ddot{x} = f(x,t)$), but only kinematics. Let $\dot{x} = F(x,t)$ be specified for an individual particle (we could then introduce the more special cases: $F(x)$ or $F(t)$). How should we describe the simultaneous motion of a large ensemble of particles? We introduce the concept of density $n(x,t)$, equations in partial derivatives, the concept of a particle flow, the divergence of the flow steady-state solutions naturally arise, here or later — the Fourier method. Completely new concepts! Together with this, we recall that at the basis are non-interacting particles with individual trajectories. And this rabbit will come out of the hat, nip the reader in the behind and awake him: we construct the equation of characteristics to solve the equation in partial derivatives — and these characteristics, so mystical in usual courses, turn out simply to be trajectories! There's nothing forgotten, nothing left out. The formal manifestations: $n(x,0 = \delta(x - x_0)$ is the motion of a δ function as a trajectory; formally, a Green function. Further, probably, we need $\ddot{x} = f(x,t) \rightarrow$ transform to x, p — phase space, two first-order equations and $n(x,p,t)$. Here is the Liouville theorem, condition of stationarity, some bases of statistical mechanics. Another branch: random mixing and a transition to a diffusion equation for $n(x,t)$ (in part, this was in the fish in Elements). [He means a random distribution of the weight of caught fish, considered in EAM. — A.M.]. A linear equation. The technique of Green functions, concept of superposition of solutions. Random pushes and friction in momentum space give a Maxwell–Boltzmann distribution for a steady-state solution. The concept of osmotic pressure, the Einstein relations between diffusion and mobility (if you don't know these, I'll explain when we meet). Problems on the first passage, giving rise to the boundary conditions. Together with the usual problems in diffusion: $n = 1/r$, $\exp(-r^2/t)$, a cylinder, on a surface — as a whole, it would be an excellent volume, completely dedicated to a mathematical description of the motion of non-interacting particles. This should very much help the reader to absorb Volume II — interacting particles: linear interaction — acoustic waves and non-linear interaction — shock waves. Volume III will be the electromagnetic field. Start writing, give me your suggestions. Come and visit, we'll settle you in the kitchen ... All my growing family send you greetings.'

We discussed this plan in detail twice, and from the middle of August 1969, I began to write the text, passing on chapters to YaB as they were ready. For various reasons, the work was subject to major interruptions, so that the main part of my work did not begin until October of the following year. As with EAM, I had to rewrite some parts after discussing them with YaB, sometimes several times. In connection with our work on this book, we met about 20 times; YaB either made comments on my text or rapidly wrote individual fragments himself, which I then edited and rewrote. In addition, YaB gave me various comments on the text together with corresponding suggestions written by him between our meetings. We worked most intensively in May and June of 1971, when we discussed the text page by page for several days in a row from morning until late evening. After our meeting on June 13, I submitted the manuscript of the book *Elements of Mathematical Physics* (henceforth, EMP) with the subtitle 'A Medium of Non-interacting Particles' to Fizmatlit on June 25. However, as with the writing of EAM, the work was far from finished! Soon, YaB himself began to have ideas about new concepts; in addition, various useful comments were made by G.I. Barenblatt and A.N. Tikhonov once they had read the manuscript, and also by A.A. Ovchinnikov, the book's editor. After two discussions of these comments with YaB, I made the corresponding revisions in December, and again submitted the manuscript to Fizmatlit. Between March and May 1972, we returned to this work again in connection with a review by V.S. Vladimirov; and, as ever, YaB had more ideas about how to improve the text. This required four meetings with YaB, and the text was only finished at the beginning of May.

YaB's preserved drafts with outlines of individual parts, comments on my text, and individual fragments carry a vivid imprint of his personality. One can see how his hand — in spite of all its swiftness, the abbreviation of words, and so forth — can barely keep up with his thoughts; the numerous exclamation marks and question marks show not only YaB's temperament, but also his deeply personal attitude toward the material. In my opinion, parts of these drafts are more instructive than the final edited text, due to their focused concentration on some main point. I think that these drafts deserve an independent study.

YaB always tried not only to communicate deep ideas connected with the material presented, but also to make the text more alive and interesting, to present memorable comparisons, and so forth. He never missed an opportunity to make a joke. I'll present as an example the problem of a random walk along a straight line. In the original text, we, following tradition, considered a drunk, who falls at each streetlight, forgetting the direction from which he came. The reviewer suggested that we change the example, so as not to 'bring to mind yet again this widespread vice.' Therefore, we replaced the drunk with a woman who, in a fit of frightful anxiety, runs from store to store looking for a French umbrella; however, YaB, with clear pleasure, added a footnote: 'Usually, the example shown here is of a drunk, but, so as not to bring to mind yet again this widespread vice ...' and so forth. Further, we talked about a set of a large number of non-interacting women lost on the x axis and named particles ...

At the beginning of September 1973, YaB wrote to me: '... I congratulate you [on the publication of EMP. — A.M.]. I've set about giving out copies, faking your signature: to Tikhonov, Ovchinnikov, Barenblatt, Kolmogorov,

Kapitsa, Pitaevskii, Lifshitz I.M., Kompaneets A.S. (I don't expect objections — the aria of Onegin)... Masses are looking for the secret in "Only completeness ..." Let's gather our wits and think [about further work. — A.M.].' The reference to 'completeness' requires some explanation. Not long before this, in *Uspekhi Fizicheskikh Nauk*, there appeared a review article by YaB, in one of the footnotes of which there was a couplet that purported to be by V. Khlebnikov, with YaB's comment 'the archival work is mine.' The first letters of the couplet made up an indecent expression addressed to one of YaB's friends; as YaB told me, 'he called me that first.' In the final text, one letter was changed (as YaB explained, his friend had taken back his words), but it isn't difficult to guess what was there originally. This episode became widely known: YaB said that even a literary critic expressed his reproach. Therefore, including genuine verses by Schiller in the preface for EMP, YaB added: 'archival work and verse translation by Ya.B. Zeldovich,' counting with pleasure on the inadequate reaction of the readers.

I'll present here an excerpt from the preface to EMP, in which the specific qualities of the book are discussed: 'The most widespread method for setting out mathematical physics is as follows: the original material is suggested using physical reasoning, after which all the studies are carried out using purely mathematical methods ... In this book, we wish to take another path. Concentrating our presentation around problems that allow a clear physical interpretation, we wish to show how mathematical concepts and methods naturally flow from transparent reasoning, how it is possible to trace more fully the connection between mathematical and physical approaches, and indicate the clear meaning behind the procedures and intermediate steps of mathematical solutions ... We have not striven for generality, but rather have tried to show the main ideas using the simplest possible material. Mathematical calculations and logical proofs play a secondary role in the book; our primary goal is that the reader should correctly understand the interconnections and analogies discussed.'

In this same preface, it says: '... Similar independent books on mathematical questions connected with hydrodynamics, electromagnetic field theory, and quantum mechanics are appearing on the horizon [on this account, I said to YaB: 'But I know almost nothing about quantum mechanics!' to which he responded: 'That will make the presentation all the fresher.' — A.M.], which together comprise elements of mathematical physics which are "physicised" in the sense discussed above. It is difficult to say whether it will be possible to realize these plans; though when we finished EAM, we weren't certain either that our work would continue ...'

Alas, the work did not continue, although some specific rough plans were made. For example, at the beginning of February 1974, I received from YaB a reprint together with a letter: '... I'll try to throw together a first approximation of a plan for the next part of EMP and send it to you separately in the next few days...' On August 26, 1974, he wrote: '... I finally submitted a request for the second volume of mathphysics, faking your signature. Hydrodynamics, acoustics, thermal processes, including the propagation of flames $T = T(x - ut)$ with definition of u as an eigennumber. We'll firm things up when (and if) the proposal passes through the editors' council ... P.S. Kompaneets has drowned, I'm very sad. P.P.S. Have you seen the reference to our books in *Pravda*? [From August 23, 1974 — A.M.] People from Dnepropetrovsk, Mossakovskii, and one other [Leonov. — A.M.], I sent them a response.' However, the proposal wasn't approved. A regrettable discussion arose about HMB between YaB and a group of Academicians who occupied key positions in Fizmatlit, provoked, in my opinion, by reasons which were far from scientific, and YaB's entire mathematical activity came to a halt. The monopolism and intolerance that are characteristic of our society played a pernicious role here, and YaB's unique plans remained unrealized. I can't forgive myself; if I had set aside other affairs and worked only on this, maybe we would have time to finish ...

From the beginning of 1974, our collaborative work ceased, so that our contacts soon began to become rarer. Of course, affairs of some other nature arose from time to time, but they became fewer and fewer. (I've already written that I was always somewhat shy of YaB, and also could easily imagine how busy he was; in particular, because of this, I always declined requests from various people to introduce them to YaB, pass some text on to him, etc.). For example, at the beginning of 1973, when I was looking for a job, YaB suggested that I join the organisation he had just left, or was just about to leave. He named several well-known scientists who worked there, then added: 'Forget those names.' He explained that he was leaving because he was tired of working for so long at a rhythm imposed from outside, and not determined by himself. I didn't follow up this suggestion. By the way, for natural reasons I never talked to YaB about his main work; however, some details, primarily of a curious nature, came out from time to time. For example, once we were talking about the fate of political literature in the period of the cult of the personality,[135] and YaB said that in a bookshop in 'Ensk' the compositions of Stalin were being sold for 1 kopeck a volume. Another time, when we got to the formula $e^{i\pi} = -1$ during the discussion of EAM, YaB told about the tale of how at some point he had had to perform complex computations by hand. Because of the responsibility involved, they doubled up on this work, the answers were compared, and each mistake per one exponent was punished with some unit of a fine (I don't remember how much), which then went toward general expenses. The question arose of what would be an appropriate fine for a sign error. Based on the formula above, they decided that the fine should be $|\pi i \log e| = 1.36$ fine units.

There were other rare meetings. On March 8, 1974, my wife and I stopped by to congratulate YaB on his 60th birthday. He drank a glass of wine with us and tried to convince us to come that evening, amusingly playing the role of a seducer: 'We'll get Myshkis drunk, and then go dancing ourselves;' but we left in the evening. About this same time, I met YaB in

[135] The official name of the period of Soviet history from 1922 to 1953 marked by the rule of Stalin. [Translator]

Chernogolovka, where he was participating in a conference and was living in a flat for visitors. He told me with pleasure that in this flat there was a large drum from a variety orchestra, and he alternated his scientific activities with playing on the drum; YaB took me to the flat to show me the drum and let me listen to its sound. On May 8, 1976, I saw YaB in the evening in a small group at his friends the Shuvalovs' (I worked in the same department as the late E.Z. Shuvalova). YaB seemed tired and his usual enthusiasm was lacking.

Our next encounter wasn't until September 23, 1980, when I, being in Alma-Ata, learned that the sixth All-Union symposium on combustion and explosions was underway, and that YaB was participating. I went to the session and saw how YaB actively made presentations, not only after nearly every talk, but also during the talks. However, when I gave YaB a compliment on his good form, he only said: *O zochen wej* — appearances are deceiving. In the evening, we went to a pantomime at the theatre, where I for the first (and the last) time met his second wife, Anzhelika Yakovlevna. YaB got me into a game where I had to pretend to guess her first name and patronymic from her appearance, but she didn't like this very much. It's possible that she was in a bad mood. Later, when in response to my puzzlement that every other presentation in the pantomime was described as a premier, YaB made a somewhat impudent comparison, and AYa commented on this rather sharply. Some of the acts had an eccentric character, and YaB, comparing this pantomime with 'Vampuka,'[136] left with AYa after the first act. This meeting left me with a painful impression.

Our last contacts were in 1986. At the beginning of May, YaB telephoned and requested that I come and see him at the Institute of Physical Problems (IPP). On May 6, I went there. YaB jokingly, but with clear pleasure, announced that now he could be considered a marshall, since he had been assigned to supervise Landau's famous seminar after the death of I.M. Lifshitz. He had asked me to stop by as the possibility had arisen of publishing HMB in Estonian, but he would like to somewhat expand it, taking into account EAM, so was I interested, and so forth. In addition, the question of publishing in English a book combining the new HMB (this last book, substantially reworked and supplemented by I.M. Yaglom and in part by YaB himself, was published in 1982 by *Nauka*) and EAM, updated of course, was being discussed with a Western publisher; this had to be discussed with I.M. Yaglom. When we parted, YaB gave me a reprint with the inscription: '... in the hope of establishing contact, at least Moscow–Tallinn–Moscow.'

On May 11, I called into the IPP again with a plan for expanding HMB. However, it was as if YaB had been changed. He considered the publication of HMB in Estonian not to be worthwhile due to the narrow circle of potential readers. With regard to the publication of a combined HMB–EAM, he said that physics and mathematics were different now to how they were,

[136] A parody opera of the early 19th century, which ridiculed opera clichés.

say, 30 years ago, and the appearance of his name on such an elementary text could be misunderstood. (It is possible that I'm not conveying his thoughts precisely, since he expressed them indirectly.) YaB even suggested that Yaglom and I publish such a unified book on our own, and that he write a recommendation for it in the preface. On this note we parted. I never saw YaB again. I.M. Yaglom continued for some time to try to convince him to change his position, but without success.

In conclusion, here are several remembrances relating to various times.

As is known, in the 1960s, it was very common to collect signatures in support of various protests. At some point in one of our conversations, I talked to YaB about the difficult situation of a person who is asked to sign a petition in connection with a trial about which he is not sufficiently informed. YaB said that he was in such a position, and that, when signing the petition, he also wrote that he was concerned not so much about this specific trial, as about the need for openness in processes of this kind.

At some point we were talking about the Jewish problem, and YaB talked about a group of young people who were trying to revive Jewish customs, right down to their clothing, keeping the Sabbath, and so forth. I expressed doubt about the reasonableness of this, but YaB talked with certainty about people's right to behave in this way. In general, he was sensitive to national feelings. For example, when we considered a problem in EAM about a saddle point, I wrote about a path in the mountains leading from one *kishlak*[137] to another, but YaB thought this could be a possible source of offence (of the sort 'Let two Jews ... and so forth' he said), and in the final version of the text, we wrote about two villages in a hilly area.

Once I asked YaB whether the fact that he didn't have a higher-education diploma had ever posed problems in his life. He answered that yes, it had been a problem, until he became an Academician.

About the conflict associated with Professor A.A. Tyapkin; I first saw this name presented in a very unflattering light in a text about Lorentz transformations written by YaB for EMP. (Subsequently, I learned that Tyapkin was a physicist who had become widely known for his doubtful methodological presentations.) However, after discussions, we decided that such a reference was inappropriate in a book of this sort. Later, YaB told me that he had left the editorial staff of *Uspekhi Fizicheskikh Nauk* because of Tyapkin. Tyapkin had submitted an article to the journal in which he claimed to refute the bases of quantum mechanics. Judging from YaB's tale, the editors, having decided that they would in any case be forced to publish the article under the pretext of freedom of discussion, decided to accept it, adding a 'counter-paper' with a detailed refutation. YaB was indignant ('They haven't even been threatened with flogging yet, and they're already taking their pants down'), and decisively objected to the publication of Tyapkin's article.

[137] Traditional name for a village in Central Asia. [Translator]

I remember an episode connected with our meeting in Almaty. As everyone knows, one of the main accusations levelled at HMB by some mathematicians was that the main concepts of mathematical analysis are introduced without setting out the theory of limits. However, not long before our meeting in Almaty, a book was published by Academician L.S. Pontryagin (one of the activists fighting against HMB), dedicated to an introduction to mathematical analysis, in which he writes in the preface: 'Some physicists [read here Ya.B. Zeldovich — A.M.] consider that the derivative can be described without limit theory. I agree with them.' YaB and I discussed whether we should send a congratulatory telegram to Pontryagin on this account.

I'll now finish this article. Of course, it would not have been appropriate to publish much of what I've written here while YaB was alive, and it was never intended to be published. It is possible that some things here are of interest only to me. However, I think that now, when it has become especially clear what an eminent personality lived among us, any details about him are necessary and important.

Part VI

Astrophysics and Cosmology

As I remember him ...

A.G. Doroshkevich

'If you want to be happy — be happy!' For me, this saying of Koz'ma Prutkov has close associations with YaB. In the 27 years over which I knew him, I can't remember YaB in a depressed mood. It seems to me as if he was always accompanied by an atmosphere of good luck, success, and happiness. I remember his constant activity, discussions of scientific problems — his own (most often) or others', discussions of plans, perspectives. The appearance of YaB in the department soon got everyone moving. His constant active discussion with researchers and colleagues, his impetuous and often unexpected transition from one topic to another, from work to relaxation and vice versa — these features of YaB's style allowed him to intensively and successfully carry out scientific research up to his last days.

I went to work with YaB at the Institute of Applied Mathematics of the USSR Academy of Sciences in the period when he was just beginning to do research in relativistic astrophysics (though this term only appeared much later) and the theory of strong gravitational fields. This circle of problems was completely new to him. That made it all the more interesting to observe how effectively YaB absorbed specific problems in astrophysics, as he impetuously became a scientific leader and created a powerful school in this area of science.

YaB's extremely high and sound creative potential guaranteed freedom of discussion (in modern terminology — *glasnost*), and a benevolent attitude toward critical comments. I remember 27 years of continuous arguments, not only about scientific questions ... To this day, I'm surprised that our collaborative work (and arguments) didn't end much earlier!

One could say a lot about the peculiarities of YaB's approach to both scientific and daily problems, recalling his numerous and always pithy statements about the essence of questions and methods for their solution. This inexhaustible theme is far from fully reflected in his articles, books, and presentations. However, I would like to introduce just two examples here — in my view, they are the brightest ones.

One day in 1969, YaB told the department about one of his new results — a non-linear approximation theory for gravitational instability. At that time, this problem had become topical, and many researchers had attempted to construct such a theory. Solutions of some special problems had been published in scientific journals, several phenomenological models had been constructed (including some very useful ones), the linear theory had been excellently developed and studied in detail, but further than that ...

And then we were presented with a new theory, in which gravitation played a secondary role, and the conclusions were so far from those expected that it was difficult to believe in this theory. Yes, the mathematical part was flawless, but what relation did all this have to gravity? It seemed improbable that gravitation did not play an appreciable role in the non-linear stage and inhomogeneities developed in a kinematic regime!

For nearly a year, we in the department examined the new theory from various angles, got used to it, elucidated its possibilities, and learned how to apply it. Abroad, this work of YaB was acknowledged only seven years later, in 1977, when the large-scale structure of the universe, predicted essentially in YaB's first work in this area and confirmed later in numerical simulations, was detected in the observed distribution of galaxies. Subsequently, YaB's theory served as a point of departure for the development of a Lagrange representation theory (parts of a catastrophe theory) and an inertial instability theory, and today lies at the basis of all theories for the formation of the structure of the universe.

More than once, listening to talks by YaB on various questions connected with his theory, I came to the conclusion that a large role was played in its development by a beautiful analogy with the motion of free particles (this simple problem later also proved to be useful in the theory of inertial instability). However, I still didn't ask YaB what reasoning he had used to write his famous equation for the motion of particles describing the development of a gravitational instability in linear and non-linear regimes. And he already clearly understood all the main physical assumptions underlying this theory in 1970!

A second qualitative shift in cosmology associated in my memory with YaB occurred in the spring of 1980. At this time, an intense situation had formed: observational estimates of fluctuations in the temperature of the relict radiation[138] proved to be appreciably lower than predicted by existing theories. Numerous discussions of this problem by both theoreticians and observers had all led to dead ends. It appeared that the bases of cosmology were in danger. It was necessary to urgently supplement cosmological models and modernize the theory developed by YaB, which provided a good description of many non-trivial observations, but could not explain why radio astronomers did not detect the predicted fluctuations in the relict radiation. It was possible to remove the difficulties by including various hypothetical processes, but they all seemed very exotic ...

[138] More often referred to as the cosmic microwave background. [Translator]

In the spring of 1980 at the Institute of Theoretical and Experimental Physics in V.A. Lyubimov's group, the first series of experiments on measurement of the mass of the neutrino was completed. After returning from a seminar at ITEP, where the results of these measurements were presented, YaB said that the mass of the electron neutrino had been found to be about 30 eV, that this could be important for cosmology, and this ought to be looked into. I won't dwell on the scientific consequences of YaB's suggestion (models were calculated, then refined, and at the time of writing, 14 years later, this work is still continuing). I would like to use this example to investigate certain aspects of the psychology of scientific creativity and the evolution of the world view of a great scientist.

Up until 1980, all physicists firmly 'knew' that the neutrino was a massless particle, and that there were serious experimental, and more importantly, theoretical, arguments to support this conclusion: that the 'heresy' of a massive neutrino is in contradiction with everything, has no chance of success, etc. The problem was not so much in physics as in human psychology.

The introduction into cosmology of 'hidden mass' in the form of weakly interacting massive particles not only provided a new model for the universe, but also led to serious changes in views of the physics of the microworld (now, questions associated with cosmology and microphysics are the subject of cosmomicrophysics). In 1980, I didn't really understand why YaB considered it so important to work on cosmology with massive neutrinos. After all, in 1972, he had (in my presence) unambiguously rejected this possibility during a discussion with the American physicist S.A. Bludman. The situation in both physics and astrophysics was probably different in 1980: the necessity and unavoidability of fundamental solutions was keenly felt, new ideas were carried in the air, and the seminar at ITEP was like a detonator. One way or another, YaB decided to try to solve a 'little' problem — taking into account the mass of the neutrino in cosmology. As a result, the 'neutrino' model of the expanding universe appeared, immediately obtained international recognition, and strongly influenced modern cosmology.

After a brief period of euphoria, a difficult psychological reorientation began. YaB had earlier taken a position in support of 'Occam's razor' (don't introduce extra complications: everything that is not absolutely necessary is excluded), but now, long conversations were conducted — and not only at seminars — about rightfulness and the need to introduce various 'exotic' hypotheses, the relation between 'permitted' and 'forbidden,' justified and excessive ... There was serious work on liberalizing the principles of one's approach to physics, to methods for constructing physical theories, to science in general. This work continued more or less intensively, and may have facilitated YaB's successes in creating his theory for the birth of the universe (in which there is no dearth of new hypotheses). YaB writes a little about this in his autobiographical afterword to the second volume of his selected works. For me, it was highly educational to observe YaB's work. After all, at that time, he was approaching his seventieth birthday, and, as one would have thought, should have firmly established views. Nevertheless, he sharply changed his point of view, which requires no small spiritual strength.

During constant contact with physicists on the scale of YaB, criteria and evaluations often shift, and many unusual events are taken to be ordinary. Only time establishes their true value. After YaB's move to the Institute of Physical Problems of the USSR Academy of Sciences, our contacts became more sporadic, and I suddenly realized my habitual need for his presence. However, there were still valuable possibilities to talk with him about various problems (not only scientific ones). Now that possibility is gone ...

We met several times a year up to the end of his life. YaB was interested in results, plans, and perspectives; he criticized and helped with specific problems. In his last two years, he successfully worked to organize research in cosmomicrophysics. His active scientific and organizational work continued up until his last day. The end came unexpectedly ...

In my memory remains the unique charm of YaB's original personality and a sense of his energy, success, and happiness.

The beginning of work in astrophysics

I.D. Novikov

I am convinced that Yakov Borisovich was not merely a great scientist; he was a phenomenon in physics. As far as I know, everyone who had contact with him in science considered this to be extraordinary good fortune. Why? What was so unusual in him as a scientist?

One can list many things — his fundamental knowledge of very different fields of physics, and, after he became interested in astrophysics, his knowledge in this field; his legendary physical intuition; his mathematical technique; his completely non-standard ideas connected with literally every question with which he was concerned; and finally, the absolute trustworthiness of his conclusions.

'In our scientific and technological age Ya.B. Zeldovich is one of those rare and valuable properties that every university and research establishment yearns for: a great physicist who is equally at home in laboratory discussions of experimental techniques, in the rarefied atmosphere of nuclear theory and elementary-particle theory, or in technology-oriented studies of shock waves and other explosive phenomena. To all these disciplines and others, Zeldovich has made fundamental contributions,' wrote the well-known American astrophysicists K.S. Thorne and W.D. Arnett.[139]

[139] K.S. Thorne and W.D. Arnett, Foreword to the English translation of the book *Relativistic Astrophysics, Vol. 1* by Ya.B. Zeldovich and I.D. Novikov, Chicago: University Press, 1971.

However, his most characteristic feature was an unlimited, child-like love of physics.

I can formulate this quite briefly: in science, he was a Master.

Our first meeting in the autumn of 1962 was not quite ordinary. I had just finished my postgraduate course at the Sternberg Astronomical Institute of Moscow State University (SAI) under the supervision of A.L. Zel'manov, and had started my further work there. My speciality was the general theory of relativity and its application to astronomy. At that time, this was a rare speciality, since astronomical objects with such strong gravitational fields that it was necessary to use Einstein's theory instead of Newton's to describe their properties were still unknown. Neither pulsars nor black holes or quasars had been discovered yet; it is now hard to imagine astronomy without these. Only cosmology — the science of the entire universe — required the general theory of relativity.

At a conference in Tartu, my wife — Eleonora (Nora) Kotok, also an astronomer — heard Yakov Borisovich talking about general relativistic effects during the catastrophic compression of a body. Knowing the sceptical attitude of some astronomers toward general relativity, and especially toward prospects for its application in astronomy, Nora decided to ask Yakov Borisovich whether he felt it was worthwhile for her husband to work in such an exotic field. In their conversation, she mentioned my Candidate of Science dissertation, in which I considered the possible existence of a so-called 'half-closed world' — an enormous clump of matter whose gravity strongly distorts the space within and around it, nearly enveloping it in a closed 'cocoon,' with this 'world' connected to our universe through a narrow spatial tube. As it happened, Yakov Borisovich was also investigating the possible existence of a half-closed world. He asked Eleonora how he could contact me, and when he returned to Moscow, he came directly to the SAI. Our first conversation was long, and gripping for me. I saw that my companion immediately understood everything (I have not encountered such a person either before or since). Before our meeting, I had heard of Yakov Borisovich only a few times. Not much earlier, he had turned to astrophysics as an object of new scientific interest. Earlier, Yakov Borisovich had worked at the closed 'object,' and it was quite natural that we young astronomers didn't know anything about him. This was probably why I didn't feel timid before such a famous man. Once our discussion had begun, no fear or awkwardness could arise, since Yakov Borisovich was interested in the truth, in science, and there was no question of any 'inequality' between him and the person with whom he was talking.

At that first meeting, I was also impressed by his illustrious intuition — his striking ability to sense a completely non-trivial result before performing complex calculations or proofs. If he had serious physical arguments in favour of some conclusion, he often omitted a rigorous mathematical proof of the result. For example, to show that a half-closed world could be connected with our universe through a tube, I constructed an exact, rather unwieldy, solution, assuming that the orifice was filled with matter with a variable density. Yakov Borisovich solved the problem, as they say, with one sweep of his pen, 'joining' the two known solutions for empty space. To my question about whether the proof showed that this was, indeed, an orifice, and not a double passage to the same space acting in opposite directions, he replied, 'Well, it's obvious.' However, it wasn't obvious to me, and Yakov Borisovich asked: 'And how did you set about it?' I told him about the matter filling the orifice. After reflecting for a second, he said: 'That's reasonable. I'll be an official opponent at your Candidate of Science defence.' And in conclusion, he proposed that I work in his newly founded astrophysics group at the Institute of Applied Mathematics, headed by M.V. Keldysh — then president of the USSR Academy of Sciences.

At that time, I didn't yet realize this was a turning point in my life. It seemed impossible that I should go and work with Yakov Borisovich, since I was already working at the SAI, which I considered my home, and I felt responsibilities to my teacher, A.L. Zel'manov.

However, Yakov Borisovich kept me. He took me to the Keldysh Institute and, after the seminar, introduced me to the president: 'This is that very same young man about whom I told you. I would like to take him into my group.'

'So what's the problem, take him,' answered Mstislav Vsevolodovich.

'The trouble is that he is assigned to SAI.'[140]

'Well, call Ivan Georgievich (I.G. Petrovskii, then the rector of Moscow State University) and come to an agreement.'

Here, Yakov Borisovich, in a somewhat crude manner, told part of the well-known joke: 'Mstislav Vsevolodovich, as they say, Papa can do it, but the bull is better ...'

The president laughed and answered him in kind: 'You know, Yakov Borisovich, if you use the bull on every occasion, he won't last for very long. Well, what can I do with you, I promised you I would talk to Ivan Georgievich. Let's go to my office.'

In his office, Mstislav Vsevolodovich called Ivan Georgievich on a 'vertushka;'[141] he ordered me to come immediately and himself directly wrote an order for my transfer to Keldysh's institute. Thus began the 16-year period of my work under the supervision of Yakov Borisovich.

The first amazing fact I encountered was as follows. Yakov Borisovich, as I have already mentioned, had only recently turned to applications of general relativity in astrophysics. In his words, he had become acquainted with general relativity from the well-known textbook of L.D. Landau and E.M. Lifshitz, which he considered an unsurpassed classic. Many students had come to understand general relativity using this text, including myself. The wisdom of this science seemed to me to be unshakeable. However,

---

[140] According to the rules, a young specialist who had finished his post-graduate studies was required to work at an institute to which he was assigned for three years.

[141] Direct classified government telephone line. [Translator]

Yakov Borisovich, in the course of his 'acquaintance' with the topic did several pieces of research that immediately became classics. This occurred several times in succession: often his 'acquaintance' with a new field produced important new results. Usually, Yakov Borisovich completely solved the problem immediately, using very original methods. However, there were cases when the truth came to him after investigations undertaken in various directions. For example, one of Yakov Borisovich's interesting astrophysical works of that early period was his study of the possibility of a cold beginning of the development of the universe. In the first half of the 1960s, it was not clear whether the expansion of the superdense matter of the universe occurred at a low, essentially zero, temperature (a cold universe) or, as George Gamow believed, that the temperature was initially enormous (a hot universe). The main problem with the cold universe theory was thought to be that the expansion of a superdense, cold plasma of protons and electrons leads to the formation of neutrons, and then, as a result of nuclear reactions, to the unavoidable transformation of all matter into helium. However, the current universe consists primarily of hydrogen!

To eliminate this contradiction, Yakov Borisovich proposed that in the beginning, in addition to protons and electrons, there were also neutrinos. In the presence of neutrinos, it is impossible for protons and electrons to react in dense matter with the formation of neutrons, so helium cannot form either. Before Yakov Borisovich, no one had considered the hypothesis of a cold universe with neutrinos. Psychologically, this was probably due to the fact that neutrinos can freely escape from any clump of dense matter. Thus, in clumps of cold matter of finite size, it is impossible to stabilize protons in this way. However, neutrinos can't escape from the universe! Such stabilization is possible — and this was the 'simple' (as is everything that is truly interesting) idea of Yakov Borisovich.

Zeldovich's hypothesis made it possible to mark a path for the solution of other difficult problems in cosmology as well. E.M. Lifshitz rightly described this idea as exceptionally clever.

After this work, it became clear (at least to those of us working with Yakov Borisovich) that in order to elucidate whether the hot or cold universe model was correct, it was necessary to try to detect 'relict radiation' — weak electromagnetic radiation at centimetre and millimetre wavelengths that remained from the hot beginning and cooled with the expansion of the universe.

At that time, our group consisted of only three people — our supervisor, A.G. Doroshkevich, and myself. Yakov Borisovich did everything he could to encourage us to work actively in cosmology. We were still quite inexperienced, and often couldn't say 'yes' or 'no' or anything at all intelligible about the many suggestions and results that were given to us almost daily by our supervisor. We tried to sort out one question, and he would hand us the next one, then another, and so forth, at a constantly accelerating rate. His temperament was incredible. Yakov Borisovich couldn't imagine that someone in his group couldn't work as quickly and as much as he himself did. He demanded from us prompt critical discussions of new ideas: 'Either agree with me or criticise.'

Before my arrival in Zeldovich's group, I began to calculate the spectrum of the electromagnetic radiation from all sources in the universe. I later continued this work together with A.G. Doroshkevich. We first heard about the 'hot universe' theory from our supervisor, and calculated whether it was possible to detect relict radiation against the background of the known sources. It turned out that this was possible at centimetre and millimetre wavelengths. Yakov Borisovich was very supportive of this work, and recommended that we publish it in the journal *Doklady Akademii Nauk*. No one, either in our own country or abroad, had paid attention to the possibility of searching for relict radiation. No one, that is, except our supervisor — and the existence of relict radiation would contradict his hypothesis. A year after our publication, relict radiation was detected by chance by the Americans. Yakov Borisovich immediately appreciated the value of this discovery, and in subsequent years, became one of the founders of the modern astrophysics of the hot universe.

I remember another case that characterises Yakov Borisovich's attitude toward young scientists. At some point, in our group we were all reading the work of W. Fowler on the equilibrium of supermassive stars. Yakov Borisovich was sceptical about a number of his conclusions. However, A.G. Doroshkevich, one of the biggest sceptics that I've ever met, and a person who was very meticulous in science, scrupulously redid all the calculations and came to the same results as Fowler had. He told this to Yakov Borisovich, who initially waved him away with a certain irony; however, the next day, he arrived and first thing raised Doroshkevich's hand as a sign of his victory. And then our supervisor explained the fine point that had initially prevented him from believing the result. This was extremely useful for us.

In those years, the scientific interests of Yakov Borisovich lay almost completely in astrophysics. He closely acquainted us, scientists just starting our careers, with well-known physicists whose research touched on astrophysics in one way or another. I remember meetings with V.L. Ginzburg, E.M. Lifshitz, M.A. Markov, and I.M. Khalatnikov. Our discussions with A.D. Sakharov were especially remarkable. At the end of the 1960s, he often spent time in our group discussing new ideas in cosmology. The logic of Andrei Dmitrievich's reasoning was completely different from Yakov Borisovich's style — more abstract, detached, and formal. Specific physical problems were closer to Yakov Borisovich. I should say that often, I was also attracted to more abstract questions — 'other universes,' the time after the 'absolute future,' and so forth. I gained a great deal of satisfaction from the fact that Sakharov considered his so-called multi-sheet model for the universe to be, as he wrote, 'based on ideas of I.D. Novikov'.[142] At the suggestion of Yakov Borisovich, Sakharov was the opponent of my Doctor of Science dissertation in 1971.

Yakov Borisovich taught us to continuously follow the scientific literature, and tried to have us work in very different areas of the science founded by him — relativistic astrophysics. He was impatient when he got new ideas. It sometimes happened that he would telephone in the early morning (well before six), excitedly exclaiming: 'I have an idea, come without delay.' What a joy it was to work with him!

Our first joint article was devoted to an analysis of the motions of bodies in the vicinity of a black hole, and was published in 1964. After this, we published 48 joint works, among them several books. Yakov Borisovich didn't tolerate a moment of complacency or self-satisfaction. Even during his vacation, he wrote us excited letters from the south demanding a critical evaluation of the state of affairs, and expressing some new ideas.

I have briefly touched in these notes on only a few episodes characterizing Yakov Borisovich to some degree — as a scientist and teacher in the years of our collaborative work. Our collective was unusually friendly and happy together long after its creation. Everything was done openly and with general consent. Gradually, new people came to us.

In the first years, we discussed many questions freely with our supervisor, including non-scientific ones. However, at the end of the 1970s, when a conflict arose in connection with one (non-scientific) question, Yakov Borisovich decided that everyone who did not agree (the whole group, with one exception) should leave. This breakup was tragic for us and, as far as I know, for Yakov Borisovich too.

Many years later, he and I were talking as we sat in the garden of the University of Rome during a foreign trip. Yakov Borisovich asked about my children as if they were very young, though they had grown up in the meantime.

In 1987, the life of this extraordinary person who led us into a great science was curtailed.

Fifteen years and more

G.S. Bisnovatyi-Kogan

ACQUAINTANCE

In 1963, my friend V.M. Chechetkin and I decided to study astrophysics. At that time, we were in our last year of study at the Aeromechanics Department of the Moscow Physical Technical Institute (MPTI) and had to travel to the base in Podlipki, very far away; we knew only one astrophysicist,

---

142 A.D. Sakharov, *A Multi-sheet Model of the Universe*, Preprint of the Institute of Applied Mathematics of the USSR Academy of Sciences; I.D. Novikov and A.D. Sakharov, *Relativistic Collapse and the Multi-sheet Structure of the Universe*, Preprint no. 7, 1970, p. 17.

S.B. Pikelner, who, at Valerii's suggestion, we visited at the Stern- berg Astronomical Institute (SAI).

'I don't feel very well now,' said the extremely delicate Solomon Borisovich, 'so I can't take you on as postgraduate students. But Academician Zeldovich is getting together a new group at the Institute of Applied Mathematics, so I'll try to help you make contact with him.'

Upon receiving the telephone number a week later, we called I.D. Novikov and A.G. Doroshkevich, arrived at the IAM, and roughly nine months later became postgraduate students of Yakov Borisovich Zeldovich. My collaborative work with YaB had begun, and lasted nearly 15 years, up to our sad parting of ways in 1979.

THE SEMINAR AT THE SAI

I first saw YaB at one of those seminars at the SAI that soon became famous and were named JAS (Joint Astrophysical Seminar). He talked in his somewhat muffled voice about the expanding universe, neutrinos, quarks, and stiff stars — all this was completely new to me. Moreover, I couldn't imagine how it was possible to learn all this in one lifetime. Having received a very applied education, I was especially struck by the figures: 10^{33} g, 10^{47} erg/s (at that time quasars had just been discovered), and so forth. Conversing at some point with Doroshkevich, I asked him:

'Andrei, how can you work under such conditions? It seems you can just make up anything — go and check.'

'Don't worry, you can't go too far afield,' answered stern Andrei. 'It's enough to show the boss, and he immediately leads you out to clear water.'

'You mean he knows everything?' I asked sceptically.

'Yes, he knows everything that you can imagine.'

That was roughly what Andrei said after working with YaB for about five years. In the subsequent 15 years, coming constantly to YaB with various ideas, I heard his impatience many times: 'keep at it.' This meant that the idea was raw, required further work, or was still quite unsuitable. It often happened just that way.

THE EXAM

The postgraduate entrance exam for the speciality 'Properties of matter at high temperatures and pressures' was based on Landau and Lifshitz's *Statistical Physics*. The committee consisted of chairman Ya.B. Zeldovich, I.D. Novikov, and A.G. Doroshkevich.

The supplementary question was 'What is the adiabatic index for a non-relativistic electron gas?' Turning my face to the board after 15 minutes, having covered the board with writing several times, I obtained the answer — 5/3. YaB's face did not exactly shine. He gave me the next question: 'What is the adiabatic index of a gas of relativistic electrons?' Again, turning my face to the board and shaving five minutes off my derivation due to my previous experience, I obtained the answer: 4/3. YaB's face clouded even more. What was the problem, I was puzzled, it seemed I had answered correctly ...

'Well, I can't give him a five,'[143] said YaB; here I pull a long face, and turn into a mute question mark. 'These things should be known by heart,' he concluded. I opened my mouth, tried to say something, but didn't have time. Another member of the committee, Novikov, got involved. 'Yakov Borisovich, why should we give him a four, after all we want to accept him,' he says half-questioningly. 'Yes, well, probably,' answers YaB uncertainly. 'So in that case, we'll give him an 'excellent' in order to avoid difficulties with his matriculation.' 'You think so? OK, we'll give him the five, but only on credit, mind (here he was already talking to me) — all basic things one must know by heart.'

Since that time, I've learned many things by heart, but each person has his own list of basic things. YaB's list was very long. For example, I never saw him look in an address book when he telephoned somebody.

THE BOTTLES

It was the beginning of my second year of postgraduate school. I had adapted a bit, and began to argue with YaB; the first time, the question concerned obtaining a solution for a model of a star with steady-state outflow. YaB wrote a solution, and I didn't believe it. He patiently explained it to me once, twice, but in vain. It seemed to me that I understood better. And Novikov supported me: 'Let's listen to him, it seems that he is talking sense,' said Igor. YaB was somewhat shocked. 'We'll bet a bottle,' he says, 'I'll bet a bottle of cognac, and you a bottle of mineral water, but with a label stating why you lost.' 'Fine,' I agreed, looking forward to my victory.

The first question from YaB the next morning: 'Where's my bottle?' 'But I think that I'm right,' I announced impudently, having verified my ideas with long calculations and attained an equality between the number of equations and number of unknowns. 'Well, you know —' YaB was speechless, and went into his office. Attentively studying YaB's manuscript once more, I began to feel a slight trembling in my knees, and my impudence gradually evaporated, giving way to despair. Coming into the room after some time, YaB announced seriously: 'If you are going to continue in this spirit, you'll strongly degrade yourself in my eyes; you should think about whether we can work together further.' 'I've already sunk to the bottom,' I said. 'No,' answered YaB, 'a downward descent is a bottomless abyss, so there is always reason to stop.'

On the next day, I understood everything, and brought in the bottle. This was my first bottle; it became clear that it's better not to bet with YaB on science: the probability of your being correct is negligible, and it's much more

143 The Russian marking system is based on a scale from five to one, with five being 'excellent' and below three usually being 'unsatisfactory.' [Translator]

useful to think things through once again. Bottles of water with labels from various people gradually filled an entire shelf in the cupboard in our room.

DOES A NEUTRON STAR HAVE A MAGNETIC FIELD?

1967, the General Assembly of the International Astronomical Union in Prague on the eve of the Prague Spring. YaB talks about the radiation of a neutron star during accretion (a splendid piece of work done together with N.I. Shakura, which became Kolya's [Shakura's] undergraduate thesis). Soon afterward, in Moscow, I became acquainted with Alik Fridman from Novosibirsk, who was working in plasma physics. We decided to supplement the Zeldovich–Shakura work and give the neutron star a magnetic field. Having made estimates indicating the importance of the field's influence, we wrote a short paper and, toward the end of the year, brought it to YaB. He listened, but didn't like all of it.

'Why should the neutron star have such a magnetic field?' he asked. 'But even if you compress the Sun, there will be a very large magnetic field due to flux conservation,' I asserted. 'Well, maybe, but the structure of your collisionless shock wave is completely unconvincing, and therefore the spectrum is not believable' (we were considering synchrotron radiation from electrons with a relativistic Maxwellian distribution). Indeed, we couldn't perform serious computations on the structure of a collisionless shock wave with relativistic velocity; as far as I know, even now, this problem has not been solved entirely. Attempts to convince YaB that the work deserved to be published, at least because of the idea of a magnetised neutron star and rough estimates of the pattern of accretion onto it, were not successful. The paper lay in this state until the spring, when a report of the discovery of pulsars appeared in *Nature*, with its reference to 'little green men'. And again, YaB at first didn't accept the idea of a magnetised neutron star, citing the principle of maximum simplicity, and constructed a theory for the radiation of pulsars based on radial oscillations of a white dwarf. I think that only after the discovery of a pulsar in the Crab Nebula did he finally reject this idea.

When I came to him with our paper, he ordered us to publish it without delay. I later used this case several times as a trump card in our arguments. When I was very sure of something, and YaB expressed doubts, I would say, 'And remember, you didn't believe in neutron-star magnetic fields.' After this, YaB usually softened and answered: 'Yes, indeed, I wasn't right that time. Well, publish it if you want.' However, this happened perhaps two or three times during the entire time of our collaborative work.

AFTER THE BREAK-UP

At the end of 1978 and the beginning of 1979, our group, founded by YaB and consisting primarily of his students, unexpectedly fell to pieces: nearly all his researchers working at the Space Research Institute were forced to move to another department. I remember that, for me (and I think for all of us), this was like a bolt from the blue. A disagreement about a problem unrelated to science arose between YaB and most of the group, but none of us dreamed in our worst nightmares that it could all end like that.

In spite of all this, my personal relations with YaB were preserved, and I never felt enmity from him. Although our discussions of scientific questions virtually ceased, I could count on his help with some everyday and other problems. For example, in July 1987, he was able to get me into an extremely interesting conference in the Federal Republic of Germany dedicated to the five months of research since the explosion of Supernova 1987A in the Large Magellanic Cloud, although there was only one month between my receiving the invitation and the departure date. At that time, due to bureaucratic complications, such a trip could not have taken place without YaB's help.

From 1986, our scientific contacts were resumed somewhat. Memories of the events of the winter of 1978–1979 always made YaB sad. 'I wasn't completely right then,' he told me at one of our meetings, 'because I didn't correctly understand the motives behind the behaviour of those on the other side of the argument. It seemed to me then that everyone was making a choice between me and someone else, but now I think it wasn't really like that.' I assured him that, indeed, it was not like that at all, and if we had known that he would take it that way, none of us would have started that argument. It is now clear that not only he was mistaken — the behaviour of the other participants in the argument could also have been more tactful.

YaB suggested that we revive our collaboration, and I joyfully agreed. He prepared for my transfer to another division of the Space Research Institute, but the practical realization of the plan was dragged out. The move was made only after his untimely death.

When we were young

R.A. Sunyaev

I distinctly remember March 1965. I was a student in my fifth year at PhysTech,[144] living out a dream of working in theoretical physics. Olya Zeldovich — a bright spark in the group of L.G. Landsberg in A.I. Alikhanov's laboratory, where the department head K.A. Ter-Martirosyan had sent me — said to me, 'Why are you tormenting yourself? Give my dad a call, he's forming a group of theoreticians. He'll like you.' Every time I pass through the entry gates of ITEP, I am filled with gratitude to all the people mentioned above, and also to my teachers at the Institute for Theoretical and Experimental Physics, L.B. Okun', G.A. Leksin, N.A. Burgov, and many, many others. It may be that I spent the most carefree years of my life there,

[144] Moscow Physical Technical Institute. [Translator]

in the dormitory for the students and graduate students from PhysTech, in the library, in the experimental-accelerator hall. There, at ITEP, I.Ya. Pomeranchuk shook my hand (for some reason) at the seminars, and I saw V.B. Berestetskii — one of the authors of *Quantum Electrodynamics*, which since then has become a primary handbook for me.

But the main thing is — that day. I believed Olya. She was the only person who did not hide instruments and soldering irons from the students, and who was kind and friendly in, according to our judgement, untroubled times. Therefore, without asking her who her father was and not suspecting that that day would change my fate, I called the telephone number she gave me. The next day, at the appointed time, I stood at the entrance to a building at Miuss;[145] there were no signs, but I later discovered that the Department of Applied Mathematics, headed by Keldysh himself, was located there. I remember that the lithe Nora Kotok brought me my pass, and I climbed the stairs to the second floor. An important-looking fellow in glasses gave me a rather thorough interview (who were my parents, where were they from, my entire biography). At that time, I took him to be a high-level '*chekist*,'[146] and decided that this place was doing serious work (he later turned out to be a quite harmless junior researcher who was simply, as always, collecting all the information that was accessible to him).

During the interview, a smiling man in a shirt and tie flew into the room; his shirt buttons were undone at his stomach, revealing a healthy crop of hair. I could tell from the appearance of the '*chekist*' that he was an even bigger cheese. He looked at me and said, 'Let's go and have a talk.' Our conversation was held in a luxurious (according to my standards at that time) office (which turned out to be the office of I.M. Gelfand). I was given three problems to work on, and after a month, it had been decided that I would work on my master's thesis with this altogether somewhat strange man, who was incapable of spending three minutes in a quiet state. A few years later, I learned from him that they were not enthused at ITEP about giving me leave, and that he called L.B. Okun' and asked if it was worth insisting on it. I am very grateful to LB that he did not give a negative answer, though I now understand very well the colossal difference between us, both in character and in our perception of the world.

I remember that day so clearly because it completely changed my fate. Until that day, the main authority in my life was my father, Ali Sunyaev — a man with a very complex fate, whom I love without limit and whose memory is sacred to me. My meeting with Yakov Borisovich Zeldovich completely changed my life. From the time I met him to 1970, my life was full of festivity. Nearly every day, he called me in the morning (or I called him, when I spent the night in Dolgoprudnyi, at PhysTech) to fix a time for our meeting that day. To this day, it is difficult for me to come, at the invitation

[145] Muscovite slang for Miusskaya Square, where the IAM is located. [Translator]

[146] An officer of the KGB; refers to the original name of the KGB, the ChK. [Translator]

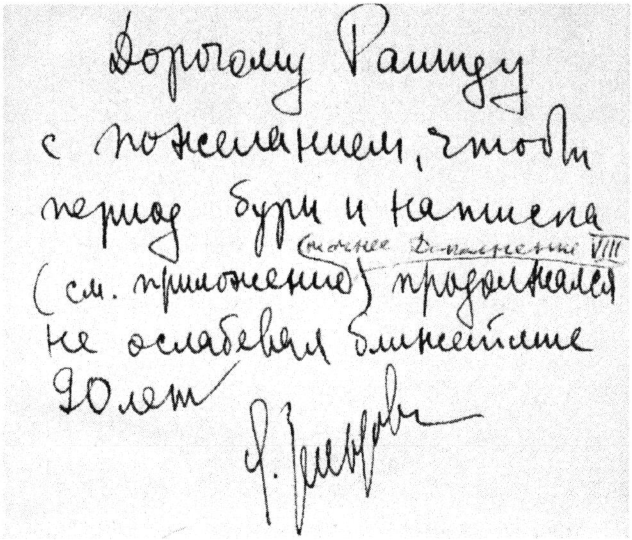

Inscription written by Ya.B. Zeldovich to R.A. Sunyaev on a copy of the book 'Relativistic Astrophysics' published by Nauka Publishers in 1967. The inscription reads:

To dear Rashid,
with wishes that the period of *Sturm and Drang* (see Addition VIII) continues without weakening for the nearest 90 years.

Ya. Zeldovich

of his daughters, to the building on the former Vorobiev Road, to the room where he worked and to which he invited everybody — where I came several thousand times. In the kitchen, his wife Varvara Pavlovna would feed the 'boy from the dormitory,' or he himself would cook frankfurters for two, and then (after 1977) we would drink tea with Anzhelika Yakovlevna. Later, when I was a Corresponding Member of the Russian Academy of Sciences, his third wife Innessa Yur'evna received me in that same kitchen.

But the main thing was not that they accepted me, a boy from Tashkent, into their home like a decent person, nor that I — a night owl from birth — was called nearly every morning for almost a quarter of a century by a man whose nature was like a lark, and whose position was completely incommensurate with my own. The main thing was his enormous interest in life; his thirst for novelty; his search for truth and essence; his lively interest in everything unusual; his colossal ability to work, multiplied by his need to teach and excite work in others, which I still find amazing; his surprising readiness to accept another's point of view; his joy at others' results, that someone was able to do something important earlier and better, which at times seemed like it must be affected. His patience with groundless criticism at seminars was striking. When, 22 years ago and spoiling for a fight, I asked him why he was so patient with the fact that they were saying both quietly and aloud at the seminars that his 'pancake' theory would die like the dinosaurs, he calmed me down, saying that he was hesitant to square his shoulders, since he didn't know how grievously he might injure those who were criticising him. It could be that this property was exploited by the brightest personality among his astronomical surroundings — I.S. Shklovskii, who caused him more than a few bitter minutes. I was always struck by how his activity, constant readiness to be taught, talent for absorbing new information, absence of fear to confess that he didn't know something, and instantaneous mastery of new material frightened acknowledged masters. Solomon Borisovich Pikelner, Abram Leonidovich Zel'manov, and Dmitrii Yakovlevich Matrynov were, perhaps, the few exceptions to this rule.

I arrived in YaB's group at a time that was happy for him, and indeed for all astronomy. Many remember the seminars at the Sternberg Astronomical Institute (SAI) in 1965 to 1968, when they were supervised jointly by YaB, V.L. Ginzburg, and I.S. Shklovskii; when Solomon Borisovich Pikelner was prepared at any minute to step up and explain something to the physicists; when D.Ya. Martynov, S.I. Syrovatskii, B.A. Vorontsov-Vel'yaminov, B.V. Kukarkin, N.N. Pariiskii, Ya.A. Smorodinskii, P.N. Kholopov, A.L. Zel'manov, A.G. Masevich, and M.U. Sagitov sat in the front rows at nearly every seminar; S.A. Kaplan sometimes came in from Gor'kii;[147] the young L.M. Ozernoi, I.D. Novikov, N.S. Kardashev, V.S. Imshennik, V.G. Kurt, Y.N. Efremov, V.F. Shvartsman, and A.M. Cherepashchuk actively participated; E.M. Lifshitz, I.M. Khalatnikov, or A.D. Sakharov occasionally appeared. And I remember the conference hall at the SAI, packed to the limit, with doors open on both sides, behind which clustered the latecomers, for whom there weren't enough seats. And many young people, who later became the flower of our astrophysics.

I remember the first seminars after the discovery of relict (cosmic microwave background, CMB) radiation, when YaB talked, as if it were something obvious, about the dipole component as a means for measuring the velocity of the Earth, and about the presence of a reference system that could be defined by this radiation. I remember how I first heard from him at these seminars about the quadrupole component in the case of an anisotropic universe, and the unavoidable existence of angular fluctuations of the background. Only a year later, reading fresh issues of *Nature*, *Physical Review Letters*, and *Astrophysical Journal Letters*, did I understand that there was a level starting from which things became evident, and cosmologists of this level in Cambridge, Princeton, and Moscow were required to seize this simultaneously. And then I realized for the first time that to be alongside YaB meant, as a minimum, to be at a world-class level.

I also remember that YaB expressed regret at the seminars that he had held to the cold model of the universe until the discovery of the relict

[147] A large city in central Russia, since then again given its original name, Nizhnii Novgorod. [Translator]

With American astrophysicist J. Ostriker. Moscow, early 1980s.

radiation. It was striking how he immediately acknowledged the correctness of G.A. Gamow's point of view, and delightedly praised it, which was not so simple at that time. In general, YaB worked so that the names of A.A. Friedmann and G.A. Gamow and their gigantic contribution to cosmology were appropriately reflected in the scientific literature. He loudly bemoaned his own mistake, when he tried to explain (including in a published paper) the newly discovered phenomenon of radio pulsars as the effect of processes on white dwarfs.

At that time, he told me that that mistake was the result of his many years of work on weapons research: it was sometimes necessary during the course of a single night to make a decision about what to do the next day. There had to be only one final option, and it had to be realistic, simple, reliable, and the most economical. It was not permitted to make mistakes. Nature, as he convinced himself time and time again, could allow itself things that seemed at that time absolutely unexpected or improbable, but, from the point of view of physics, there were entirely natural solutions. And he was glad when he could, on the fly, adopt the ideas of T. Gold and F. Paccini about the nature of pulsars and the articles of P. Goldreich, W. Julian, J. Gunn, and J. Ostriker who were still little known at that time. He also treated the work of V.L. Ginzburg and V.V. Zheleznyakov on mechanisms for the radio emission of pulsars with great attention and respect. In the last years of his life, he, a man who was absolutely certain that neutrinos should have zero mass, radically changed his point of view and often repeated: everything that is not forbidden is permitted. Not everyone can undergo such a radical change of perspective and begin work supporting a new philosophy.

Yes, it was a different time then. Right up to 1973, YaB's talks at sessions of the Division of General Physics and Astronomy of the USSR Academy of Sciences at the Lebedev Physical Institute attracted a full conference hall; at times, there were hundreds of people standing in the entryways. The words black holes, neutron stars, accretion, X-ray sources, supermassive black holes in the nuclei of galaxies, and quasars cast a spell over the imagination of physicists; his reviews in *Uspekhi Fizicheskikh Nauk* and books on relativistic astrophysics with I.D. Novikov attained colossal popularity. It would seem that, with the confirmation of accretion theory by the Uhuru X-ray satellite, he should have worked on details of the theory. But he left this field to his students, while he himself went off to entirely new things: to the problem of extracting energy from a rotating black hole, together with one of his best-loved and most talented students, Alesha Starobinskii (you can read in Stephen Hawking's book *A Brief History of Time* that this work, in *Soviet Physics JETP*, awakened this English giant to the possibility of black-hole evaporation); to work on the magnetic dynamo in astrophysics, with Sasha Ruzmaikin and Dima Sokolov; and to what became his main research topic over the last twenty years of his life — the origin of the large-scale structure of the universe, or, as many like to call it, the theory of the formation of galaxies. Most of all, he worked on this theme with Andrei Doroshkevich and Sergei Shandarin.

I would like to write about one difficult conversation I had with him. In 1982, the question arose of the organisation of an experimental department for high-energy astrophysics at the Space Research Institute. The very powerful and authoritative I.S. Shklovskii proposed me as a candidate to head this department. After a conversation with R.Z. Sagdeev, toward whom YaB felt very positively (and not only because of his old friendship with D.A. Frank-Kamenetskii), he called me in and began to persuade me, repeating that to refuse the director would automatically imply that I should leave the institute, and asking which I would choose — SAI or the Institute of Applied Mathematics. I clearly understood that accepting the offer would mean that I would have to leave theoretical astrophysics, where, as it seemed to me, things were not going badly for me; that I would not be good as the supervisor of the collective, a manager; that I didn't have any experience as an experimentalist. It was all absolutely clear, and I categorically turned away from this new and evidently ill-fated affair.

Then, YaB began to talk about himself, which is why I have presented this story, which was not very pleasant for me and again changed my fate completely. He said that many times he had radically changed not only the theme of his research, but also essentially his speciality: he had nearly been a chemist, and in the end had become almost an astronomer. And not only the turns of fate were responsible for this.

In his words, it is difficult, but interesting to master ten percent of the information and specific methods in any field of the natural sciences, but this is essential in order to begin independent work, or at least to calmly get oriented. Further, the path from ten to ninety percent understanding is pure pleasure and genuine creativity. And to go through the next nine percent is infinitely difficult, and far from everyone's ability. The last percent is hopeless. It is more reasonable to switch to a new problem before it is too late, and have the joy of continuous creation. I don't know if he indeed thought that this was true or simply adopted this method of persuasion, but the conversation was memorable. Many stories he used to teach me were amazingly pragmatic. He apparently thought that they would have a stronger effect on me. There was the feeling that he adopted an individual teaching method with each of his close students, in accordance with the impression of that person he received at the very beginning.

He then said that he had already 'sold' me, and had given agreement to the offer in my name, but he promised to help. And he kept his word until the end of his life. He received funding for ten positions from the then Head of the Ministry of Medium Machinery,[148] E.P. Slavskii, went with me to establishments in Frunze,[149] Leningrad, and Moscow that supplied and developed equipment, called the administrations of various ministries, got from them what was needed, met with the supervisors of foreign collaborations and with the supervisors of enterprises that developed satellites. His participation nearly always helped, everybody knew his name. It was only after his death that I fully realized what it meant to prepare a space experiment without having a powerful escort.

And today, with some successes with the orbiting observatories in the 'Kvant' module and the 'Granat' satellite behind me, I can say that without him, this country would not have any capability in modern experimental X-ray astronomy. It is a joy that he was able to see the arrival of data from 'Kvant' on the X-ray radiation from the supernova in the Large Magellanic Cloud, due to the radioactive decay of cobalt transformed into iron. He was vividly interested in this. And not once did he tell me that the detection of the neutrinos from this supernova, predicted by him and O.Kh. Guseinov, was much more important, although I understood this perfectly well myself. He was the first to suggest to me that the Rayleigh–Taylor instability was probably operating in the shell, mixing the radioactive cobalt.

There are few who know how much he did for the success of the 'Granat' spacecraft, which enabled detailed investigation of the central regions of our Galaxy and discovered first microquasar GRS 1915+105 with superluminal motion of radiojets; about his participation in the birth of the perspective 'Spektrum X–Gamma' project, intended for studies in cosmology and extragalactic astronomy, and capable of becoming our country's first national observatory in space, whose data will become, must become, as is usual in

[148] Minsredmash — the Soviet ministry that oversaw the atomic industry.

[149] Then the capital of the Kirgiz Republic, now named Bishkek, the capital of Kirgyzstan. [Translator]

the West, accessible to all observatories, institutes, and universities in our country.

He helped in the development of the 'Relict' experiment, designed to measure the large-scale anisotropy of the relict (CMB) radiation, carried out by I.A. Strukov and D.P. Skulachev with the 'Prognoz' satellite; he did all he could so that the first top-class space experiment in the field of cosmology in this country received the wide international recognition it deserved.

Like many PhysTech students, I read all the translated literature about the creation of the American atomic bomb, the tormented reflections of J.R. Oppenheimer, how the pilot who dropped one of the first nuclear bombs on a Japanese city subsequently lost his mind. With our physicists it was a different story: they understood why they were working and were very much idealogically well-grounded. Nobody lost his mind.

It was not comfortable to ask direct questions. YaB himself was silent on this topic. The most he did was laugh it off, or tell anecdotes about the policeman who had rehung road signs in the closed city in accordance with the actions of a physicist (read YaB) who had broken traffic rules.

Several times, he told me that he advised his son to learn by heart some everyday jokes, which would make it possible for a young man to always have something to say in a group of friends without making comments about topics that were worrying to everybody. (It would be interesting to ask B.Ya. Zeldovich if he was given this advice). I'm certain that such indirect comments were made around him after his latest unsuccessful attempts to get me approved to go to some conference abroad.

As a person who had been denied travel abroad, he was long certain that if he were ever allowed to go to some major conference in the West, something important would come of it. However, over almost 20 years, as he said, he had only the orbital velocity. He was allowed to travel only to Eastern Europe, at that time firmly building Socialism.

In 1967, he was able to get me — his postgraduate student — approved for a group of young scientific tourists going to the General Assembly of the International Astronomical Union in Prague. To this day, it has not been possible for me to attend another IAU Assembly. And that was my first and next-to last trip abroad before more than 12 years of refusal for foreign trips. Naturally, I remember this meeting well. I remember the mood of the people, carried in the air, which was unusual to us, and clearly heralded the imminent arrival of the Prague Spring. But I also remember YaB, busy 100% of the time, leading behind him a group of his students to meet various celebrities (and they were scheduled one after the other). At these meetings, I first saw Edwin Salpeter, Margaret and Geoffrey Burbidge, Herbert Friedmann, the young and very active Riccardo Giacconi, George Field, and Dennis Sciama. Professor Sciama was memorable for also being accompanied by a student, who later became well-known — Martin Rees. YaB managed during a short meeting to say something about his latest work, ask several specific questions, primarily on experimental data, and prompt a few words from his young student, who spoke English with difficulty. He led us to a swimming pool, where he, as always, swam pleasureably and talked about the film he had seen the day before. At intervals, he could be seen sitting on some sofa and rapidly writing several lines in his large, expressive handwriting, usually giving them straight to I.D. Novikov for their newest book.

It is surprising, but in Moscow, he behaved essentially in the same way: he was never late to appointments, of which he had many each day, he regularly went skiing or walking, wrote in school notebooks nearly every free minute. Everything he wrote he tried to give away immediately for inclusion in a book or paper, or simply to be used as food for thought. Sometimes, this was a prepared text, sometimes only formulae or estimates with questions between them. I have kept many such sheets. However, he rapidly understood that I was 'not a writer' and was very disorganised. As a result, I received primarily questions and formulae.

It was striking how he could instantly concentrate, sit a bit more comfortably, fall silent, and rapidly write. Then, completely unexpectedly, he would come out of this state and begin a conversation on a new topic. If the papers were not given to someone on the spot, they were intended for somebody else, most likely for a book.

It is difficult for me to understand how he managed to find time to be present at the birthday parties, thesis-defence celebrations, weddings, and housewarmings of all his students and young colleagues. After all, he was always extremely busy. Apparently, at some point he decided once and for all that such social activities were very important.

On the other hand, there are few who would have done so much to help with finding work positions, and even with '*propiskas*.'[150] Everyone knows that it has always been almost useless to try to get a Moscow '*propiska*' for a young scientist who has finished his post-graduate education. Who to Zeldovich were Kolya Shakura, who came to Moscow State University from a Belorussian village; the PhysTech student Sasha Ruzmaikin from Central Asia or his wife and fellow student Toma from Melitopol in Ukraine; Volodya Lipunov from Kiev, and many others, who didn't then have support from anybody else, but have since all acquired international reputations in their fields? Who to him was I, a Tatar student from Tashkent, without relatives or acquaintances in Moscow except for my friends in the dormitory, who had stubbornly tried to explain to him at our first meeting that astronomy was completely uninteresting to me, and that I was sure I wanted to study elementary-particle physics?

What force made him go and ask about positions for all of us, about *propiskas*, what made him work in order to get us rooms in communal, and later individual, flats, to dance at our weddings, to express joy at the birth of our children? I think that few of us fully acknowledge that all this was far from easy. But it was done somehow simply and naturally. And it was rare that someone had to ask for help. Most often, he would report on the first unsuccessful approach in connection with the problem, and then request that we remind him about the need for another attempt a few days later.

[150] See footnote 82 in the contribution by S.S. Gershtein (p. 160). [Translator]

Ya.B. Zeldovich with colleagues (left to right: Joachim Truemper, Director of Max-Planck Institute; Rashid Sunyaev; Ya.B. Zeldovich; Vladimir Zharkov, Institute of the Physics of the Earth, USSR Academy of Sciences; and Keith Moffat, Cambridge University). Space Research Institute, Moscow, 1970s.

In 27 years of active work in astrophysics and cosmology, YaB raised at least a dozen splendid Doctors of Science in theory, whose names are widely known. Many of these have worked completely independently for many years, and have virtually no contact with him. At the same time, I think that today, they nearly all acknowledge that the best and most cited of their works were carried out either jointly with YaB or under his influence. And he was generous, shared ideas and news, asked to be included in new problems, never hid his results, always acknowledged the rights of others to priority.

How did he find time for all of us? Each of us had to be given a problem that fitted his character and interests, each had to be looked after and inspired, each given papers with references to new articles related to his theme. Each day there was a microseminar with news, and a short conversation with each of us that had to have enough content so that we would not be annoyed to sometimes wait for hours. And work, work, work.

Perhaps this is a good time to interrupt myself and tell about one of my conversations with him. It surprised me when some of his students stopped working with him and left, sometimes conducting themselves in a fashion that was, in my opinion, far from good. He once gave me an answer to this question, which was clearly unpleasant for him. He said that it was necessary to spend some two years on any undergraduate or post-graduate student before he begins to produce results; then, this person can help with collaborative work for another two or three years, and there are opportunities to expand the research front. Further, some people tire of working each day at such a tempo, some lose interest, and some become fully independent and strong people. It is important to feel this moment and fully free them from guardianship when the time comes. And teaching young people is certainly a sacred responsibility, a joy, of huge use for oneself. I'm afraid that I haven't

COSPAR (Committee on Space Research of the International Council of Scientific Unions) and the Russian Academy of Sciences introduced in 1989 the Zeldovich Silver Medal for young (under 35) scientists to mark their outstanding achievements in space sciences. Every two years eight best young scientists of the world are awarded the Zeldovich Medal in eight different fields of space sciences from space biology and climate research to planetology and space astrophysics. The medals are presented during the General Assemblies of COSPAR, which take place every second year in the presence of 1500 professionals. 52 young scientists have received this medal to date.

quite conveyed word-for-word the thoughts expressed at that time by YaB. I remember only being left with the feeling that perhaps I had been too importunate with my questions, and that his gentle and quiet answer was that of a tired and bitter person, who did not have a drop of cynicism in him that could involuntarily creep into my retelling of his words.

How he 'educated' me in connection with the fact that I was in no hurry to write up some ideas — which then seemed to me inaccessible to experimental verification — about the future (10^{30} years later) of the universe, with decaying protons and explosions of neutron stars that 'lost weight' in the process. He was delighted by the very possibility of the explosion of a neutron star that was losing mass for some reason. Having despaired of ever forcing me to publish this work, he referred to the idea in his editor's supplements to the Russian translation of the splendid book *The First Three Minutes* by Steven Weinberg. But what an earful I got in the meantime. The most honourable accusation was of 'Heisenberg syndrome.' In YaB's words, the most horrible thing is when someone publishes one work of high quality, and is then wary of publishing something else that is not quite so perfect, according to his own estimates. YaB believed it was far from shameful to publish 'student research' at any age. And as an example, he presented L.D. Landau (an unquestioned authority), who actively published and continued to collect the 'integral' his whole life.

For some reason, all these arguments, and especially the references to famous names, had virtually no effect on me at the time. But now, more than ten years later, it seems to me that he was right in this as well. The initially somewhat wild idea of explosions of neutron stars has been reined in preparation for development, and a number of very interesting detailed calculations have been published. It is a question of internal discipline — it's important to conduct a reasonable piece of work to its end, publish it, and rapidly turn to the next problem. That was his style. Hesitations (is it worth it — isn't it worth it?) are of little use. At the same time, one can't live without one's own criteria. Several years ago, choosing papers for a two-volume tome of YaB's selected works, I was struck that it was virtually impossible to find a single paper that had not found further development in the work of other authors, both at home and abroad. He didn't spend time on unpromising problems to no good effect.

The great importance he placed on experimental data is remarkable, as is how he was gladdened by research of his students that made precise predictions for experimental tests.

And it may be that he perceived any problem differently than we young people did. After all, he had a huge amount of experience under his belt, excellent knowledge of methods developed over many decades in very wide-ranging fields of science — from problems connected with the propagation of flames to the physics of elementary particles. He had the sharpened acumen of an independent researcher who has more than once known the success of solving problems that no one before had even attempted to attack, and whose solutions would subsequently become widely acclaimed.

Approximating the truth, or the Zeldovich ansatz

A.D. Chernin

On March 8, 1965, Zeldovich's birthday, the few and very young students and colleagues of Yakov Borisovich at the Institute of Applied Mathematics (IAM), which then included myself, when I had a temporary fellowship, merrily expressed the wish that he complete the problem of galaxy formation by the end of the year, and simultaneously resolve the mystery of the quasars. Both themes, as is known, deeply engaged Zeldovich. Quasars were discovered in 1963, and the problem of the origin of galaxies had been waiting for a solution since the middle of the 1920s, when it was first established that the extragalactic nebulae were gigantic stellar systems. In our birthday wish, the only joking and non-serious lines were, perhaps, the deadlines we set; in its content, the wish corresponded well to the goals of Zeldovich, which were subsequently realized. Zeldovich turned out to be among those who first suggested that black holes generated the energy in the nuclei of quasars. Today, there are few who doubt that this is, indeed, the case. And the largest contribution to modern galactic cosmology belongs to Zeldovich — his famous 'pancake' theory.

The 'pancakes' began with a short note that YaB showed us in the summer of 1969, still in manuscript form. This gave an exact non-linear Newtonian solution describing the development of gravitational instability in an expanding universe. The solution is one-dimensional and flat (for perturbations), and represents a special case (a single arbitrary function of a single Cartesian variable). To this day, it remains impossible to find the three-dimensional exact analytical solution. At that time, Zeldovich made a brave assumption: the time behaviour of the perturbations in all three directions was the same as that for his exact solution. Zeldovich called the corresponding three-dimensional formula, written 'by hand,' an approximate solution.

When he discussed this at the JAS (Joint Astrophysical Seminar), in the auditorium at the Sternberg Astronomical Institute which was over- flowing, as always at his talks, I asked: 'If the solution is approximate, to what is it approximate?'

'To the truth,' was the instant response.

And this was absolutely correct: this was elucidated much later, when numerical computations performed first by A.G. Doroshkevich and then also by many others demonstrated that the approximate solution (perhaps it would be more accurate to call it the Zeldovich ansatz) not only provides an accurate qualitative picture of the non-linear three-dimensional process, but is also quantitatively close to the 'truth.'

How could he guess this? And people believed that human intuition can only act on linear problems ...

Returning to the distant but unforgettable 1960s, with memory bringing alive the best and brightest, one is gladdened (and surprised) by how much inspiration, ardour, merriness, and also softness and coziness there was around Zeldovich, in an atmosphere that he created with patience and tact. For example, it was touching each time he took care that a still very green visiting junior scientist, who the academic bosses wouldn't notice if they were looking straight at him (this is even more usual today), could get a room in a hotel: he would call some important bureaucrats in the Academy and talk with them firmly about the issue. Here he demonstrated a direct sensitivity to the naive young people who had been charmed by him, together, I think, with a certain pedagogical system.

YaB's gentle attitude made a considerable impression on me the first time I came to see him. After I returned to Leningrad, I told my teacher L.E. Gurevich (YaB said and wrote more than once that he considered Lev Emmanuilovich one of his first teachers) that Zeldovich even seemed shy to me. Lev Emmanuilovich did not comment on this statement, as he always did when he listened to nonsense.

Someone who is writing memoirs can't fail to talk about oneself; to the justifications and excuses expressed on this account by writers of memoirs more esteemed than mine, I'll add only a word for physicists: a writer is like an instrument, without which we can't manage if we wish to study an object. 'That's nice, and should be expressed elegantly,' said Zeldovich, 'sit down and write.' And he dictated my first scientific work, of which he was not a co-author, from beginning to end. This paper reported a new exact solution for a Friedmann cosmological model, taking into account both matter and radiation. Zeldovich was interested at that time in a cosmology with a cold beginning, and we took the radiation to be an isotropic background of ultra-relativistic neutrinos, and possibly gravitons. YaB had already published papers on this topic — one with Ya.A. Smorodinskii and another with S.S. Gershtein. No reference was made to a hot beginning in 'my' text; we believed that Gamow's theory of a hot Big Bang had been soundly disproved by the newly published work of two of Zeldovich's coworkers, A.G. Doroshkevich and I.D. Novikov. YaB presented my talk at Landau's seminar, which had then been meeting for three years without Lev Davidovich. That was in the winter of 1965.

And then in September, I received in Leningrad a letter from YaB: 'It looks like the cold model was a mistake. The Americans have measured the background radiation. For now there are only rumours, not yet in print.' The discovery of the radiation predicted by Gamow (which I.S. Shklovskii immediately began to call relict radiation) became the biggest event in the entire history of cosmology since the time of Friedmann and Hubble. Soon, YaB, with ardour and temperament, was reading lectures about the relict radiation of the hot universe in packed auditoria. The cold model was thus lightly left to the history of science, or perhaps 'put into the sidings.' However, 15 years later, Zeldovich again returned briefly to it, though on completely new grounds.

In 1966, I travelled with YaB and B.V. Komberg to the Moscow PhysTech at Dolgoprudnij, where his lecture created extraordinary excitement. Going through the corridor to the lecture hall, we heard the students rushing to get places, loudly passing on the word: 'Zeldovich is here! He's going to talk about the relict! Yes, everyone's going!'

I remember that we went there in a big black car, a ZIM. YaB asked the driver to stop at a village grocery store outside Moscow, and bought a loaf of black bread, and we rapidly and merrily ate it during the rest of the drive.

At that time, one had the feeling that cosmology was really taking off. It seemed that things were only beginning (this was only partly true), and the main progress was just about to happen (this was also only partially the case).

The sense of things taking off was not only felt by young scientists just starting out (by the way, there were only a few young people studying cosmology in our country at that time, and nobody suspected that 15 to 20 years later there would be what Shklovskii called the 'self-stampede in relativistic astrophysics'). Zeldovich himself was then over 50, but in his energy and passion for science, he was younger than the youngest. And young scientists adored him completely and boundlessly.

His jokes, and stories 'about Zeldovich,' were repeated many times in both capitals,[151] and he was a living legend. For example, he suddenly arrived at the IAM with all his 'stars' on his chest: 'The traffic police took my license, I have to go and rescue it.'

This story had a continuation. Coming into the room where we were waiting for him, he hung his suit jacket on the back of a chair and left again for a moment; and the cheekiest of us managed to put the 'heroic' jacket on himself for a moment!

In his approach to young people, it stood to reason that he expected work from them that was, so to speak, at the level of the best international standards. This exactingness was taken to be natural. The question didn't even arise 'How are we any worse than they?' In the area in which we work ourselves, we may even be better. This was not an attempt to intentionally apply pressure, and the high demands went together with a relaxed atmosphere.

Visiting Zeldovich with my latest new work, I either told him about it with chalk in my hand at the board in his home office, or, as happened more often with time, gave him a preliminary text (a draft of a draft, as he said), which I received back with oral or written comments no later than the next morning. After thinking over these comments, I could ask questions during a short or more prolonged conversation, discuss it all with him, and sometimes agree with him on the final text with references (no small importance was attached to this last step). I regret that so few of these letters with comments from YaB — usually on school notebook pages — have been preserved. In them, there were rarely questions, and more often assertions — exact, brief. Sometimes, they included an element of irony or joking. For example: 'Oh, how much discussion there will be. We are the swan, the crayfish, and the pike![152]
Greetings! YaB
1/IV
(Symptomatically)'

This was in connection with the book *Hydrodynamics of the Universe*, which YaB, Doroshkevich, and myself undertook to write. It seems that the 'draft of the draft' had already been composed, but the book was never finished.

Almost every conversation with Zeldovich created an almost physical sensation of a leap in one's understanding of science. And not only science.

'Let's not be naive,' he would say to a very young scientist, seriously looking him right in the eye, and somehow, with a push, expanding his field of view a little.

Both at that time and later, I noted the special clarity and depth that sometimes arose in his eyes, full of life and warmth, like in some portraits by Rembrandt. It was unforgettable ...

We were delighted with Zeldovich's scientific acumen, the bright flares of his physical imagination and intuition, which we had the good fortune to

[151] Moscow and Leningrad (St. Petersburg), as Russians traditionally call their two largest cities. [Translator]

[152] An allusion to the tale by I.A. Krylov, 'The Swan, the Crayfish, and the Pike.' [Translator]

observe at close hand. It seemed that no problem could not be solved by a talent of such scale; solutions that were hidden from others were clear to him. That was most often the case: he was always ahead of everybody. However, there were also exceptions. He lagged behind Hawking with the quantum evaporation of black holes by only a short time, I think a matter of months. He took up inflation, alas, after Guth, and in this case, there were earlier papers by the Leningraders E.B. Gliner, I.G. Dymnikova, and L.E. Gurevich in which the embryo of what would later become inflation theory was already present. Initially, YaB sharply rejected the Leningrad work; later, he told us that 'at the time, he didn't appreciate them.'

How did he himself view all this in reality? His self-evaluation should in all fairness be very high. I'll allow myself to suggest that, in his attitude toward honours and awards too, there was no casualness, indifference, or, especially, irony. He knew what he received them for.

'The Russian soldier saved us.' Behind these words, which I heard from him more than once, there stood definite uneasy reflections and anxieties. It could be that this was a compact formula found to denote a complex mixture of ethical, behavioural, and political concerns. Or it may be that it was like an invocation in the hope of co-ordinating and bringing into agreement the higher landmarks as he understood them, his own life experience, and the positions of people with whom he dealt.

He deeply felt and revered the power and strength of the state, and the gravities of those aided by power. He himself didn't object to being called a statesman.

At the monument to Kurchatov in Moscow (whose most obvious feature is naturally the beard), I asked whether he had indeed been so imposing. After a pause, YaB answered seriously: 'He immediately addressed generals as *ty*.'[153]

At some point long ago, Zeldovich asked a question in the form of a merry joke: 'Well, how's your new boss? He doesn't disturb your work?' And in reality, what could disturb a theoretician, except for his own ignorance, stupidity, or laziness — and even then, only up to a point?

YaB sometimes had to call big bosses, and himself be 'a boss in charge of people.' I never knew what he was like in this role. My transfer to the IAM, proposed by him in 1966, didn't take place for some reason; later, I came to appreciate the advantage of a distance of 650 km. In YaB's letters were scientific news, comments on my work, reviews, and reasoning about ideas and results that had attracted his attention. Occasionally, he wrote not only about physics.

In one letter — not in the glorious 1960s, but in another epoch, at the beginning of the 1980s — he wrote, 'Wise and benevolent tranquillity! This is a very rare quality nowadays; I value it very much in my old friends Ovsei Leipunskii, the late David Albertovich Frank-Kamenetskii... I don't have enough wise tranquillity..., this makes me value it all the more in others ... .'

[153] The informal form of 'you' in Russian. [Translator]

Tranquillity was impossible for him; his style was energetic, boiling, directed — and then again sometimes not very; this was the means of his existence in life and science. Health in all areas is necessary for this, however ...

He was very strongly, extraordinarily strongly concerned with the theme that was reflected in his last work, which came out in *Uspekhi Fizicheskikh Nauk* after his death, in the summer of 1988. Why not react toward this with 'wise' tranquillity? But no, he boiled, was indignant. The peals of his thunder were carried all the way to Leningrad. Late in the autumn of his last year he called, talked loudly, firmly, sharply; he was in an evil temperament, and didn't take in answers. I listened, surprised (and it was said that he was becoming more mellow with the years), trying to object and, in general, not giving way. The conversation didn't end well. Soon, R.A. Sunyaev happened to be in PhysTech and agreed to convey to YaB my explanations about the matter. YaB called again, his unreasonableness easily dissipated, and this conversation, which (who would have thought it at the time?) proved to be our last, ended on a warm note: YaB invited me to visit him, discuss some things, and so forth ...

On the day after the funeral at Novodevichy Cemetery, Rashid [Sunyaev] told me that my name with an exclamation mark, written by YaB's hand, remained in the corner of the board on the wall of his office, which is so well known to all of us. So that he wouldn't forget to tell me or write to me about something? Whether it was something serious or a trifle, who knows.

With regard to theoretical cosmology, one can say that is in origin a Leningrad science. In Petersburg–Petrograd–Leningrad, the founder of the theory of the expanding universe, Aleksander Aleksandrovich Friedmann was born, lived, worked, and died. Georgii Antonovich Gamow, a student of Friedmann at Leningrad University, took cosmology into his hands, and became the author of the hot universe theory.

Zeldovich was born in Minsk, and his bronze bust is installed there. However, in terms of his scientific origin, he is, of course, a Leningrader. It was here he studied physics, at Leningrad PhysTech he found his first mentors and companions in physics, to whom — as is well known — he always remained faithful. Many continued to associate him with Leningrad when he lived and worked in Moscow: the best Muscovites are Leningraders, someone said at one of the PhysTech jubilees at which Zeldovich was present.

Together with Friedmann and Gamow, Yakov Borisovich Zeldovich won by his labours a place in the science of the universe. It is on the backs of these three Leningrad giants that cosmology now stands.

Zeldovich's magnetism

Alexander A. Ruzmaikin

For about 20 years, I had the pleasure of working on cosmic magnetism together with Yakov Borisovich. YaB magnetized me at first glance, when he appeared wearing an ordinary sweater at one of the Moscow PhysTech meetings and, in a very simple and powerful manner, explained that the universe must have been hot at the beginning of its expansion. However, his first response to my desire to work with him was cool. 'I am not obliged,' he said, 'to take every student.' And then, after a pause, he added: 'Can you figure out what is going on with muon neutrinos at the beginning of the expansion?'

Thus, I got my first problem to work on. This opened a way of communicating with YaB. It seems to me that it would have been impossible to communicate with him in any other way than through solving problems. He himself would get completely caught up, fully concentrating on finding a solution. I've never met another person who was able to sort out and solve a problem so fast, a problem against which you had beat your head in vain for several months.

Yakov Borisovich had a rare instinct for the new. In this context there is an interesting story about his response to a work by George Batchelor, a famous British scientist, an expert in fluid mechanics. It had been known for a long time that any scalar quantity, such as the scent of a perfume in the air, smoke, or a hot spot, is rapidly destroyed through turbulent diffusion. In his 1950 paper Batchelor claimed that a vector (the magnetic field in a conducting flow) can survive the dissipative action of turbulence and can even be amplified by the turbulence. The authority of Batchelor was so great that his non-rigorous reasoning passed the 'Landau threshold' and had been published in the Landau and Lifshitz *Electrodynamics of Continuous Fluids* (1957 edition). In a short note, Zeldovich pointed out that Batchelor's arguments are not convincing, moreover, argumentation is wrong in the case of two-dimensional turbulence.

The attack against such an expert as Batchelor was remarkable in and of itself. But Zeldovich's short publication was not purely critical, it also contained several new results. In particular, he pointed out the possibility of magnetic field growth in a limited time interval and indicated that conducting turbulence behaves as a diamagnetic material. These results are now on a golden list of basic principles of turbulent magnetohydrodynamics and are widely used in studies of the origin of magnetic fields in planets, stars, and galaxies.

Another of Zeldovich's qualities was his vivid and often pictorial way of presenting his scientific results. For example, a great deal suddenly became clear in the complex problem of a turbulent dynamo when he introduced his now famous 'stretch–twist–fold' picture of magnetic field transformations by turbulent motions. It is worth remembering an anecdote of how he came up with this idea.

It happened in Poland at the international symposium on fluid mechanics in 1971. Someone stated that self-amplification of the magnetic field in a highly-conducting fluid is not possible since magnetic lines are frozen-in (carried by the fluid motions), which implies the conservation of any initial magnetic flux. YaB jumped in, asked the sceptic to remove his belt and, while the man was holding up his pants, turned the belt into a ring, twisted the ring into a figure of eight, and then lay the two rings of the belt into each other demonstrating the doubling of the flux through the cross section of the doubled ring. This topological trick is now widely used in the science community as an illustration of the fast dynamo action.

In his pictorial style YaB was closer to Faraday than to Maxwell. This view is supported by the way YaB told a story of a dispute between I.E. Tamm, a Nobel laureate in theoretical physics, and the famous Soviet electrical engineer V.F. Mitkevich. This story, in a politically correct way, is written in our book (S. Vainshtein, Ya.B. Zeldovich, and A. Ruzmaikin, *Turbulent Dynamo in Astrophysics*, Nauka, Moscow, 1980). Mitkevich, following Faraday, treated magnetic fields with the help of magnetic lines, like threads or ropes. Tamm pointed out that the concept of magnetic field lines is not useful because it depends on the frame of reference. Via the Lorenz transformation, the magnetic field can be converted into an electrical field, so that for one observer, two points will be connected by a magnetic line while for another observer this might be not the case. Tamm wittily remarked that the question of whether a magnetic line joins two points is similar to the question of whether the geographic meridian passing through Moscow was green or red. To this remark, Mitkevich immediately replied that he (and apparently the audience) was absolutely sure that the Moscow meridian was red. In his comments on this dispute, YaB (a theoretical physicist himself) paid full respect to the rigorous, Maxwellian approach of Tamm, but defended the engineer Mitkevich. Tamm, said YaB, is completely correct in vacuum, but in a highly conducting plasma the concept of magnetic lines is quite realistic and very useful.

Let me make a few remarks about the social character of YaB. In his once very popular book *How to Make Friends and Influence People* Dale Carnegie advises avoiding arguments if you want to incline people toward your viewpoint. Surprisingly enough, YaB inclined and influenced people in the exactly opposite way, by arguing with them! He argued in person, in writing — especially whilst on holiday, when you would expect him to be resting — and by telephone. Telephone contacts with YaB had nothing in common with a standard phone dialogue. It was more like being a receiver in an electrical circuit, with YaB serving as a source of a strong current, carrying a large charge with each word. Many young (and not young) scientists found their best results as a result of telephone arguments with YaB. He most often called early in the morning, and rarely, except in his last years, in the evening. Being charged up by these telephone calls for the rest of my life, I've developed a great dislike for common telephone chats.

In serious arguments, the loser had to pay. In such situations YaB bet a bottle of cognac against a bottle of mineral water. The loser was required to write a witty label on the bottle. In most cases, the label ended up on the mineral water bottle. Over the many years at the Keldysh Institute of Applied Mathematics where we worked, an entire cabinet was filled with the labelled bottles. YaB lost, to my knowledge, only once, and brought a bottle of the best cognac without delay.

He liked to do things by himself. Any collaborative work carried his substantial contribution via ideas, proofs, and direct writing. Even relatively unimportant work, like writing referee's reports and reviewing PhD theses he always did himself. To those who tried to put in front of him a prepared paper to sign he explained: 'I'm not disabled yet.' He wrote very fast, in a sweeping manner, but the result turned out to be accurate and right to the point.

Many met YaB on the skating rink; he also skied, swam well, and enjoyed driving his Volga car. However, it was a risky venture to relax with him, even when away from work. I remember an episode when he and I had a 'relaxing' walk along the Vltava river during a break in the European Physical Congress in Prague. YaB suddenly asked me where the river flowed. After a short guess, I indicated a direction. In the next morning, after having carried out a simple experiment, YaB festively proved that I was wrong.

The magnetism of Zeldovich depended on how close you were to him. He was accessible, polite, and gentle to people with whom he had no close contacts. But those who approached him at working distance experienced extraordinary stress which eradicated idleness, pompousness, and complacency. And the maximum effect of this energy, intensified rather than smoothed, was felt by Zeldovich himself.

Yakov Borisovich Zeldovich: some personal reminiscences[154]

Leon Mestel

For me for far too long, the name of Yakov Borisovich Zeldovich, applied mathematician, astrophysicist, and cosmologist, had belonged along with the legendary Lev Davidovich Landau, Solomon Pikelner, Vitali Ginzburg, Iosif Shklovskii and others to that extraordinarily talented group of disembodied voices from behind the Iron Curtain. We eventually met for the first time during the 1967 Prague IAU, and then two years later at a Symposium on Interstellar Gas Dynamics, held at Miskhor near Yalta in Crimea. Even after the thaw, bureacracy continued to interfere with his wish to travel, and our subsequent interactions had to be by mail, until our paths crossed again at the 1982 Patras IAU, and finally at Sukhumi and Moscow in 1986. But we

[154] This article was not included in the original Russian version of this book, published in 1993. [Translator]

always found plenty to talk about, not least on our contrasting family histories, my grandparents having emigrated from the Ukraine to England, some grand-paternal cousins to the United States and grand-maternal cousins to the Ottoman Empire.

He being such a polymath, most astrophysicists would find that a Venn diagram of respective interests would look like a slice of raisin bread. Zeldovich and Novikov on *Relativistic Astrophysics* is of course a classical source for anyone dabbling in high-energy astrophysics and cosmology. The Sunyaev–Zeldovich effect has become a veritable gold-mine. I still recall with gratification Yakov's verbal acknowledgement of the Lin, Mestel, and Shu paper as a harbinger of the galactic 'pancake' picture. We shared a fascination with cosmical magnetism in all its manifestations, and I continue to find new insights on the dynamo problem when returning to his book with Ruzmaikin and Sokoloff. Our very first discussion, at a session in Prague on stellar magnetism, was on the obliquely rotating fossil field model for early-type stars — not contradicting but rather complementary to the contemporary dynamo model for solar-type stars and planets. He usually had something original to add to any scenario, e.g., at Miskhor, where he and Pikelner pointed out the crucial importance of conductive heat transfer between the different phases in the Field–Pikelner picture of the interstellar medium.

Discussion with him was always rewarding, a genuine two-way flow of ideas, whether on science or public affairs. A loyal but far from servile citizen, he appreciated any hint of an irreverent attitude towards authority and dogma, such as the borrowing of Marxist clichés — 'a classical model contains the seeds of its own transformation into a quantum model.' His intellectual legacy to the astrophysical world is manifest. He leaves with us also the lasting memory of an engaging personality.

He lived science

L.P. Grishchuk

There are many people who worked with Yakov Borisovich for longer than I did, and who knew him more closely. But somehow, I feel like having my say, telling what he was like. It is possible that others will do this better. There is no need to emphasize that these notes carry a deeply subjective character, and refer only to isolated areas of his life and creativity.

In my opinion, Yakov Borisovich lived and died young. I never saw him old, dimmed, helpless. He loved his work up until his last day. It is difficult to believe that everything done by him in astrophysics was accomplished in a relatively short 27-year interval, and that he first began to work in astrophysics at the age of 50.

Both content and style are important in scientific work. The scientific content has a life independent of the author, in his professional publications. The style is better learned from contact with the author himself, or from the testimony of his co-authors and other witnesses. I'll talk about some research in which I participated. It is concerned with cosmology and gravitation, and includes a series of methodological publications.

It seems to me that Yakov Borisovich's strength was in the breadth with which he enveloped a problem, his deep physical intuition, his understanding of all aspects of events. He mastered mathematical techniques, but was not moved by long and complex calculations. I remember that when he saw 50 pages of calculations for one of my papers, he said that it was long overdue for publication. I more than once was struck by his ability to ask precise and deep questions about calculations, whose details he was not familiar with.

One of our first serious conversations concerned my postgraduate student work on singularities in the general solution to the Einstein equations for a dust-like medium (a medium without pressure). What joyful amazement I felt when I discovered that someone else could find their way through that tangle of formulae. The problem at hand was how general solutions with singularities were. The question of singularities — infinite values for the matter density, curvature-tensor invariants, and other physical quantities — was widely discussed at that time. Important results were obtained by E.M. Lifshitz, I.M. Khalatnikov, and their colleagues. At that time, it was already known that caustic singularities arise in the special case of a non-rotating, dust-like medium. In this case, we can choose a co-ordinate system that is simultaneously synchronous and co-moving with the medium. Due to the theorem that says the metric determinant of a synchronous coordinate system must vanish, sooner or later, the element of the co-moving volume vanishes, and the matter density, accordingly, becomes infinite. In the spirit of hoping that such singularities were not a general property of the solutions, it was proposed that including all degrees of freedom in the initial data — i.e., adding rotation — would lead to the disappearance of the singularity. However, I showed that, in the general case, that is, in solutions admitting the maximum set of arbitrary functions describing the initial data, there is indeed a singularity for a dust-like medium. Its character is such that it forms caustics. As elements of the medium move, the distance between particles along some direction (the local rotation axis, if there is rotation) decreases to zero, and the volume element is transformed into a 'pancake' with infinite curvature and density.

I felt inspired when I discovered that Yakov Borisovich understood and appreciated this result. In a conversation with A.L. Zel'manov and myself, he said that this sort of treatment would disprove hope that singularities were not a general property of solutions of the gravitational equations. Of course, at that time, I was worried about the unavoidability of singularities and the generality of this solution, and I wasn't thinking about any specific astrophysical applications. I don't know if this work influenced Yakov Borisovich's subsequent activity in connection with the non-linear theory of the growth of perturbations and the 'pancake' theory for the formation of large-scale structure in the universe in the post-recombination period. It is possible that he did take into account the unavoidability of caustics.

Yakov Borisovich lived science, and continuously thought about his work. Not a single interesting question was put on the back burner. He strove with a sportsman's ardour to be the first to give the correct answer. In some cases, disagreements with regard to some specific assertions arose among his young followers. If each stood his own ground, Yakov Borisovich would suggest that they bet a bottle. In the overwhelming number of cases, he won, in which case the loser bought a bottle of mineral water and attached a note in which he explained the essence of the argument and repented of his errors. It stands to reason that I also experienced such losses. The useful side of these arguments was that they forced one to carefully analyse all arguments and mobilize all resources. In some cases, this facilitated the development of good research.

One of my first joint papers with Yakov Borisovich also began with a disagreement. I discovered that, using certain observational data, one could draw conclusions about the structure of the universe on scales exceeding even the Hubble radius (i.e., on scales that were not directly observable). Yakov Borisovich presented arguments against this point of view. As a result of our discussions, we achieved total clarity. In the paper, we showed that the observed isotropy of the temperature of the microwave background (relict) radiation places strict constraints on possible deviations from uniformity and isotropy, even on scales exceeding the Hubble radius by factors of 50 to 100. True, we assumed that possible perturbations were random, and not specially correlated. There is little to say based on observations about scales that are larger than these, since there are some perturbations that could be significant, without affecting the isotropy of the temperature of the relict radiation. Many years later, Yakov Borisovich returned to this work and praised it. At that time, it was elucidated that the universe, having undergone an inflational expansion, could be very large and complex, and its structure on scales exceeding the current Hubble radius is not only of academic interest.

Yakov Borisovich worked easily and rapidly. If he was caught up with something, the work was taken right through to the end. We agreed on the main results of a paper on gravitational instability in a multi-component medium on the telephone. Intuition didn't let Yakov Borisovich down. As he predicted, in this case there exists only one instability mode. I relied on an analogy with several coupled oscillators, and for some time thought that there could be just as many instability branches. Generally speaking, this is correct, but not in this case, where we considered the components of a medium at rest bound only by gravitational forces. I then joyfully discovered that similar problems have been encountered in a number of very different areas of physics.

As Yakov Borisovich himself emphasized, with the years, he became more tolerant of 'unnecessary' hypotheses. He said that attitudes toward such questions had changed somewhat in science as a whole. Earlier, the rule was to try and preserve a theoretical scheme as long as possible, but subsequently, it became generally acceptable to act as if everything which is not forbidden is allowed. (Yakov Borisovich applied this term in science even before it became popular in connection with economic activity.) I remember in one of our early joint papers, Yakov Borisovich proposed to everywhere replace the word 'metagalaxy' with 'universe.' I cautiously wrote metagalaxy, having in mind the observable part of the universe, and in this way wishing to avoid getting tangled up with the complex concept of the universe. Yakov Borisovich said that he suspected that these were fears in the presence of philosophers, and a concession to them. He proposed that we should not be afraid to speak directly of the universe. In part, this reflected our specific understanding of how the world was constructed. It would seem simplest and more natural if it were the same everywhere as we see in our area. In this case, the characteristic size of the world (the radius of its spatial curvature) could be not much larger than the Hubble radius. And a few years later, we wrote a paper about the 'new' universe, about what questions arise if the universe is indeed very large, inhomogeneous, and closed; many arguments had appeared in support of this. At first glance, unavoidable contradiction arises. In our vicinity, the universe is uniform and, most likely, spatially open or spatially flat, as indicated by observational data. In this case, the dynamical fate of this region of the metagalaxy is unlimited expansion. How could this be reconciled with the proposed closed nature of the universe as a whole, and with the necessity that sooner or later, it must be compressed? It turns out that caustics inevitably arise. Particles from 'outer' regions of the universe will necessarily penetrate the local region considered, and these particles are able to change the dynamical character of this region. This example is a good illustration of how our cosmological views are gradually changed and refined.

Yakov Borisovich's ideological influence on the development of those directions in which he worked was enormous. Perhaps this is especially true of his work in astrophysics and cosmology. Yakov Borisovich understood his importance in this regard. In the celebrations for his 70th birthday, he said that he felt he resembled a certain character from some play (I believe it was German). Together with his comrades, this character works on transporting soil in wheelbarrows. The role of this character is that he is not afraid from time to time to turn over his wheelbarrow, clamber up onto it, and make a passionate speech to his companions about how to live and work further. In the field of gravitational research with which I'm most closely acquainted, I can think of at least three global subjects where the enthusiasm and the first papers of Yakov Borisovich himself and research with his participation substantially influenced the development of these areas and their transformation into 'fashionable' research topics. I'm thinking of the creation of particles in a gravitational field, the origin and magnitude of the cosmological constant, and the quantum birth of the universe.

The fate of first works, especially qualitative ones, is often not enviable.

At a conference, 1980s.

Although they remove important psychological barriers, they often contain few mathematical details. Researchers very quickly get used to new ideas, and begin to treat them as self-evident. Further, the work is formalized by other authors, who add many concrete results and make a couple of references to the first works as 'early' or 'naive,' after which they drop out of the literature altogether. I remember hearing R. Feynman talk about this at a conference in America, with humour, but also ill-concealed bitterness. The reason for this was that some young and energetic inventors of asymptotic freedom in quantum chromodynamics had referenced one of Feynman's papers using the adjective 'naive.' I believe that Yakov Borisovich preferred to know everything and participate in everything, even if sometimes on a qualitative level, rather than scrupulously and methodically develop work in a single field to the end and have his name associated only with that.

In his last years, Yakov Borisovich attached enormous importance to the quantum birth of the universe from 'nothing.' He believed that relativistic astrophysics and classical cosmology were already settled, and living independent lives; they could now manage without him. And here he felt a great mystery, and he couldn't stand not to obtain at least some preliminary results. This desire to concentrate on the most important issues was also manifest under other circumstances. When he attentively listened to a report about some new experimental project and discovered that it could take 10 to 15 years to realise it, he would say, with some sadness, that this was clearly not for him. He once asked me (probably as a joke) to not let him write reviews and popular articles, since they were distracting him from his main work.

There is no need to say that the question of the quantum birth of the universe is a new and extremely complex problem. Adequate concepts and mathematical apparatus are still being developed, usual gravitational methods are being compared with superstring theory, and so forth. I even felt that in several of our collaborative works on this topic, the two of us could impart somewhat different meanings into the same sentence. But nevertheless, Yakov Borisovich was glad that he had the opportunity to construct a complete cosmological theory that included descriptions of both the current and quantum–gravitational stages in the development of the universe. It seemed attractive to us that there was no reason why the universe couldn't spontaneously be born as a 'sphere' with the Planck parameters, then be encompassed by an inflationary expansion. It is thought that deviations from a uniform and isotropic solution, including the structure observed today, could develop from unavoidable quantum fluctuations of fields against the background of the inflationary expansion. Now, this research direction has been transformed into an 'industry' with many branches. Not only the birth of our large universe is discussed, but also the numerous births of small microuniverses. A second approach is currently especially popular. This approach calculates the contribution to physical processes occurring in our universe caused by the fact that Planck-scale microuniverses can separate and join with our universe via tunnelling. It's a pity that the development of such investigations, partly stimulated by Yakov Borisovich, must now occur without him.

Yakov Borisovich never made light of the services of others; on the contrary, he was delighted by vivid personalities, presented them as examples, tried to put them in the forefront. Twenty years ago, during preparations for the fifth international conference on gravitation in Tbilisi, he reminded us several times that we must not forget to invite the young and then little-known American physicist K. Thorne. Yakov Borisovich had never met him, but had read and appreciated several of his papers. Thorne's name was on our list, but at a crucial moment, we all forgot about him. Yakov Borisovich was extremely upset and didn't let the matter lie. Having then just finished my postgraduate studies, this event was etched into my memory. Yakov Borisovich greatly valued the English physicist Stephen Hawking, and called upon him for his teaching. He once told me as a joking reproach, 'You see, Hawking has already become a member of the Royal Society of London.' Many times, Yakov Borisovich spoke favourably of the work of A.A. Starobinskii, A.D. Linde, and others.

He once made a curious comment about a reference of mine to another work, which considered closely related problems, but from another point of view and much less extensively. He said that it was not enough simply to refer to the paper, and that if I wanted that author to be pleasant and have a good attitude toward me, I should move the reference to a more prominent place, at the beginning of the paper. And that is what I ended up doing.

Yakov Borisovich was an honest partner in collaborative work. He participated in various ways, and his co-authors never felt that they didn't receive their fair share. In any case, that was my experience. He wrote directly and clearly, almost immediately. This delighted but also grieved me, since I always had to work long and hard on my own sentences, and all the same, they turned out wordy and rough.

For Yakov Borisovich, the most important thing was the truth; here, he didn't allow any compromise. In my opinion, he was rather soft and even-tempered in personal relations, with a good sense of humour. I don't know if this was really the case, but perhaps he simply expressed a deeper wisdom in such questions. He considered it essential to express himself clearly and definitely in a scientific argument, but never found biting remarks necessary; perhaps he wasn't indifferent to the pride of his opponent. He and I wrote several papers together on the fundamental issues of the theory of general relativity. From the scientific point of view, the question was absolutely transparent, but the situation was complicated by accusations directed toward us of 'mistakes,' 'lack of understanding,' that 'all reasoning on this question in this paper ... has neither physical nor mathematical meaning,' and so forth. Yakov Borisovich didn't feel the need to respond in kind; he joked that since Einstein and the mathematician Klein had also been accused of making mathematical mistakes, it would be a sin for us to complain — we had obviously fallen in with good company. His humour was also evident in business conversations on this theme. When we discussed the ancient question of whether it was possible to introduce a non-gauge-invariant tensor in place of the energy-momentum pseudo-tensor of the gravitational field, Yakov Borisovich instantly described the situation using a Ukrainian saying: 'Danila didn't die — an illness got him.' True, in the published text, this was replaced by the more poetic phrase: '... drive Nature out the door, and she will come in the window...'

He was sincere and firm in his evaluation of the actions of others, but always remained decent. In confidential conversations about situations where, at the very least, there could be mutual grievances, I never heard from him such sharp words about his opponent as those addressed toward him. It was deplorable for me to read about some 'unseemly deed of Zeldovich.' I'm sure that this was a result of misunderstanding; Yakov Borisovich was not capable of such acts.

It seems to me that Yakov Borisovich also behaved very decently during the years of A.D. Sakharov's ordeal. It is known that he felt great respect and honour toward Sakharov. At the seminars in Zeldovich's group at the Institute of Applied Mathematics, I saw how he dropped everything and everyone and ran into the other room when Sakharov telephoned. Yakov Borisovich's respectful attitude toward Sakharov is also clear from the details of his scientific publications. Yakov Borisovich didn't really share his thoughts about Sakharov's social activities; it is possible that he even had another opinion on these matters, but it was clear that he tried to support Sakharov when he could. For example, he told the story of how he had written a preface for the Russian translation of S. Weinberg's book *The First Three Minutes*, in which he mentioned the scientific accomplishments of Sakharov. Of course, the censors threw out any reference to his name without obtaining the consent of Yakov Borisovich, and he had to don his 'formal uniform' with all of his stars and engage them in serious discussion. It was only then that the original text was reinstated. It is also known that Yakov Borisovich hid and left Moscow when they wanted to obtain his signature on angry, damning letters. One can imagine just how much the organisers of these affairs would have liked to get his signature.

It is good that Andrei Dmitrievich had many friends in his last years; everyone felt flattered to be in his company. However, to correctly judge today's events, it is necessary to understand the events and atmosphere of those years. I have some personal experience in this regard. First, I was forced to defend the need to reference a scientific (!) article by Sakharov (which was referenced by the entire world) in one of my own scientific papers. (By the way, in an earlier version of this paper that Yakov Borisovich read, this reference was missing, and he brought this to my attention.) Further, the reference didn't go unnoticed, and I was subject to an onerous, moralizing conversation. Subsequent unpleasantness followed swiftly.

The position and behaviour of each person are evaluated through comparison. It was rumoured that Yakov Borisovich wrote a joking self-evaluation related to the events of that time whose meaning was that he 'is not Salieri'...

Yakov Borisovich was not only our leader and teacher, but also our immediate boss — he headed the department of theoretical astrophysics at the Sternberg Astronomical Institute for about six years. As a boss, he never strove to regulate the work of the researchers, but expressed certain demands, especially in matters such as attendance at scientific seminars. He formulated large and interesting problems, which at first seemed unrealistic. He loved from time to time to invite us to lectures he gave to the students. It was not uncommon for him to lose a factor 2 during a lecture. But his lectures were always lively and interesting. He always spoke about new and difficult things in a popular, accessible, but not superficial way. And I have compared him many times with others speaking or writing about complex topics. Often, their attempts to be accessible consist of nothing more than the use of simplified turns of speech and pathetic intonations which are intended to create an illusion of simplicity, but which are not exact or deep. Yakov Borisovich's numerous review and popular science articles were of enormous importance for the education of young people.

Perhaps Yakov Borisovich's most characteristic feature was that it was pleasant to have contact with him, both at work and in everyday life. His democratism and the simplicity of his relations attracted people. People constantly turned to him with questions, asking for consulations or advice. It is impossible to imagine Yakov Borisovich separated from people, an important boss in a fancy office. He was not indifferent to his high awards, but did not make a point of showing them off; they were never a barrier to conversation. I liked very much the great modesty and naturalness of his home, and his warm relations with his children. He knew literature and poetry well, and loved to joke. However, with regard to television, he said that he couldn't imagine Einstein sitting in front of a telly.

I had occasion to observe Yakov Borisovich at the wheel of a car. I can't say that he scrupulously observed the traffic rules. Several times, I was a witness when he was stopped for violating some rule. With a guilty expression, Yakov Borisovich unexpectedly handed over his certificate as a Hero of Socialist Labour together with his driver's license, after which they let him go peacefully.

It is well known that Yakov Borisovich loved sports. At home, from time to time, he proposed that we throw about a medicine ball in order to limber up, or called me to his balcony to work out on the apparatus. I remember that it was quite difficult for me to carry out the exercises he was doing.

Yakov Borisovich could preserve his dignity even in unpleasant everyday circumstances. I was once with him at a sanatorium in Uzkoe. We were going to walk to Profsoyuznaya Street and call into a shop for vodka on the way. (Yakov Borisovich himself almost never drank, but he kept spirits for guests.) Everyone knows what it is like in alcoholic beverage departments, and what type of people one can meet there. We likewise didn't manage to avoid such 'contacts.' I was already preparing myself for a fight, but, with a few words and a great deal of dignity, Yakov Borisovich smoothed it over. For me, this was a valuable lesson.

People often turned to him for help. In some cases, he would get out his letterhead, which listed all of his titles, and directly write by hand letters to important official persons. Several times, this was done in connection with me, and it worked without fail.

Like many others, I felt enormous respect and love for Yakov Borisovich. Nevertheless, I remember with shame that there were actions on my part that were regrettable — tardiness that forced Yakov Borisovich to wait, excessive stubbornness in arguments, inadmissible forgetfulness, and so forth. They sometimes warranted more serious reproaches than his sad look and the two or three words that accompanied it. However, I feel that he taught me a lot with this sort of attitude, for which I'm deeply thankful.

Now, that he has gone forever, for some reason I miss most of all his early-morning telephone calls.

I have read what I've written and reflected. Is there anything here that I would say differently in a conversation with Yakov Borisovich himself? I think not.

The teacher

A.G. Polnarev

I was acquainted with Yakov Borisovich for more than 20 years, and was his undergraduate student, postgraduate research student, and then colleague. It seems to me that any attempt to reconstruct in detail his multifaceted and contradictory image is inevitably doomed to failure. First, there is a great danger of subjective evaluations. Second, 20 years is a long time, during which any person can change a lot, all the more so that time affects one's memories. Third, it is very difficult to separate the influence of circumstances and history on the behaviour of a person from that person's nature. Therefore, I will not even try to lay out my many memories. My most vivid remembrances of Yakov Borisovich are associated with him as a teacher, whose talent and charm determined my entire life's course, for which I will always be infinitely grateful to him.

It was 1967. A packed auditorium. The students at PhysTech looked at the blackboard as if bewitched — Yakov Borisovich was giving a lecture on relativistic astrophysics. And at times it seemed to me, a second-year student, that I could reach out and touch the universe with my hand. For several students present at that lecture, that was the day that determined their fate — a day when their choice was made.

1971. 'You can work on whatever you like, but you'll be responsible for gravitational waves,' — these were YaB's words, followed by the formulation of the problem that ultimately became the theme for my thesis. Apparently quoting one of the classics, Yakov Borisovich said more than once that a teacher could not teach his student anything — his task was to create the right environment and atmosphere for the student to teach himself. A great many students had the luck of falling into this environment and feeling that same atmosphere, permeated with energy, scientific intuition, and the prodigious breadth of Yakov Borisovich's knowledge and interests (qualities that a person receives directly from God).

Now, when I myself am in the position of working with students, I return each time in my memory to my unforgettable conversations with the Teacher, and I feel that I was issued a lucky lottery ticket that will never fall to my own students.

1972. When trying to get into postgraduate school, I became convinced that the problem of the fifth point[155] can not always be solved simply. In spite of the existence of all the necessary recommendations, and my having satisfied all formal and informal conditions, I was dejected to find that I was not on the list of PhysTech graduates who were approved to take the entrance examinations for postgraduate school. When he learned about this,

[155] In all Soviet forms, the fifth point was the applicant's nationality, or ethnic identity. Certain nationalities were widely discriminated against. The 'fifth point' gained its notoriety primarily in relation to the treatment of Jews. [Translator]

Yakov Borisovich gave me an order (that is how he put it — it was an order) to keep him abreast of the situation right down to the tiniest details of all possible conversations in the dean's office and the postgraduate office of PhysTech. This I did through a whole series of complicated developments, including conversations with the assistant dean of the postgraduate section. Yakov Borisovich instantly reacted with telephone calls, including calls to the rector. The affair ended with him being required to write a note on his own letterhead, guaranteeing the defence of my thesis in time (in three years) and a place for me to work afterwards. At that time, this was complete nonsense. This is probably why, after all ended well and I was finally admitted to postgraduate school, Yakov Borisovich smiled sadly and said that I was the champion in admission to postgraduate studies in 1972. 'In what sense?' I asked. He answered curtly and without a smile: 'The worst sense.'

In spite of the 'guarantee' dragged out of YaB by the bureaucrats, he later (but before my defence in time) gave me a medal made by him personally from cardboard and foil. The medal depicted a goat, and was given to someone about whom it had been possible to say at one time 'getting anything useful out of him is like getting milk from a billygoat.' My only comfort was that the medal was transferable. I don't know where it is now.

Now, when I have to help undergraduate and postgraduate students, I silently think to myself: 'What a pity that you weren't lucky the way I was.'

One remembers many things when one thinks about Yakov Borisovich, and alas, not only happy moments. But the more time passes from the moment of his death, the more distinctly I feel that incomparable emptiness, because someone has left who knew nearly everything and was interested in a circle of problems ten times larger than most people, and who — because of his gift — attracted people who sought in him the objectivity of scientific opinion. He also made mistakes, but time has presented me with a filter that leaves in this world only his rich bequest. So much of YaB remains in this world and in science that I'm sure his name will influence the choice of path for many, many generations of scientists to come.

How one book on cosmology was born

M.V. Sazhin

I had the pleasure of working with Yakov Borisovich on a book entitled *Cosmology of the Early Universe*. It was finished not long before YaB's death. In September 1987, he made the last corrections to the proofs. I would like to write about this work.

Initially, the book was conceived as a collection of Zeldovich's lectures for fourth- and fifth-year undergraduate students in the physics department of Moscow State University. The course was not long, only six months, and was dedicated to the new cosmology theory — the inflationary universe. The appearance of the term 'inflation theory' dates from January 15, 1981. It is a rare case in the history of science when the birth of a new field can be pinpointed with accuracy to within a day. This was the title of A. Guth's paper in the *Physical Review*. YaB immediately realized the fundamental value of this direction for cosmology as a whole, and the possibility of now constructing a unified cosmological picture — from the singularity to our time — and actively developed the ideas of inflation further (I would like to note here that many of the principal ideas of inflationary cosmology had been expressed by YaB and his students before the work of Guth; acknowledgement of this fact is provided by the abundant citations of YaB's papers in both the first, pioneering, paper of Guth and in numerous subsequent articles on inflation theory). Therefore, YaB considered it necessary as early as one year after the appearance of the theory to begin to give student lectures laying out the theory's main ideas and results.

The synopses of the lectures were put together quickly. In fact the synopsis for one lecture had already been prepared by the beginning of the next. However, it soon became clear that such a modest volume of material would not be sufficient. This was true for two reasons. First, the theory was being developed very rapidly, and each week, preprints with new results arrived. It was necessary to understand these and include the material in the book. Second, YaB attentively searched for (and achieved) clarity in the material already incorporated. He included more and more new material, and, before my eyes, the raw synopsis was transformed into a simple and accessible text, as formerly disconnected paragraphs, sections, and chapters became connected and acquired a new sound.

For me, there was much in his manner that was new. On the one hand, it was necessary to achieve complete clarity and simplicity (so that it would be understandable to the 'man on the street'); this involuntarily drove things to one side, due to the necessary explanations, digressions, and so forth. Together with this, it was important to get across the strict main idea, which, like a Christmas tree, was hung with facts, proofs, and examples.

YaB's amazing ability to work was remarkable. In front of me lies one of the first typewritten versions of the second chapter with his comments. There are many of them. I look at them dazzled. He read and worked on every sentence. However, this was due not to mistrust of his co-authors, but rather to his drive to work out everything himself, to write everything such that there remained no unclear, cloudy places. Here, I should add that, at the same time as writing this book, he actively continued to work on and create new theoretical papers.

He was demanding of both us and himself. Even when the proofs came, he began to actively refine and improve the text. Such self-criticism was not only expressed in his work on this book.

During his lectures, he told the students about the difficulties that are encountered in the work of each researcher, although not all researchers are willing to admit them, even to themselves. These difficulties are associated with unrealized ideas, plans, and articles. Someone overtook you, someone published work earlier. With the current saturation in science, this is not surprising. As they sometimes say, ideas are 'carried in the air.' He talked about the psychological difficulties that he encountered along his way — it was a living history of science! Two such episodes have especially remained in my memory. One is associated with the work of Hawking and the other with the work of Guth.

Roughly 20 years ago, YaB and his students developed questions in connection with the birth of particles in a gravitational field, especially as applied to the non-steady-state field of the early universe. This stimulated many works, in particular, those of Hawking on the evaporation of black holes. YaB had also proposed writing a paper on this, but didn't get round to it. A similar situation arose with the work of Guth. YaB and his students came very close to similar ideas, but didn't manage to formulate them quite like Guth had. Therefore, YaB called for the students not to put off publishing good ideas that would get developed in their future work.

This attitude toward himself and toward his creativity inspired students much more strongly than the regalia and positions of other, more 'venerable' scientists. This was precisely why there were always young people around YaB. They liked his simplicity and the complete absence of haughtiness.

These lines are being written slightly more than a year after the unexpected death of Yakov Borisovich. It is only now that we are beginning to understand the extent of the loss. New problems arise, new questions are asked — and often they 'hang in the air.' It was possible to go to YaB with any difficulty — be it scientific or everyday. He dealt with scientific problems using his intuition and deep erudition; his sense of human decency helped him find solutions in the sea of everyday problems.

We have lost not only a splendid person, but an entire epoch in astrophysics.

He founded a school of relativistic astrophysics

A.M. Cherepashchuk

The creativity of Ya.B. Zeldovich determined the fate of many young scientists. At the very beginning of my path through astronomy, I had the pleasure of becoming acquainted with his work on the physics of stars, of meeting him and discussing this problem many times.

He possessed an amazing and rare talent — to illuminate classical fields of science with a new light. With his arrival in astronomy, the physics of stars and stellar systems took on qualitatively new development. Studies of the late stages in stellar evolution, prediction of the radiation of neutrinos during the collapse of a star, the theory of neutron stars and black holes, the accretion of material onto relativistic objects as a source of gigantic energy release, the physics of close binary systems with relativistic components, and many other ideas and results obtained by Zeldovich in the 1960s at the threshold of the era of X-ray astronomy determined our level of understanding of the nature of these fundamentally new objects subsequently detected by instruments on board spacecraft. His work laid the foundation for a new science — relativistic astrophysics — and was so gripping, interesting, and new that it determined the main direction for my entire life's research — the physics of highly evolved close binary systems.

Zeldovich appeared at the Sternberg Astronomical Institute (SAI) of Moscow State University at the end of the 1950s and the beginning of the 1960s. At that time, we — students of the astronomy division of the physics department of Moscow State University — did not yet suspect that a new era in astronomy was at hand, which would subsequently be described as the second revolution in astronomy. True, the first Soviet satellite had already been launched, and the cosmonaut Yurii Gagarin had been into space. At the SAI, the radio astronomy department of I.S. Shklovskii was already actively operating, and we often heard the word 'satellite' in G.N. Duboshin's lectures on celestial mechanics. Nevertheless, at that time many of us didn't realize that a 'golden' era in astronomy would soon arrive.

Under the supervision of my teacher D.Ya. Martynov, I was then working on studies of close binary stars, in particular, binary systems with Wolf–Rayet star components. We were interested in the problem of extracting the maximum amount of information about the masses, radii, and temperatures of peculiar stars from optical observations of eclipsing binary systems. Together with A.V. Goncharskii and A.G. Yagola, students of the mathematics division of the physics department, we enthusiastically worked on this inverse — and generally speaking, ill-posed — problem. We managed to 'overcome' the ill-posed nature of the problem by introducing *a priori* information about the monotonicity of the desired functions for the brightness distribution over the stellar disc, and to construct a regularising algorithm in accordance with work by A.N. Tikhonov.

Then, at the end of the 1960s, we excitedly listened to talks by Zeldovich and his students at the Joint Astrophysical Seminar (JAS) and read his papers. I should point out that Zeldovich's appearance at the SAI was inseparably linked with the foundation and operation of the JAS. For us — students, postgraduate students, and junior researchers — each of these seminars was like a holiday, since the most interesting, topical, and often preeminent results were reported there. The entire astrophysical scientific society of Moscow, and sometimes of the country, came to this seminar. And how many brilliant Soviet and foreign scientists actively participated in the JAS!

Three works by Zeldovich that were approved at the JAS made especially strong impressions on me. First and foremost was 'Fate of a Star and the Release of Gravitational Energy During Accretion' published in 1964. This paper laid the foundations of relativistic astrophysics, predicted the powerful release of gravitational energy at X-ray energies during the accretion of material onto relativistic stars, and indicated the important role of binary stars in investigations of neutron stars and black holes. For me, this work was a real discovery, since the theme of my undergraduate thesis was connected with studies of close binary systems with peculiar components.

The second work (published in 1966 jointly with O.Kh. Guseinov) was 'Collapsed Stars in Binaries,' where a program for searching for neutron stars and black holes in binary systems was first formulated. As a postgraduate student specializing in close binary systems, I was very excited by this paper. It became obvious that binary stars were a powerful instrument for studying not only normal but also relativistic stars.

Finally, the third work by Zeldovich (produced jointly with I.D. Novikov in 1966) was 'Physics of Relativistic Collapse,' which directly pointed toward the optical star in a binary system as a source of material whose accretion onto the relativistic object could lead to the powerful release of X-ray energy. It became clear that neutron stars and black holes in binary systems could be directly observed from their surrounding X-ray halos, and we excitedly read publications (primarily in foreign journals) which were still rare at that time, about investigations of the X-ray sky using instruments on board rockets — in particular, the discovery of the X-ray source Scorpio X-1 and its identification with a 12th magnitude optical star.

All these new ideas and results were published in full in 1967, in the well-known book by Zeldovich and Novikov, *Relativistic Astrophysics*. This became for us a real encyclopedia on the most current problems in astrophysics. At that time, discoveries in this area were happening one after another. New terms could constantly be heard at the JAS: quasars, relict radiation, pulsars, cosmic masers, and so forth.

Concluding my description of the period preceding my meetings with Ya.B. Zeldovich, I cannot fail to mention three works by his students, which have passed into the science hall of fame:

- the defence of V.F. Shvartsman's Candidate of Science dissertation in 1971 on the accretion of material onto single neutron stars and black holes
- the work of N.I. Shakura (1972) on the disk accretion of material onto relativistic objects in binary systems
- finally, the famous work of N.I. Shakura and R.A. Sunyaev on the theory of accretion onto relativistic objects in binary systems (early 1973).

These immediately preceded the era of X-ray astronomy, the beginning of which is usually associated with the launch of the American X-ray satellite 'UHURU' in 1971 when X-ray astronomy first stood up on a firm observational base. In 1972, a group of American scientists under the leadership of R. Giacconi published an article about the discovery of the first eclipsing X-ray source, Centaurus X-3 — an X-ray pulsar with a rotation period for the neutron star of 4.8 s and an orbital period for eclipses of X-ray source by the optical star of about 2.1 days. Several months later, the same group reported the discovery of X-ray eclipses in the source Hercules X-1 with a period of 1.7 days, with the period of the X-ray pulsar in the system being 1.24 s. Using a collection of photographic exposures from the photograph archive of the

Ya.B. Zeldovich and his wife, I.Ya. Chernyakhovskaya, at a reception at the American Embassy during the International Cosmic Ray Conference in Moscow, 1987. On the left is Maurice Shapiro (Naval Research Laboratory, Washington D.C.), on the right is Jim Cronin (Nobel Prize Winner, 1980).

SAI, N.E. Kurochkin identified the Hercules X-1 system with the known irregular variable star HZ Herculis. The period and phase of the optical variability of this star coincided with the period and phase of the X-ray eclipses and with the period of variations in the radial velocity of the X-ray pulsar Her X-1. This identification was made independently by the American astronomers J. and N. Bahcall. The collected X-ray and optical data for this object and new ideas about the accretion of material in binary systems led us to conclude that the main reason for the optical variability of this system was reflection — that is, heating the surface of the optical star by the powerful X-ray radiation of the accreting neutron star. In 1972, in the *International Bulletin of Variable Stars*, Yu.N. Efremov, N.E. Kurochkin, N.I. Shakura, R.A. Sunyaev, and I published an article interpreting the optical variability in the HZ Herculis system. The results of this work were presented at the JAS in the spring of 1972.

This was my first presentation at the famous Unified Seminar, and my first encounter and scientific discussion with Zeldovich. During my talk and our subsequent discussion in the foyer at the entrance to the SAI seminar room (a precious place to many), he expressed strong interest in our results and recommended that we publish them as soon as possible. With the help of B.V. Kukarkin, we managed to quickly — in two months — publish the article in the *IBVS*. It became clear that Zeldovich had not hurried us in vain: roughly six months later, at the end of 1972, a related letter by J. and N. Bahcall appeared in the American *Astrophysical Journal*.

My second encounter with Zeldovich occurred at the Institute of Applied Mathematics (IAM) in the summer of 1972 when Sunyaev, Lyutyi, and I presented work on the interpretation of optical variability in the object Cygnus X-1 — a black-hole candidate. I remember Lyutyi, who arrived at the SAI from our Crimean station in the spring of 1972 and was showing the results of his photometric observations of this object — more precisely, of the optical companion in the X-ray binary system. He tried to use his data to find the orbital period of the binary system, which was difficult due to large physical variability and non-continuity of the observations. Literally the next day upon coming into the SAI library, I saw a fresh issue of the journal *Nature* with a radial-velocity curve for the optical star in the Cygnus X-1 system on its cover and the headline 'Cygnus X-1 is a Binary System.' This was the famous discovery of L. Webster and P. Murdin of a spectroscopic orbital period of 5.6 days in Cygnus X-1 and determination of the mass function of the system. I rapidly communicated this information to Lyutyi. It turned out that his data correlated very well with the American ones.'

Lyutyi, Sunyaev, and I reported results interpreting the light curve of the Cygnus X-1 system as the effect of ellipsoidality of the optical star in the gravitational field of the relativistic object and estimated the mass of the No. 1 candidate for a black hole at Zeldovich's seminar at the IAM in the summer of 1972 and at the JAS in the autumn. He decisively supported our work which, with the help of S.B. Pikelner, jumped the queue of the *Astronomicheskii Zhurnal*[156] and was published at the beginning of 1973. A year later, we published in this same journal a paper qualitatively interpreting all types of optical variability observed in X-ray binary systems. All these studies brought us into close contact with Zeldovich. He supported our work and helped us with valuable advice. In 1974, at Zeldovich's suggestion, we made a presentation at the Bureau of the Astronomy Council of the USSR Academy of Sciences about our results and the perspectives for optical studies of X-ray binary systems and prepared a programme of co-ordinated observations of all known objects of this type, which we sent to all observatories in the country. Thus, under the influence of Zeldovich and with his direct support, a new direction in ground-based astronomy arose — optical astronomy of X-ray binary systems.

I should also mention the event of the century — the explosion of a supernova in the Large Magellanic Cloud in 1987 and the detection of a pulse of neutrino emission from it, in accordance with the theoretical predictions made by Zeldovich and Guseinov in 1965. The X-ray emission of this supernova was detected on board the Soviet Mir orbiting complex with the 'Kvant' astrophysics module created under the leadership of Sunyaev.

[156] The English version of this journal during the Soviet era was *Soviet Astronomy*, and is now *Astronomy Reports*. [Translator]

In subsequent years, I had many opportunities to meet Yakov Borisovich and discuss various scientific problems. The most important results on X-ray binary systems went through approval at the JAS under the direct scrutiny of Zeldovich. At the beginning of the 1980s, the department of relativistic astrophysics, under the leadership of Zeldovich, was founded at the SAI, and I had the opportunity to attend their working seminars. As a rule, the discussion of a talk was ended with some words from Yakov Borisovich about important scientific news, and the relaxed and informal atmosphere made it possible to ask very naive questions without shyness. An excellent pedagogue, Zeldovich collected around himself a cluster of talented young students, who later became internationally-known scientists. In the last years of his life he worked especially enthusiastically in the field of cosmology and the theory of the large-scale structure of the universe. In 1984, when I was already the head of the stellar astrophysics department of the SAI, I listened to a semester course of his lectures on cosmology for students in their last two years in the physics department. He conducted this course brilliantly and sometimes, interrupting his presentation, posed questions to the listeners to test their level of understanding of his lectures. I was also 'hit' with several of these questions.

I could write a lot about various meetings and conversations with Zeldovich. I will dwell on only two episodes that are etched especially strongly in my memory.

In 1984, I asked Yakov Borisovich to put my talk 'Search for Relativistic Objects in X-ray Quiet Binary Systems' on the schedule for the JAS. By that time, many data had been accumulated indicating that the number of neutron stars and black holes in quiet (non-X-ray) binary systems should exceed by a factor of several hundred the number of relativistic objects in X-ray binary systems. In our department, we had collected a large amount of observational material on investigations of such quiet X-ray binaries. We studied them using a method described by Zeldovich in 1966 based on the motion of the optical star in a binary system. For me, it was a great honour to hear Zeldovich's evaluation of our work at the end of the talk: 'It's good to hear that the search for neutron stars and black holes is being undertaken at an industrial level at the SAI.'

My last conversation with Yakov Borisovich took place on the telephone at the end of November 1987, two weeks before his death. He called and proposed that I present a talk at the JAS on the theme 'Highly Evolved Close Binary Systems.' I began to prepare intensively, however I was destined not to meet him again in a scientific discussion — several days later, he passed away. I gave the talk after his death at the first session of the Astrophysical Seminar that is now managed by a collective organisational committee, which continues the tradition of the JAS and forms a scientific memorial of Ya.B. Zeldovich.

I have always felt indebted to Ya.B. Zeldovich for the positive influence that he exerted on my life and I am proud that my scientific fate touched on the creativity of such an eminent scientist. His contribution to science, and especially to astrophysics, is enormous. Many of his ideas and results are still to be understood and acknowledged — the scientific creativity of Ya.B. Zeldovich was often far ahead of his time.

One of Yakov Borisovich's great dreams was to obtain direct observational evidence for the real existence of black holes in the Universe. At the end of his life, in 1987, only two black hole candidates in X-ray binary systems were known: Cyg X-1 and LMC X-3. Today, 15 years later, two dozen black holes in X-ray binaries with masses between 4 and 15 times greater than the solar mass have been discovered as well as about a hundred supermassive black holes in the nuclei of galaxies, with masses 10^6 to 10^9 times greater than the solar mass. Just recently black holes of intermediate masses (10^3–10^4 times of solar mass) have been discovered in the centres of globular clusters and first determinations of masses of single stellar-mass black holes have been realised from observations of gravitational microlensing effect. A new branch of astronomy was born — black hole demography. A considerable contribution to the discovery of black holes has been made by the Russian X-ray space observatories Mir–Kvant and Granat, under the supervision of Academician R.A. Sunyaev, a former student of Yakov Borisovich.

It is very significant that the observational manifestations of accreting black holes and neutron stars in close binary systems are different from each other, in full agreement with Einstein's general relativity theory. Yakov Borisovich was a firm supporter of general relativity theory and believed in the correctness of this most beautiful physical theory.

It is a great pity that he is not alive now, when his scientific ideas and predictions have been confirmed.

On first-name terms with physics

M.Yu. Khlopov

His coffin in the hall of the building of the Presidium of the Academy of Sciences, which was so full it couldn't hold any more; uncovered heads in spite of the –20°C temperature at the funeral gathering at Novodevichy Cemetery — all this happened, but it seems impossible to connect these facts with the death of Yakov Borisovich. All the same, you still expect his morning telephone calls. Passing the gate of the Institute of Physical Problems, you involuntarily glance at the window of his office to see if he's there. Unconsciously you try to find him at scientific seminars and conferences. The bright, lively impression of his presence remains. For everyone who knew him, the words 'death' and 'Zeldovich' are incompatible.

He always lived science, and his life is continued in it. He seemed in science unlimited and all-encompassing. MAAS — Member of All Academies of Sciences[157] — that was how he sometimes presented himself. He realized his genius and greatness, but his mischievousness and ardour removed the import of this self-perception. He didn't consider himself a '*Vy*' in science.[158] He spoke informally, on a first-name basis, with science and with himself. In his approach to the solution of a scientific problem, the vivid and integral image of the phenomenon being studied was what attracted him, and formulae and computations seemed only elucidations and justifications for some intuitive solution which he had already 'envisioned.' 'After all, he sees it' — an approving evaluation of some scientist would sound on Zeldovich's lips, with a hint of a kind of jealousy if Zeldovich didn't share this 'vision' at that moment. Results that could not be reduced to clear physical pictures which could be explained on the back of an envelope, to use his expression, 'jumped away' from him. He could not accept formally correct but physically incomprehensible, dry, mathematical derivations. Such work bored him and provoked open fits of yawning.

Our collaborative scientific work occurred in the last decade of his creative life, and was related to problems at the junction between elementary-particle theory and cosmology; in the 'Autobiographical Afterward' to his selected scientific works, he characterises these problems as follows: 'Life goes on, and cosmology goes deeper into the area where physics is far removed from experimental verification. The new generation of theorists talk not about the first three minutes or seconds, nor about nuclear reactions in plasma. Processes on the Planck scale, over the Planck time, with the Planck energy are discussed ... Field theory considers 5-, 11-, and 26-dimensional spaces. Under laboratory conditions, these will necessarily imitate our usual (3+1)-dimensional space–time, and the extra dimensions are hidden, rolled up, leaving traces only in the systematics of particles and fields. Twenty-year-old kids come and immediately, without the burden of previous work and traditions, sink their teeth into the new topics. Do I look like a mastodon or archaeopteryx among them?

Humanity, like never before, is on the threshold of amazing discoveries. The idea of all-unifying theories presents itself more and more vividly ... It is probable that precisely cosmology will provide a testing chamber for the verification of the new theories.'

He was in that group of leaders who directly guided the dramatic development of natural science over the last 50 years, and his personal perception of this development is most fully expressed in our work on a book about the drama of ideas in physics, which carries autobiographical impressions of his personality. We began to write it in the spring of 1982. Our collaborative work on the encyclopedia *Physics of the Cosmos* (he as the editor,

[157] In Russian, the corresponding abbreviation is *ChVAN*, which sounds like a play on the word *chvanlivost* (boastfulness) — a quality totally lacking in Zeldovich. [Translator]

[158] *Vy* is the formal form of 'you' in Russian, and indicates respect for the person addressed, especially when capitalised. [Translator]

I as one of the contributing authors) provided a direct push for this. A number of fundamental questions in the theory of elementary particles, which were not directly related to the topic of the book but played an important role in the development of modern theoretical astrophysics and cosmology, aroused in us a desire to trace how the understanding of elementary particles, charges, and fields in the history of physics were augmented as science developed.

In spite of our fairly clear general plan, work on the book stretched over five years. In one way or another, its content touched on virtually all the basic questions in theoretical physics, and it was difficult to reconcile accessibility in the presentation with minimum loss of scientific rigour.

The appearance of the book *α, β, γ, ..., z* by L.B. Okun' presented us with unexpected help. Our task was simplified — it was no longer necessary to describe modern concepts of elementary particles in detail, and we could concentrate on the evolution of these concepts. Working in this vein, we understood that the initial title for the book — *Particles, Fields, Charges* — could be no more than a subheading for the final full title — *The Drama of Ideas in Comprehending Nature*. However, here we came upon another difficulty — the necessity of evaluating current tendencies in the development of physics. Turning to the past in order to understand the current situation was one stimulus for work on the book. It was interesting to get inside the skin, let us say, of our contemporaries Dirac and Fermi and to try to sense how they reacted to the idea of anti-particles, to the concept of the creation and annihilation of elementary particles. At the end of the 1980s, this type of 'living through' the physics' past was psychologically important for an understanding of its present and perspectives for its future.

During work on the book, the dialectic of the development of scientific ideas came alive before our eyes. I'll present one story told to me by Yakov Borisovich in connection with a discussion of magnetic lines of force. The question of whether or not this concept, which is convenient for electrodynamics, has any physical meaning became the subject of a heated discussion between V.A. Fok and the electrical engineer Mitskevich in the 1930s. Fok, having actively refuted the physical meaning of this concept, said that the question was meaningless, as was the question, 'What colour is the meridian?' To this, Mitskevich, convinced of the fundamental importance of force lines, spitefully responded: 'It is strange that Comrade Fok doesn't know that our Soviet meridian is red.' 'It may seem,' Zeldovich concluded his story, 'that, in this discussion, Fok embodied the purity and uncompromising nature of current science, while Mitskevich played the role of a simple demagogue. However, the truth is more complex, because magnetic force lines have evident physical meaning for a magnetic field frozen in plasma.'

On the threshold of 1984, he and I talked about Orwell's book *1984*. Unexpectedly, Yakov Borisovich expressed his lack of delight with the book. 'Why?' I wondered. 'The painfully wretched fantasy of the author — socialism without queues,' answered Zeldovich. Reflecting on this evaluation, I was struck by its depth: queues are the very essence of a distributory system.

Another example is connected with *The Inspector* by Gogol. 'I figured out,' Yakov Borisovich told me, 'that the 'fathers' of the city should have been veterans of the Patriotic War of 1812. And having risked their lives for their native land, it doesn't seem shameful to them to take a piece of cloth from a merchant, for whose sake, essentially, they risked their lives. Khlestakov didn't participate in the war in his youth, and accepting the fruits of victory as his right, he cannot understand things that are elementary for the ruling class of veterans.' Since Zeldovich repeated this interpretation more than once, it seems to me that he was rather proud of it.

At the beginning of 1985, Zeldovich visited an exhibition by the artist Sel'vinskaya, leaving in the comment book the inscription: 'It is a pity that there is no such poet among physicists.' When I stopped by to see him at home several days later, he was surrounded by copies of Volume I of his selected scientific works, given to him by the library of some research institute. I flatter myself with the vain hope that, signing a copy for me, he consciously corrected his inscription in that comment book: 'Dear Maksim Yur'evich, to a poet (or dramatist) among physicists ... this book stolen from a research institute with risk to life and reputation, with hope for our third volume, from YaB.' 'Then you can compose a tale about how you and I stole this copy from the library. You stood on guard and I nicked it. You can think up some spine-chilling details — the squeak of the floor boards at the most crucial moment, or how we were nearly caught in the act, and so forth,' he said, handing me the signed copy.

The third volume to which he referred was our *The Drama of Ideas in Comprehending Nature,* which came out after his death. It indeed was a logical conclusion to the collection of his works, as a volume of philosophical reflections completes the collected works of every major scientist.

The genius of Zeldovich is immortal. A classic textbook polish shines from the fragments of his living memory. He is too alive, and I cannot believe that he is no more. 'He was a man, a man in everything, and such as he I'll never meet again.'

'Bonjour tristesse' Reminiscences from meetings with an outstanding scientist

R.M. Bonnet

Before I met Yakov Zeldovich for the first time in June 1980 on the occasion of the COSPAR meeting in Budapest, I had hardly paid attention to his name, which I first read on the cover of the famous *Relativistic Astrophysics,* whose formulae at first sight rebuffed me until I found them and the text around them to be astonishingly clear. I also knew of him because of his association

P. Dirac medal for work on theoretical physics, 1985.

with my great friend Sunyaev, through the famous effect which carries both their names, and which allows in principle to determine the distance to clusters of galaxies through a combination of X-ray and submillimetric observations, thereby allowing an unbiased determination of Hubble's constant.

We had no reason to know each other: he was a famous theoretician and I had spent most of my scientific life playing with screwdrivers, diffraction gratings, mercury lamps and oscilloscopes. By chance, however, in June 1980 we found ourselves attending the COSPAR meeting, sitting in the same dusty auditorium at Budapest University, in fact on the same bench at the top of the auditorium, side by side, listening to a series of lectures on future projects in space physics. At that time I had become fairly excited by the prospects of helioseismology from space, and at the meeting several papers reported on recent results obtained from the South Pole, from which a long-term series of measurements and power spectra could be obtained. Zeldovich showed a keen interest when, in the coffee break, I found myself explaining to him how promising helioseismology was for probing the solar interior. It was one way, I added, to study the origin of the apparent neutrino deficit.

I had pronounced a magic word. In 1980 the first results of experiments conducted in the USSR on neutrinos seemed to confirm that they had a mass

Meeting Pope John Paul II during a trip to Italy (The Vatican, 1986).

and could oscillate between three types of particles, offering a possible explanation to the neutrino deficit. Zeldovich was keen to report on the extraordinary results, and was enthusiastic about helioseismology. Our conversation continued, and very soon we were engaged in a discussion of the outstanding questions in solar physics, the field which, due to my past work, remains the most familiar to me. I was very happy to find an interlocutor so interested in acoustic and magnetic heating of the chromosphere and of the corona, solar wind propagation, and the activity cycle. I was amazed to see the famous relativist Zeldovich leaning with interest towards such specific issues in solar physics.

That experience was pleasant indeed and it marked the start of a continued, although short, warm relationship. Less than three years later in April 1983, I was invited by E. Trendelenburg and our mutual friend R. Sagdeev to Samarkand, where the Soviets had organised the most incredible farewell party one could imagine in honour of Trendelenburg, who was leaving ESA at the end of that month after 20 years' activity (this is probably not the right word when talking of Trendelenburg) at ESRO and then at ESA. I had been elected as Trendelenburg's successor and he had kindly invited me to share these moments of celebration with him and his closest collaborators. These moments are unforgettable in many respects. Sagdeev had booked the local Communist Party VIP hotel for all of us. Trendelenburg was driven in a bulky, black, official Chaika car preceded by the police and followed by an ambulance. Policemen were placed at each major traffic intersection in order to lead the way. For three days we had the vivid impression of being part of a royal train, being the guests of the highest local political, as well as scientific, authorities. Yakov Zeldovich was one of Sagdeev's guests, together with R. Sunyaev, N. Kardashev, V. Barsukhov, and several others.

These three days of continued eating, drinking, and dancing offered me another opportunity to get to know Zeldovich more closely. We found ourselves once again sitting on the same bench, but this time in the well-kept and sunny garden of the Communist Party hotel in Samarkand, discussing recent progress in helioseismology, neutrino physics, and ... French literature. I admired the fact that he could read Zola and Hugo in French, not to mention Françoise Sagan, of whom he was very fond and able to quote without difficulty. (His favourite was *Bonjour Tristesse*.) My interest in cosmology had grown considerably since our first encounter, and I was more than happy to learn that he was in the process of preparing one of the plenary evening lectures for the General Assembly of the International Astronomical Union, to be held in Patras a few months later, on modern cosmology. I found the topic crystal clear as explained by the brilliant and glittering Zeldovich. He knew I was going to leave active research by entering ESA and he expressed the strong wish that I could remain actively involved in space science, advice that I have tried to follow faithfully during all these years at ESA.

In the evenings we usually left the field of science for that of fantasy and dancing, he with his wonderful wife and me with the local Uzbek beauties, and vice versa ... The nights were too short, so were the days. At the end of this exhausting period we went to Moscow and I flew to Paris. He remained in Moscow. We had become close friends.

While at ESA I had several other chances to meet Zeldovich on the occasion of my regular visits to IKI and, when we could not meet there, since he knew of my visit, he never failed to invite me to his office at the Landau Institute for Theoretical Physics of the Academy, a fairly shabby place, very close to the huge and ugly Gagarin monument on Leninsky Prospect. At the Institute he was Head of the Theoretical Department and occupied Landau's own room. He had been invited to the Institute by the great physicist Kapitsa in the 1980s.

I certainly cannot forget our discussions on cosmology and neutrinos once again, in his very small office, fairly dusty, which was perhaps kept in this state in memory of the early days when Dirac was invited to the office and had many discussions there with Kapitsa and his colleagues. On the blackboard, one of those wooden boards on which it is nearly impossible to write, with those large Russian pieces of chalk, Dirac had written a few equations. Zeldovich had preserved those from any ignorant attempts to erase them, by means of a rectangular piece of plexiglass. He was proud of being the keeper of this precious autograph.

It was in Landau's room that Zeldovich told me of his early involvement in weapons, and I learned of his ability to manage hardware in addition to theory. In fact, his very early scientific works were on the theory of combustion and detonation, and in several other areas of chemical physics. I had heard from Sagdeev and Sunyaev that he had been involved in the development of the Katyusha and the famous Stalin Organs. I wanted to know from him and, in Landau's dusty room, Zeldovich explained to me that, as originally conceived, the organs were dangerous, not for the enemies but for those who fired them. They had to be redesigned and Zeldovich himself proposed the mechanical modifications which eventually made them the most dreadful weapons against the Nazis.

In November 1984, on the occasion of the annual meeting of the Inter-Agency Consultative Group in Tallinn, I met Zeldovich again. He was sad and lonely: his wife had passed away a few months before, and we remembered the happy times in Samarkand. One good way to avoid too much sorrow was to sit down in a small room and to let him explain the latest developments in cosmology. At no time ever before had negative pressure, positive energy density, inflation theory, pancakes, cosmic strings, and many other ingredients of cosmology and quantum physics become so easy to understand. After having enjoyed science, we joined Sagdeev, Galeev, and others in the hotel sauna where beer was served!

A memorable event cast some new light on this exceptional character. This was on November 7, 1986 in Rome, the very day of the October revolution, but this was not the reason. The IACG had met a few days earlier in Padua and had been actively discussing its future activities after the great success of the space missions to Halley's Comet, all of which were miraculously successful! The Pope had invited the whole delegation to the Vatican and a solemn audience, with a presentation of the scientific results to the Pope, had been organised in the Sala Reggia, in the presence of all the cardinals and the ambassadors. Zeldovich was part of the Soviet delegation. At the end of the presentation all the participants were allowed to shake the Pope's hand or to kiss his ring. When it came to Zeldovich, he extracted from his pocket a copy of his book *Nuclei, Particles and the Universe* and presented it to the Pope, saying loudly enough for his neighbours to hear: 'When I was younger I thought that science and cosmology were able to explain the origins of the universe. Today I am not so sure!' Surprising Zeldovich! The Pope thanked him warmly, took the book and put it under his left arm with only the word universe left uncovered. What an image!

The last time I saw Zeldovich was on the occasion of the forum organised by Sagdeev in Moscow on October 4, 1987 to celebrate the 30th anniversary of the launch of Sputnik-1. Less than two months later, a telex from Sunyaev informed me that Zeldovich died suddenly of a heart attack.

I cannot forget this man, small in height but a giant in modern astrophysics. In a paper published in *Nature* in February 1988, the late A. Sakharov, who had worked with him closely between 1948 and 1968, wrote a long article in which he called Zeldovich 'A man of universal interests.' No sentence characterises more effectively this outstanding man. My all too short companionship with Yakov Zeldovich can only confirm this impression. When I learned from Sakharov's paper that Zeldovich had started his career as a laboratory technician at the age of 17, I was even more impressed by him. His brain was always on the alert, ready to seize any challenge, be it on matters of science, art, rockets, sport or girls. He was a man of eternal youth who could communicate easily with all those around him. This is a sign of personality. To me, he was a great friend who, after many years, is still present in my memory, and whose death I will always consider to have come much too soon and to have created a great loss. Not only for me, for his friends, his students, but above all for science.

Ya.B. Zeldovich and large-scale structure: the impact in the United States

Adrian L. Melott

This is a largely personal account of the influence of Yakov Borisovich on the development of our understanding of large-scale structure. It reflects my own perspective, and is not intended as a comprehensive review.

My encounter with the ideas of Yakov Borisovich Zeldovich began in 1980. At that time many of us were excited about the possibility of a neutrino mass around 30 eV providing most of the mass in the universe. I realized, as had many others, that the so-called 'pancake' theory would be important for the formation of structure in that case.

Pursuing this interest led into numerical models for structure formation. Due to the time lag between publications in our countries, I began work which was nearly identical to Doroshkevich *et al.* (1980). This important paper showed that a cellular structure, not (as many had assumed) isolated 'pancakes,' are a natural outcome of the kind of picture of galaxy formation presented by Zeldovich.

I remember that at that time in the USA there was a great deal of scepticism about the reality of emerging evidence for very flattened structures on supercluster scales. There was a general feeling that they were some kind of observational artifact. The fact that they arose in pancake models did not particularly lend credence to that theory here.

I remember very clearly a noted astronomer pointing to a picture of the galaxy distribution in the CfA survey and pronouncing 'there are no filaments *here*.' I think most of us now accept the strong flattening of the local supercluster and accept that it is a dynamical effect. The modern continuation of the CfA survey has given a spectacular confirmation of the existence of large voids surrounded by flattened structures.

I must say that I think there has been a reluctance to give Yakov Borisovich his due here. He developed a theory which *predicted* this kind of structure. He did this at a time when there were only hints of flattened structure locally and no evidence from large redshift surveys. There has been a tendency to emphasize the novel, unexpected quality of these observations, and all the structures, filaments, bubbles, great walls — but not to acknowledge them as structures predicted from theory in 1970, with later numerical refinements. As time passes, I believe the significance of the prediction will be remembered.

I spent the last winter and spring of 1983 in Moscow. I remember one conversation rather well. I remember discussing some numerical modelling results with him, and I remember his demand to *see* the *derivation*. Perhaps one of his shortcomings was an inability to understand the nature of a numerical experiment. The visit proved extremely valuable, although in the political climate of that time we had to be extremely devious to arrange informal meetings with members of his group.

The era of *glasnost* began only in the last years of his life. It is unfortunate that his personal contact with people in the West was limited by this historical coincidence.

Evidence has steadily increased that pancake-like processes are of general importance. We studied numerical models of cold dark matter, and found that they had percolation properties rather like those of the pancake models. Visual evidence began to accumulate that flattened structures were important in Cold Dark Matter.

We undertook systematic studies to determine under what kinds of initial conditions Zeldovich pancaking is important. We found that there is a more or less continuous range of types of clustering as the initial power spectrum is continuously varied. Pancake-like structures arise not only when the power spectrum has a short-wavelength cutoff, but also when there is power on all scales — if the long wave power has sufficiently high amplitude.

We did some explicit statistical tests by cross-correlating different evolved models with the same phases in the Fourier components of the initial conditions, but different amplitudes. This showed that the somewhat controversial filaments in hierarchial clustering models trace out the structure seen in the corresponding pancake models. The amplitude of the cross-correlation increases with large scale power in the initial conditions. This shows clearly that the filaments in hierarchial models reflect power in their initial conditions, and are not accidental patterns or numerical artifacts of the lattice.

It appears that the 'Cold War' in large-scale structure studies is over. Twenty years ago, the hierarchial clustering picture and the pancake picture were seen as radically different theories for the formation of structure in our universe. Now we see them as merely different emphases in viewing the clustering process. The observational data seem to suggest that something intermediate has been operative in our universe. It has neither the regular, sharply defined structures of the pancake model nor the isolated disconnected lumps of extreme hierarchial models. Our final picture will include a synthesis of these.

In the spring of 1990, we held a workshop in honor of Zeldovich at the University of Kansas. It was the first workshop to be held under the 1989 agreement between the US National Science Foundation and the Academy of Sciences of the USSR. Thirty-two people from five countries attended.

Our common theme was our individual modern extension of work by Yakov Borisovich. It is a tribute to his breadth and creativity that so many different topics were discussed by Soviet citizens, Americans, and others. There is talk of continuing some kind of joint astrophysics and cosmology workshop, perhaps every two years.

During our evening banquet, there was a round of stories about him. The one I remember most vividly concerns his conversation with a colleague about physics. The colleague told him that he was working on a theoretical model in which time ran in several different directions simultaneously. Yakov Borisovich got down on the floor, on his knees, and begged the man 'not to work on such a stupid thing.'

He had a fine sense of what is important and physically reasonable. That is why so many of us are still working on these things, and were willing to travel large distances to gather to talk about them.

I am grateful to the IREX Foundation, the Shternberg Astronomical Institute, Moscow State University, and the Institute for Space Research for support of my 1983 visit to Moscow. I also thank the US National Science Foundation (Grant INT-9000598), the National Aeronautics and Space Administration (Grant NAGW-1793), the University of Kansas, and the Academy of Sciences of the USSR for support of our 1990 workshop in Lawrence.

Zeldovich and modern cosmology[159]

P.J.E. Peebles

INTRODUCTION

Landau's contributions to astrophysics were deep, as witnessed by the papers at this conference on neutron stars and supernovae, but not nearly as broad as those in other fields, such as condensed matter and particle physics. The explanation I have heard is that Landau felt that astronomers were always changing their stories on him. Whether or not that is apocryphal, it does reflect the situation: if you want to do astrophysics it is one of the conditions you must learn to live with. Yakov Zeldovich, whose memory we are also celebrating at this conference, was a master at doing physics on this slippery ground.

---

[159] This article was not included in the original Russian version of this book, published in 1993. [Translator]

My last chance to talk with Zeldovich was at the IAU Symposium No. 130, on the large-scale structure of the universe, at Balatonfured, Hungary, in June 1987. We see one aspect that was hard to miss — a deep appreciation of life and all it has to offer. It did not take much further study to discover that here was someone who loved doing physics, was generous in his help to others, and a wonderful physicist. Through all my career in cosmology I could be sure that if Zeldovich was not hard on my heels it was because he was racing far ahead.

My first slightly personal contact with Zeldovich was in 1965, through a letter he wrote to congratulate Bob Dicke for his key role in the discovery of the thermal cosmic microwave background radiation (CBR), in initiating a search for the effect at Princeton and in interpreting the actual discovery by Arno Penzias and Bob Wilson at Bell Laboratories. Zeldovich was hot on the trail of the CBR. In that same year he had published an account of the behaviour of blackbody radiation in an expanding universe and of the connection between the present radiation temperature and the production of helium in the hot Big Bang. He knew that the Bell Labs' Holmdel telescope was capable of detecting this radiation if the temperature were 1 K or more. Unfortunately he misinterpreted the reports on the operation of this telescope as meaning that the background temperature would have to be below 1 K, whereas the instrument actually was indicating an anomaly that Penzias and Wilson convincingly showed is due to extraterrestrial radiation.

The discovery of the CBR and the demonstration that the spectrum is fairly close to thermal marks the beginning of the modern era of cosmology. The CBR gives us tangible evidence that the universe really did expand from a state a good deal more dense than it is now, for the universe is observed to be optically thin at radio wavelengths and so could not have caused radiation from known sources (stars, radio sources) to relax to near statistical equilibrium. There is still the possibility that the radiation is starlight thermalized at a redshift as 'low' as $z \sim 100$, but the demands on the energy source and grain stability are so severe that the more conservative and I think more reasonable interpretation is that this radiation existed back to extremely high redshift.

Since the universe is not exactly homogeneous and isotropic the CBR must deviate from an exactly isotropic blackbody spectrum, and these deviations are an invaluable probe of how structure in the universe developed. Zeldovich and Rashid Sunyaev were among the leaders in developing the theory of how one relates observations of the CBR to physical processes in the early universe. I think they were the first to emphasize the importance of bulk plasma motions as a source of small-scale anisotropy of the CBR, an effect that is now receiving wide attention in the case where galaxies formed at high redshift. Following the work by Ray Weymann, they showed how hot electrons would tend to upscatter CBR photons, lowering the background temperature in the Rayleigh–Jeans part of the spectrum and raising the temperature at shorter wavelengths, and they showed how this effect could be used as a diagnostic for the state of the intergalactic medium at high redshifts and for the intracluster medium at low redshifts. The Sunyaev–Zeldovich effect may have been observed in the Berkeley–Nagoya CBR spectrum measurement.

The study of how the CBR departs from an ideal heat bath is part of a larger problem, to understand the origin of the large-scale structure of the universe–galaxies in groups and clusters and superclusters. This structure must have developed under the action of gravity, electromagnetic forces and perhaps other significant stresses out of departures from exact homogeneity that originated in the very early universe in a state of such extreme energy density that the classical Friedmann–Lemaître cosmological model is not a useful approximation. The search for the principal actors and the outlines of the scenario by which large-scale structure developed out of the physics of the early universe and evolved up to epochs accessible to observation has been one of the main themes of modern cosmology and one of the main subjects of Zeldovich's work in this field. It is appropriate therefore to honour his memory by giving a progress report on how the search for an understanding of large-scale structure has prospered in the last twenty-five years.

MODELS FOR GALAXY FORMATION: EVOLVING FASHIONS

a) Baryon adiabatic model

When the CBR was discovered it was seen that this would be a principal actor in the evolution of departures from the ideal homogeneous expanding world model, helping fix the matter Jeans length, and, at redshifts $z > 100$, contributing a very significant pressure. In most of the early discussions it was assumed as a matter of course that the only other important actors would be baryons and their electromagnetic interactions. If that were so and if the departures from homogeneity could be described by linear perturbation theory, then the state of the universe within the horizon would be characterized by two functions to represent fluctuations in the densities of matter and radiation. Convenient combinations of these functions are the adiabatic mode, in which the local number density of photons is proportional to the number density of baryons, and the isothermal mode, in which the radiation density is homogeneous and the baryon density fluctuates. (I think Zeldovich in 1967 first used these names in this connection.) When people came to think about density fluctuations in the very early universe, where length scales of astrophysical interest would have been larger than the horizon, the word 'isothermal' came to be used to describe the situation in which the total mass density ρ (averaged over a scale larger than the horizon) is homogeneous and the baryon number is clustered. I will follow the more recent convention of calling this an isocurvature perturbation. As the short wavelength part of an isocurvature perturbation enters the horizon it turns into a combination of adiabatic and isothermal modes, the adiabatic part dissipates by photon diffusion, and we are left with the old-fashioned isothermal mode.

The first discussions of galaxy formation with the CBR assumed the isothermal mode (in my case, at least, because it is simplest). One later came to see that the adiabatic mode is in a sense generic: if at some very early epoch the baryon and photon number densities were perturbed from homogeneity by independent but comparable amounts, it would generate a combination of adiabatic and isocurvature modes, and since only the former grows by gravitational instability it would come to dominate.

The short wavelength part of a baryon adiabatic perturbation dissipates by photon diffusion, so an initially broad power spectrum of adiabatic density fluctuations ends up with a computable coherence length in the mass distribution at decoupling. Depending on the shape of the initial spectrum of density fluctuations the mass enclosed by the coherence length can be anywhere from that of a giant galaxy to that of a rich cluster of galaxies. How does a mass distribution with a coherence length evolve as the amplitude of the density fluctuations approaches $\delta\rho/\rho \sim 1$? The answer was given by Zeldovich: planar collapse produces a first generation of flattened objects, Zeldovich 'pancakes,' with size fixed by the coherence length.

What is a reasonable guess for the power spectrum of the primeval adiabatic density fluctuations? The simplest possibility would be a function that has no scales, a power law with the index chosen so space curvature fluctuations associated with the density fluctuations diverge only as the logarithm of the length scale. This would imply that the rms amplitude of density fluctuations appearing on the horizon is independent of time. This is properly called the Zeldovich spectrum because he was the one who most consistently emphasized its possible importance.

With these ideas, including pancake collapse of adiabatic baryon perturbations with an initial Zeldovich spectrum, one could build a fairly detailed picture for galaxy formation. The main problem with it is that if galaxies formed at a reasonable epoch, the amplitude of the primeval density fluctuations would violate the bounds from large-scale fluctuations in the galaxy distribution and from the isotropy of the CBR. These problems were never very seriously debated, however, because a new and seemingly much more promising idea surfaced, a universe dominated by massive neutrinos.

b) Massive neutrinos

The rise in popularity of massive neutrinos as the dominant part of the dark mass traces to four important developments. First, it was seen that in the simplest case the theory of light element production in the hot Big Bang requires that the density parameter, Ω (the ratio of the mean mass density, ρ, to the mass density in the Einstein–de Sitter model), in baryons not be bigger than $\Omega_B \sim 0.1$. A second, popular opinion fixed on the idea that the only reasonable cosmological model is the Einstein–de Sitter case, where $\Omega = 1$ (and the three-dimensional sections of constant world time are flat, with negligibly small cosmological constant, Λ). Third, in 1980 V. Lubimov and colleagues announced the possible detection of a neutrino mass at about the value wanted if neutrinos were to make up the difference between the baryon mass density allowed by light element nucleosynthesis and that needed for the Einstein–de Sitter model. And fourth, it was recognized that non-baryonic matter such as massive neutrinos could solve the problem with excessive perturbations of the CBR, for density fluctuations in the non-baryonic mass distribution would start to grow well before decoupling of the baryons from the radiation, lowering the amplitude of the primeval fluctuations.

The invention of the inflation scenario was the immediate cause of the growth of popularity of the Einstein–de Sitter model. The great virtue of this scenario is that it offers a way to understand the remarkable large-scale homogeneity of the universe: expansion during inflation could have stretched the length scales belonging to inhomogeneities to unobservably large values. It would reasonably follow that the length scale belonging to space curvature would be stretched to an unobservably large value also. The two arguments against the cosmological constant are that a cosmologically reasonable value of Λ is quite unreasonable from the point of view of particle physics, and that highly special initial conditions in the very early universe would be needed to make the epoch when Λ and ρ both make reasonable contributions to the expansion rate coincide with the epoch of life on earth. (Bob Dicke taught me this argument, which applies to space curvature as well as Λ, maybe 30 years ago.)

Gershtein and Zeldovich pointed out that neutrinos with rest mass could make an appreciable contribution to the mass density of the universe. At high redshifts neutrinos relax to statistical equilibrium with the other fields, the neutrino number density n_ν being comparable to the photon number density n_γ (assuming the conserved lepton number density is small compared to n_γ). If the neutrinos are still relativistic when they decouple from the other particles, then the present number density of electron-type neutrinos works out to about 100 cm^{-3}. If the neutrino mass is $n_\nu \sim 30$ eV, which is in the range of the laboratory estimates in 1980, then the mean mass density $m_\nu n_\nu$ agrees with the Einstein–de Sitter density. It is an attractive coincidence that this mass agrees with the phase space bound for neutrinos as the dark mass in the outer parts of spiral galaxies.

The idea that non-baryonic matter might account for the difference between $\Omega = 1$ and what is allowed by light element nucleosynthesis was discussed by many authors, as was the point that the introduction of non-baryonic matter has the beneficial effect of lowering the amplitude of primeval fluctuations because small-scale density mass fluctuations can grow before the baryons decouple from radiation drag. Also, free motions of the neutrinos while they are still close to relativistic would suppress small-scale mass density fluctuations, tending to produce a coherence length comparable to that of the baryon adiabatic case. The first generation of non-linear objects thus would be planar, Zeldovich pancakes. The assumption that the primeval adiabatic density fluctuations have the scale-invariant

Zeldovich spectrum was reinforced by the recognition that quantum fluctuations of the inflaton field of inflation would produce density fluctuations with almost exactly the Zeldovich spectrum.

A detailed model for galaxy formation based on the above ideas is given by Sergei Shandarin, Andrei Doroshkevich, and Yakov Zeldovich. Perhaps the greatest triumph for the model is the prediction that pancake collapse would tend to place galaxies in sheet-like arrangements. This prediction, made when not all of us were convinced of the reality of the effect, has been brilliantly confirmed by redshift surveys.

There are two major problems with the massive neutrino picture. First, analytic arguments and N-body model experiments indicate that if galaxies were fair tracers of the large-scale mass distribution then to match the galaxy distribution with the Zeldovich spectrum one would have to set the normalization of the primeval mass fluctuation spectrum so that galaxies form at low redshifts, which seems counter to the observations. (Observational constraints on the epoch of galaxy formation are discussed briefly in the next section.) One avoids the problem if the galaxy distribution is a biased measure of the mass distribution, mass clustering more strongly than galaxies. However, to reconcile the dynamical estimates of the density parameter, $\Omega \sim 0.2$, with the assumed value $\Omega = 1$ one usually assumes biasing in the opposite direction, galaxies clustered more strongly than mass, so the estimates of the mean mass per galaxy from the dynamics of groups and clusters underestimates the true value. I have not yet heard of any proposal for how to resolve the conflict. Another way out of the problem with late galaxy formation is to assume the neutrinos accrete around mass concentrations in an isocurvature picture.

The second problem has to do with the old question of the sequence of formation: which came first, galaxies or protoclusters? In the massive neutrino picture it would be the latter. The observations suggest to me that it is the former. For example, we see the local supercluster forming now, out of old galaxies.

The merits of these objections have not yet been adequately debated because popular attention soon swung to another candidate for dark matter, non-baryonic particles with negligibly small primeval velocities.

c) Cold dark matter

It appeared that, once massive neutrinos were seen to be an acceptable and interesting candidate for dark mass, particle physicists were glad to supply us with a host of other possibilities. Of particular interest is the simple case where the particles have negligible primeval velocities and the primeval density fluctuations are adiabatic with the scale-invariant Zeldovich spectrum. To the extent that baryons can be ignored this picture is fixed by just one free parameter, the amplitude of the primeval mass fluctuation spectrum, and the evolution of the model is described by particularly simple physics, an initially cold collisionless gravitating gas.

This CDM (for cold dark matter, a term introduced by Dick Bond) picture combines many of the attractive features of the massive neutrino model, possibly losing the pancake effect that so naturally accounts for the tendency of galaxies to lie on sheets, but adding the very important advantage that small-scale fluctuations survive and so can produce galaxies before clusters. The dramatic observational successes of the CDM N-body model experiments by the team of Marc Davis, George Efstathiou, Carlos Frenk, and Simon White have led many to conclude that here at last we may have a reasonable candidate standard model for the origin of galaxies.

There are two crises for the model. First, the scale-invariant Zeldovich spectrum suppresses large-scale fluctuations (relative to white noise). Is this consistent with observations of large-scale velocity fields and with observations of structures in the galaxy distribution on very large scales, where in the theory fluctuations in the mass distribution are anticorrelated? The debate still is in its early stages, but is well worth following.

Second, the epoch of galaxy formation in the model seems uncomfortably late. The actual epoch is not known, but is the subject of very active research; only a few points will be mentioned here. There is evidence of powerful bursts of star formation at low redshifts, but it is not clear whether this represents galaxy formation or the rapid conversion to stars of hydrogen already in place in the galaxy. The way to find out is to examine the nature and abundance of galaxies at high redshifts. The bright galaxies observed at redshifts $z \sim 1$ tend to have spectra whose blue part is dominated by young stars, but again, that can be misleading because the blue luminosity can be dominated by a small mass fraction in massive stars from ongoing star formation. The infrared spectra and luminosities of galaxies observed at $z \sim 1$ tend to look like a stellar population that formed at substantially higher redshifts, as if the bulk of the stars in these galaxies formed at high redshift. Apparently consistent with this, it is seen from the Lyman-α absorption spectra of quasars that at redshift $z \sim 4$ the intergalactic medium was already in place: between the clouds of the Lyman-α forest the neutral hydrogen density is no more than about $10^{-13}\,(1 + z)\ \mathrm{cm}^{-3}$. In CDM numerical experiments galaxies seem to be forming at $z < 1$. One could argue that the intergalactic medium at $z \sim 4$ was ionized by star clusters that formed before galaxies, though it would be difficult to see how the bulk of these stars were later collected into galaxies; and it may be possible to argue that some galaxies in the model complete formation at $z \sim 1$, though the challenge will be to explain why so many galaxies at $z \sim 1$ look old and so relatively few starburst galaxies at $z < 1$ look young. This is a rich subject; the debate will be educational.

My impression is that late galaxy formation is a natural consequence of biasing (a concept pioneered by Nick Kaiser). To reconcile the dynamical estimates of the mean mass density, that pretty consistently yield $\Omega \sim 0.2$, with the assumption that Ω really is unity, one must suppose that galaxies are more strongly clustered than is mass (so the mass per galaxy found in groups and clusters underestimates the global mean value). This biasing of the galaxy distribution relative to the mass seems fairly easy to arrange at galaxy formation. However, gravitational instability of the expanding universe causes clustering to grow after galaxies have formed, drawing together galaxies and dark mass alike, and tending to erase the biasing on large clustering scales. This is beneficial in the sense that it increases the expected large-scale velocity field, which is the route taken by Kaiser and others, but problematic in the sense that it becomes less easy to understand the low dynamical estimates of Ω. That is why I think the epoch of galaxy formation is the most pressing and exciting crisis for CDM.

d) Cosmic strings

With the successful development of gauge theories in particle physics people realized that there may be important applications to the physics of the early universe, and that relicts of symmetry changes — domain walls, strings, and monopoles — may have significant observational consequences. Characteristically, Zeldovich took one of the first steps, showing that domain walls with an interesting mass per unit area would produce unacceptably large perturbations to the cosmological model. Tom Kibble gave us the classification of these relicts of the early universe. Zeldovich remarked that cosmic strings may be good candidates for the seeds of large-scale structure.

Alex Vilenkin worked out how cosmic strings could produce galaxies. He introduced the scaling ansatz, that the string coherence length is of the order of the cosmic expansion time, from which it follows that the fractional contribution of strings to the mean mass density is $\rho / \rho_{s'} \sim G\mu$, where μ is the string mass per unit length. He found the power law form for the abundance of string loops as a function of loop mass m_l,

$$\frac{dn_l}{dm_l} \propto m_l^{-5/2} . \tag{1}$$

He found that these loops would produce galaxies of about the right mass if $G\mu \sim 10^{-5}$, about what would be expected if the strings were produced at the grand unification energy, which is an attractive coincidence. Also attractive is Neil Turok's discovery, that the abundance of larger loops would be sufficient to act as seeds for the formation of rich clusters of galaxies. Elaborations of the cosmic string model are still under active discussion.

What are the problems? I tend to focus on one point, where on the face of it the string model is inconsistent with the systematics of galaxies, as follows. Let v_c be the speed of motion of matter in circular orbit at a fixed distance $r_c \sim 10$ kpc from the centre of a galaxy. The mass within that radius is then $M_c \sim v_c^2 r_c / G$. For bright galaxies v_c typically is in the range of 200 to 300 km/s, for which $M_c \sim 10^{11}\, M_\odot$. The record is $v_c = 500$ km/s. This tells us something truly remarkable, that Nature is quite adept at gathering $10^{11} M_\odot$ within the volume of r_c, to make an ordinary bright galaxy, but strongly discourages the gathering of four times that amount into the same volume.

In the cosmic string picture M_c is expected to be an increasing function of m_l, but since m_l has a power law distribution (eq [1]) something has to suppress the collection or visibility of the accreted mass when m_l exceeds some critical value. It would be reasonable to expect to see evidence of the suppression mechanism in the morphologies of the transition galaxies produced by loops with nearly the critical value of m_l. For example, if the concentration of mass were inhibited when m_l is large we might expect to see that galaxies with large v_c have low central mass densities; and if formation of visible stars were suppressed when m_j is large we might expect to see that galaxies with large v_c have anomalous mass to light ratios. Neither effect is observed. Surface brightness runs in giant ellipticals and the supergiant cD galaxies look very similar. Also, there is quite a tight relation between v_c and luminosity. As far as I am aware this relation between v_c and luminosity extends to the brightest galaxies and the largest values of v_c. That is, whatever suppressed the visibility of the descendants of massive loops, either by suppressing the mass or the light, had to do so without affecting the correlation between mass and light, which seems difficult.

In summary, my understanding of the evidence is that the most massive galaxies look as if they were assembled by a scaled version of the formation of ordinary bright galaxies. If this interpretation were right it would seem to rule out the cosmic string picture in the form outlined above. I am hoping for some lively debates on whether the interpretation really is right.

e) Explosions

All the above models assume that galaxies were assembled by gravitational instability seeded by small initial mass density perturbations (relicts of quantum fluctuations in the inflation epoch or cosmic string loops). Another idea motivated by observations of how stars form is that explosions piled up material into lumps that then collapsed by gravity to form galaxies. The explosions would have to be seeded by primeval inhomogeneities; the important new point is that there need be no simple connection between primeval mass fluctuations and what we see now. Andrei Doroshkevich, Yakov Zeldovich, and Igor Novikov developed the first explosion model for formation of galaxies and clusters of galaxies. The idea was introduced independently by Jerry Ostriker and others. In the most recent version explosions result from energy deposited by magnetized superconducting cosmic strings.

It should be counted as a triumph here as for the pancake picture that the model predicts the tendency of galaxies to lie in sheet-like arrangements (a result of explosions that pile material into ridges). Given that Ostriker is a leading advocate for this model, one is not surprised to see that it has no manifest problems with the observations. The only possible difficulty I have been able to find has to do with a detail of the local galaxy distribution and motions. The galaxies within cosmological redshift distances $HR \sim 800$ km/s tend to lie on a sheet elongated toward the Virgo cluster. The mean motion of this local sheet relative to the local group of galaxies is no more than about 100 km/s. The dipole anisotropy of the CBR indicates that the peculiar velocity of the local group is 600 km/s, a motion that must be shared by the local sheet. In the explosion picture the sheet would be the remnant of a ridge and the motion would be a remnant of the explosion or else the effect of gravitational repulsion of the hole evacuated by the explosion. In either case one would expect that local galaxies off the sheet and in the direction the sheet is moving do not share the motion of the sheet. But the observed relative motions on and off the sheet are quite similar.

It is too soon to decide whether this is a fatal flaw; it could be, for example, that what is observed is only an accidental coincidence of motions. On the other hand, the straightforward interpretation is that the dipole anisotropy of the CBR is caused by peculiar motion due to a peculiar gravitational field with coherence length large enough to have had relatively little effect on the local relative Hubble flow. I turn now to a model designed to do this.

f) Baryon isocurvature model

It was noted in part (*a*) above that the adiabatic model is generic in the sense that if the baryon and radiation number densities were given comparable but independent fractional perturbations at some epoch in the very early universe, then the adiabatic part would grow to dominate. An independent argument came from the inflation scenario: baryons would have to be created after inflation, and it is reasonable to assume that the resulting local baryon number density (and the densities of all other conserved quantities) would be a universal fraction of the local entropy density. This would generate a pure adiabatic model. However these are not proofs, only good arguments, so it seems prudent to consider the consequences of the alternative assumption, that structure grew out of primeval isocurvature fluctuations.

A baryon isocurvature perturbation assumes negligible perturbation to the total mass density in the early universe, so space curvature is not perturbed, and with the baryon number density distributed in a clustered fashion. With this initial condition the mass distribution averaged over large scales, where radiation pressure force can be neglected, evolves so as to leave space curvature unperturbed. This means that in the present baryon dominated era mass fluctuations approach zero on scales which are large compared to the matter-radiation Jeans length, λ_J.

On scales well below λ_J density fluctuations appear on the horizon as a mixture of adiabatic and isothermal modes, the adiabatic part dissipates through photon diffusion, and we are left with the original baryon distribution as an isothermal perturbation (section *a*). If the baryon distribution had a flat spectrum we would arrive at the old picture from the 1960s, in which galaxies form at high redshift, $z \sim 30$. As was remarked in section (*c*) above, observational constraints on the epoch of galaxy formation have become a matter of lively debate. If the conclusion is that galaxies formed at high redshift, this is one way to do it. It might be noted finally that this model is consistent with the observational bounds on the anisotropy of the CBR.

There is a wavelength on the order of λ_J such that the adiabatic part of the isocurvature perturbation completes just a quarter of an oscillation before matter decouples from radiation. This has the effect of adding to the amplitude of the growing mode, resulting in a pronounced peak of the power spectrum at $\lambda \sim \lambda_J$. Maybe the peak causes pancake collapse of the galaxy distribution on the scale $\lambda \sim \lambda_J$, which might account for the observed sheet-like features in the galaxy distribution. Maybe the peak could account for the observations mentioned in the last section that suggest that the galaxy peculiar velocity field has a coherence length larger than ~ 30 Mpc.

How would this theory deal with dark matter? One possibility is that we ignore the constraint from light element nucleosynthesis, and assume $\Omega_B \sim 1$. Another, which I tend to prefer, assumes that $\Omega_B \sim 0.1$, so that either space curvature or Λ make an appreciable contribution to the present expansion rate. This has the disadvantage of an inelegant cosmological model but the advantage of agreeing with the dynamical estimates of Ω. It is too soon to say how well either case might compare to details of the observations.

g) Summary

As I have explained in the above survey, it is my impression that no picture for galaxy formation is so well founded as to be considered the standard model, but we do have some very promising candidates and a large and growing body of observations to test the candidates and drive the discovery of new ones. This happy state of affairs is the result of a lot of work by many people. It is impressive to see how so many of the fundamental ideas trace back through Yakov Zeldovich.

AFTERWORD

In the fifteen years since I wrote the previous sections there have been great advances in cosmology, but some parts remain much the same. The cosmological tests are much tighter: we have at last a substantial empirical case for the relativistic Friedmann–Lemaître cosmology. But there has been no progress on the issue of the vacuum energy density, apart from the discovery of a possibly key but still quite enigmatic clue: the evidence for detection of a term in the stress-energy tensor that acts like Einstein's cosmological constant Λ, and arguably like the vacuum energy. Fifteen years ago we were debating the merits of some half dozen models for galaxy formation. Now one of them, ΛCDM, has passed demanding tests that demonstrate it likely is a useful approximation to what happened. But aspects of the phenomenology of galaxies that we were debating a decade ago remain unexplained, leading me to wonder whether elements of the other structure formation models we were considering in the 1980s might be of more than historical interest.

a) The cosmological model and the energy of the vacuum

In the 1980s the evidence was that the CBR spectrum significantly departs from thermal blackbody. As I mentioned, this could have meant that galaxy formation deposited appreciable energy in the CBR; Zeldovich and Sunyaev were leaders in the exploration of these processes. Or the spectrum anomaly could have signified a problem with the idea that the CBR is a thermal remnant of the hot Big Bang. The USA COBE and Canadian UBC experiments in 1990 demonstrated that the spectrum actually is very close to thermal. This dramatic result gives us our strongest piece of evidence that the universe really is expanding and cooling: how else could the CBR have acquired its distinctive thermal spectrum? The considerations of how galaxy formation could perturb the spectrum still are very relevant, of course: they tell us that the process must have been relatively gentle, not producing much hot or rapidly moving plasma that would have produced a Sunyaev–Zeldovich anomaly in the mean CBR spectrum.

Until recently the relativistic expanding world picture of Alexander Friedmann and Georges Lemaître really was a working model, based on elegant physics but little empirical evidence. I didn't discuss this point much in the 1980s, nor I think did Zeldovich, because there was not much evidence to work with. Now we have a well cross-checked set of cosmological tests that so far are in excellent agreement with the Friedmann–Lemaître predictions. This is deeply impressive. Einstein based general relativity theory on laboratory physics; his only serious checks on larger scales were the Newtonian limit and the non-Newtonian precession of the perihelion of Mercury. When Einstein proposed that the universe is homogeneous — in the large-scale average — he didn't even know about galaxies, much less how they are distributed. Yet the theory and the homogeneity assumption pass demanding tests on the scale of the Hubble length, an extrapolation of some fifteen orders of magnitude from the size of the orbit of Mercury. Nature sometimes agrees with our ideas of truth and beauty.

Nature sometimes disagrees: the cosmological tests indicate the present rate of expansion of the universe is dominated by Einstein's cosmological constant, Λ, or by a term in the stress-energy tensor that has an effect similar to Λ. This requires that we happen to flourish at a special epoch, close to the time of transition from matter-dominated to Λ-dominated expansion. The coincidence seems unlikely and unnecessary; a decade ago the general — and sensible — opinion was that Λ surely is negligibly small. With a similar argument for space curvature we are led to the Einstein–de Sitter model, in which Λ and space curvature both are negligible and the expansion rate accordingly satisfies $H_o^2 = 8\pi G\rho/3$, where ρ is the mean mass density, and H_o is the constant in Hubble's law for the recession velocity, $v = H_o r$ at distance r. The problem was (and is) that dynamical estimates indicate ρ is only about 20% of this Einstein–de Sitter prediction. I didn't think the way around this problem — biasing — made sense 15 years ago, for the reasons indicated in the previous section, and so was inclined to suspect that either Λ or space

At the USA National Academy of Science, Washington. May, 1987.

curvature make an appreciable contribution to H_o^2, despite the coincidences argument. Either case seems ugly, but that is what the measurements require.

Zeldovich was not much impressed by the coincidences argument either: to him 'the genie [Λ] has been let out of the bottle, and it is no longer easy to force it back in.' I suspect he emphasized this because he had recognized a key point: if the stress-energy tensor of the vacuum were independent of the velocity of the frame in which it is measured then in general relativity theory the gravitational effect of the vacuum would be the same as Einstein's Λ. In more detail, if the vacuum is independent of velocity the vacuum energy density ρ_Λ has to be a constant, and the pressure (the diagonal space part of the stress-energy tensor) has to be $p_\Lambda = -\rho_\Lambda$. It is a good exercise for the student to check this by working out the Lorentz transformations. The less strenuous approach is to note that the metric tensor $g_{\mu\nu}$ is the same in any inertial frame — it is the Minkowski form in Cartesian coordinates — so if the properties of the vacuum are independent of the velocity of the observer the vacuum stress-energy tensor must be proportional to the metric tensor,

$$T_{\mu\nu}(\text{vacuum}) = g_{\mu\nu}(\text{vacuum}). \qquad (2)$$

This expression in Einstein's field equation is the same as the Λ term. By comparing this form to the stress-energy tensor of an ideal fluid one sees that the effective pressure is $p_\Lambda = -\rho_\Lambda$. There were at the time hints from the cosmological tests that Λ is detected. Zeldovich asked, could this mean we have a detection of the nonzero vacuum energy density?

When Zeldovich presented these considerations, in 1968, people had already known for half a century that the sum over zero-point energies of the modes of the electromagnetic field at laboratory wavelengths is absurdly large compared with what is allowed by the relativistic cosmology. Not long after 1968 people had started to work out the phase transitions of the standard model for particle physics. Each first-order transition changes the stress-energy tensor by a term proportional to the metric tensor, as in equation (2), and with constants of proportionality that again are quite unacceptable for relativistic cosmology.

In his memoirs Sakharov relates Zeldovich's report of the reaction to his first talk on the subject: the 'theoreticians took a sharply negative view of Zeldovich's ideas, which ran counter to the established tradition of ignoring zero-point energy.' That certainly was the received wisdom for many, but Zeldovich was right. As he emphasized, we know experimentally that the zero-point energies belonging to the atoms in a crystal make real measurable contributions to the total energy. Standard theory, and now accurate experiments, say the energy of a stable system determines its active gravitational mass. And in standard theory the zero-point energies of fields are equally real: they add to the mass of the universe. That is, within standard physics it is absurd to ignore the zero-point energy of the electromagnetic field at laboratory wavelengths. Might general relativity fail on the scales of cosmology? It certainly made sense to consider this in 1968. But now general relativity theory passes demanding checks from the cosmological tests. Fine adjustments to the theory would be allowed, but we are looking at a gross mismatch between what quantum physics suggests and what is acceptable in general relativity. Might the positive and negative zero-point energies of fields and particles, added to the condensate energies of the phase transitions, happen to cancel to spectacular precision? It has been pointed out that our existence does depend on such a nice balance.

This puzzle is as opaque as were Kelvin's clouds — energy equipartition and the luminiferous ether — at the start of the twentieth century. We cannot say what the vacuum energy puzzle is going to teach us, but we can be sure people will remember Zeldovich's point: the value of the vacuum energy density — defined as the sum of condensates, zero-point energies, and whatever else uniformly fills space — is empirically constrained by the cosmological tests and may even have been detected. Surely this is an invaluable if still incomprehensible clue.

b) Galaxy formation

In the last five years the community has converged on a standard model for structure formation, ΛCDM. This was driven by the wonderfully successful prediction of the angular power spectrum of the CBR temperature anisotropy (an advance that is discussed in many recent reviews).

As noted in the last section, this new standard model assumes the dark matter is a gas of nonbaryonic noninteracting particles with a negligibly small primeval velocity dispersion, structure grew out of primeval adiabatic density fluctuations, the density fluctuations are the realization of a random Gaussian process, and all this happened in a Friedmann–Lemaître cosmology with flat space sections and present mass composition in the proportion $\Omega_B \simeq 0.05$ in baryons, $\Omega_{DM} \simeq 0.2$ in the cold nonbaryonic matter, and $\Omega_\Lambda \simeq 0.75$ in the term that acts like Λ. The composition seems strange: why do we need dark matter or Λ? But it is difficult to argue with success; the fit to the CBR anisotropy is striking.

Though the ΛCDM model certainly is successful there is evidence that we need an even better approximation to how structure formed. Some of the issues mentioned in the preceding section have been resolved. For example, lowering the mean mass density from the Einstein–de Sitter value increases the mass fluctuation coherence length, relieving the problem I emphasized with large-scale clustering. Other issues, such as the redshift of formation of the large galaxies, have not much changed. But a more detailed discussion is not needed here; the main point is that this subject is being driven by observations that will teach us whether ΛCDM really needs improvement, and if so likely give us some hints to how to fix the model. There are several particularly fruitful lines of research. Optical-infrared observations of galaxies at great distance show us the star populations in young galaxies, because of the light travel time. Gravitational lensing studies probe galaxy mass structures, though mainly at more modest distances. And radio observations of the effect of inverse Compton scattering of the CBR by the hot plasma in clusters of galaxies gives us a powerful probe of the evolution of large-scale structure back to the redshift of formation of the first of the great clusters. It is appropriate here to comment on the last of these topics.

In the early 1970s the UHURU satellite detected nearby clusters of galaxies as diffuse X-ray sources. This was interpreted to be thermal bremsstrahlung radiation from intracluster plasma with temperature characteristic of the velocity dispersion of the galaxies. This is now clearly established, from the detection of emission lines from common heavy element ions. The plasma temperature is high enough that a CBR photon scattered by an intracluster electron typically gains energy — in inverse Compton scattering — and the spectrum thus is distorted by a depression from the original thermal form at long wavelengths and an excess from the original shortward of the Wien peak. The effect was under discussion in the 1960s, in America by Weymann and me, and in the Soviet Union by Sunyaev and Zeldovich. It is properly called the Sunyaev–Zeldovich (or SZ) effect: they got there first with the most.

Fifteen years ago there were indications of a large SZ effect in the mean CBR spectrum; that has gone away. There was reasonably good evidence for detection of the SZ effect in rich clusters of galaxies; now we have clear and precise measurements of the effect. The perturbation to the CBR spectrum along the line of sight to the cluster is independent of the cluster redshift (provided the telescope resolves the cluster), so SZ surveys in preparation that scan full fields (rather than the directions of known clusters) are going to give us an excellent picture of the evolution of the great clusters. In the ΛCDM model a cluster forms at a rare extreme upward fluctuation in the primeval Gaussian mass distribution, and the higher the redshift the rarer the fluctuation that can make a cluster. This means the evolution of the number density of clusters (after taking account of the general expansion) is quite sensitive to the cosmological parameters. Equally important, to my way of thinking, is that the evolution is exquisitely sensitive to the assumption that the initial conditions are Gaussian and adiabatic. For example, if cosmic string loops played a significant — though not dominant — role in structure formation it would affect the evolution of galaxies and clusters of galaxies. It will be fascinating to see what the full-field SZ surveys teach about this. But whatever the outcome, people will remember the origin of a powerful tool for cosmology, the Sunyaev–Zeldovich effect.

Encounters with Zeldovich[160]

Malcolm S. Longair

Little did I know when I set out for a year in Moscow in September 1968 that I would meet and work with Yakov Borisovich Zeldovich. It all came about more or less by accident. I had completed my doctorate in the previous year in the Radio Astronomy Group at the Cavendish Laboratory at Cambridge, where I had studied with Martin Ryle and Peter Scheuer on the astrophysical and cosmological problems of extragalactic radio sources. In 1965, I had attended the third Texas Symposium in New York and had my first encounters with Soviet astrophysicists. At that time, because of the Cold War, scientists from the USSR were very rare visitors to the West, but on that occasion, Ginzburg and Novikov had been allowed to attend the meeting. In the following year, Ginzburg came to Cambridge and I was impressed by his enthusiasm and deep understanding of high energy astrophysical processes. Moscow was already a major centre for theoretical astrophysics and cosmology and I asked Ginzburg if he would be willing to accept me as a post-doctoral fellow at the Lebedev Institute, if I were successful in obtaining a Royal Society – USSR Academy of Sciences Exchange Fellowship to Moscow. My reason for wanting to work with Ginzburg and his colleagues was to work on the implications of the number counts of active galaxies for the origin of cosmic rays and related problems. I also knew somewhat more vaguely of the cosmological work being carried out in Moscow by cosmologists such as Zeldovich, Novikov, and their colleagues. It was an exciting prospect to find out exactly what was going on astrophysically and cosmologically at first hand in Moscow. My application to the Royal Society was

[160] This article was not included in the original Russian version of this book, published in 1993. [Translator]

successful, and I spent eight months in the language laboratory at Cambridge learning enough Russian to be able to talk science on their terms rather than mine. This was not exactly the best time to visit the Soviet Union — the invasion of Czechoslovakia had taken place only a couple of months earlier.

The first port of call was attendance at the 5th International Conference on Gravitation and the Theory of Relativity held in Tbilisi, Georgia. This was my first encounter with many of the great figures of Soviet science. The most startling encounters were with Zeldovich and his young colleague Rashid Sunyaev.

As Rashid afterwards told me, the Soviet scientists were familiar with my papers in *Monthly Notices of the Royal Astronomical Society* and had expected this person Longair to be a bearded ancient, instead of which they encountered a young Scot, who even then had very limited understanding of astrophysics and cosmology. There was not much time to talk at the conference, but once I returned to Moscow, I very quickly received calls from Rashid saying that Zeldovich wanted to talk to me about the cosmological evolution of radio sources. At the same time, Rashid was very keen to discuss other astrophysical problems on which he was working.

It may strike the reader as strange, but I had very little knowledge of who Zeldovich actually was. Attendance at the weekly seminars, which he organised at the Sternberg Astronomical Institute, quickly revealed his enormous authority in all physics, cosmology, and astrophysics. It was only much later that I became fully aware of his role in the development of the Soviet atomic and hydrogen bombs during the 1940s and 1950s. There was, of course, no way in which I could have found this out in Moscow at that time.

Zeldovich wanted Andrei Doroshkevich and me to work out some different models for the cosmological evolution of the radio source population. The cosmological calculations which I had carried out in Cambridge had involved using the EDSAC computer in Cambridge, the most powerful computer of its type at that time. Zeldovich was very keen that I repeat these calculations in Moscow, but there was no way in which they could be carried out without using a computer. Zeldovich was working at the Institute of Applied Mathematics, an Institute completely closed to foreigners. Indeed, I never found out exactly where it was, since it was hinted to me that it would not be a good idea for a foreigner to be seen in its vicinity. Again, it was one of these pieces of luck that Zeldovich was able to arrange the use of a Soviet computer to carry out these calculations. The way we worked was that I would write out in detail the computations we needed to carry out and Andrei programmed a computer at the Institute of Applied Mathematics. Every week we met to see how the calculations were going and discuss the next things to be done.

Once we were well through the programme of computations, Zeldovich summoned me to his flat on Vorobiev Road. He told me to come for breakfast at seven a.m. and then we would discuss the results. This happened on several occasions. Only afterwards did my Moscow friends tell me that

Zeldovich habitually started work very early at home, and it was a common event for them to be summoned to his flat whenever he needed to discuss science with them at essentially any time of the day or night. These were memorable breakfasts. After we had eaten, Zeldovich suggested we do exercises. These were somewhat vigorous, the most potentially dangerous being when we stood back-to-back, linked arms and then flung each other up in the air in turn. Zeldovich was very short and I was fearful of doing irreversible damage to one of the great scientists of the twentieth century. It was during these breakfast discussions that his very deep understanding of cosmology and fundamental physics became apparent. As important as anything for me was his approach to mathematical physics, in which he knew instinctively how to reduce apparently difficult calculations into manageable proportions.

This work was carried out towards the beginning of my year in Moscow and it was an enormous and totally unexpected bonus. It was of inestimable value to me that my Soviet colleagues already knew my work and were keen to collaborate. In fact, although I did work in Ginzburg's group on the origin of cosmic rays, I found that most of my collaborative work was carried out with Zeldovich and his colleagues, particularly with Rashid. The work Zeldovich, Doroshkevich, and I completed on interpreting the radio source counts was the beginning of a number of new projects which Rashid and I began. These included the first estimates of the fluctuations in the cosmic microwave background radiation due to discrete radio sources in the centimetre and millimetre wavebands. We also worked on a model for the origin of the soft and hard X-ray backgrounds involving the escape of high energy particles from intense infrared sources. It turned out that these pieces of work were far ahead of their time, but the spirit of Zeldovich and his group was to follow through promising theoretical ideas to their natural conclusions, even if the ideas could not be immediately confronted with observation.

Rashid and I worked very intensively on these problems throughout the year, either at the Sternberg Institute or in his room in the flat he shared with another family. It was a long time before Rashid was allowed to travel to countries outside the USSR, and we suspected that our scientific collaboration, which resulted in five joint publications, may have caused the authorities some concern.

Zeldovich remained a guiding spirit and mentor of our joint work.

The atmosphere at the various seminars was quite distinctive. Ginzburg held his weekly seminar in the Lebedev Institute and the range of topics was vast. Ginzburg claimed that all physics, except elementary particle physics, could be included in his seminar programme. I had the opportunity of describing the work I was doing on the origin of cosmic rays at one of these seminars. Zeldovich and Shklovsky held their seminars in alternating weeks at the Sternberg Institute, which was remarkably open and at which I carried out a great deal of my work. The Zeldovich seminars were spectacular. Official copies of the *Astrophysical Journal* and *Monthly Notices of the Royal*

Astronomical Society normally arrived very late, but a few private copies of the journals were sent to distinguished astronomers and these were pored over as soon as they arrived. The period 1968 to 1969 was one of remarkable discoveries in astronomy and cosmology and these were reported as soon as possible at the Sternberg seminars. These were occasions at which vigorous debates were encouraged. Zeldovich would display extraordinary virtuosity in dealing with complex problems and in demolishing any incompetent astronomer or astrophysicist. The atmosphere was one of great intellectual excitement. Regularly distinguished scientists such as Pikelner and Kardashev, as well as visitors from all over the USSR, would take part in vigorous astronomical debates. There were often head-on collisions, frequently involving Shklovsky and the rest. It was an invigorating atmosphere.

Zeldovich was a dominant figure and rightly held in the highest regard. It was not for nothing that he boasted of having three Gold Stars to his Orders of Lenin, while Brezhnev only had two. He was personally very kind to me, on occasion inviting me to be his guest at the Academician's dining room. I vividly remember him pointing to an old figure in a corner and whispering to me, 'That is Trofim Lysenko,' the biologist who put back the development of research in biology in the Soviet Union by at least a generation.

For me, the year in Moscow was very important indeed, scientifically and intellectually. Through working with Zeldovich and Rashid in particular, I not only learned a vast amount of astrophysics, but I also became thoroughly familiar with the work of the Soviet scientists. Much of this work was little known outside the USSR and in a number of areas ahead of what was being done in the West. I found myself acting as an ambassador for making their work better known in the West. At the General Assembly of the International Astronomical Union (IAU) held in Brighton in 1970, I reviewed the work of Zeldovich and his colleagues and in the following year, at a cosmology meeting in Oxford, I gave probably the first detailed exposition of the work of Zeldovich and Sunyaev on the Kompaneets equation and induced Compton scattering. At the Brighton meeting of the IAU, Commission No. 47 was founded, and Zeldovich was elected its first President. I was elected Vice-president, largely because of my good contacts with Zeldovich. Among the outcomes of the foundation of the new Commission was the first IAU Symposium on Cosmology entitled 'Confrontation of Cosmological Theories with Observational Data,' held in Krakow in Poland in 1973.

Because of the difficulty of communications with the Soviet Union, Zeldovich asked me to undertake the organisation of the invitations to the meeting and the organisation of the Scientific Programme, as well as dealing with all the problems of ensuring that scientists of all participating nations could travel to Poland. Konrad Rudnicki did an outstanding job in chairing the Local Organising Committee. The meeting was a great success and many of the Western scientists had their first opportunity of meeting Zeldovich and his colleagues. The list of participants was remarkable, including encounters between the Soviet and Western relativists, including Misner,

Penrose, Hawking, Belinskii, Khalatnikov, and Lifshitz. Brandon Carter delivered his seminal paper on the Anthropic Cosmological Principle. Zeldovich was on top form. I recall his splendid opening address in which he reminded us of two things. First, the international language of science is 'specific broken English.' Second, when speaking, we should remember Galsworthy's remark, 'A platitude must be stated with force and clarity.'

In 1975, I made the first of a number of trips back to Moscow and the USSR. By that time, the astrophysicists with whom Zeldovich worked had moved to the Institute of Space Research and Rashid and I were able to continue our collaboration at the Institute. I was able to travel about once a year to the USSR to carry on our collaborative work. On one occasion, Rashid discovered that there is a symposium on Plasma Astrophysics in Irkutsk in Siberia, at which I was asked to give an invited presentation. This was another opportunity to meet with Zeldovich and Rashid. One of the memorable events of that meeting occurred when Zeldovich was invited to a Siberian breakfast by A.A. Fridman, the conference organiser. Zeldovich asked if Rashid and I could come as well, and so we were driven early in the morning to Lake Baikal where Zeldovich and Fridman had their early morning swims. We saw fishermen nearby who disappeared with their catches while Fridman organised a splendid breakfast of vodka, caviar, and other delicacies. Then, we all left the lake and found the fishermen preparing a huge fish soup in a very large pot over an open fire. Potatoes and a variety of vegetables and herbs were added to the brew. It was delicious. Only years later did I learn that the fish were omul, a species only found in Lake Baikal and regarded as a supreme delicacy. Even then, it was a protected species, but not for Zeldovich and his party! At the end of the meal, Fridman whispered to me, 'The last time we did this, it was for Fidel Castro!'

Still Zeldovich was not allowed to leave the USSR, and it remained very important to organise IAU meetings within the Eastern bloc, to which astronomers from all countries could be invited. Jaan Einasto agreed to host a major IAU Symposium on the large-scale structure of the universe in September 1977. I was asked to be Chairman of the Scientific Organising Committee, and to edit the publication of the proceedings of the meeting jointly with Einasto. This was another successful meeting, largely thanks to the huge efforts of Einasto. It was the first meeting at which the importance of structure on the very largest scale in the universe was fully appreciated. Zeldovich was again on spectacular form.

At last, at the 1982 IAU meeting at Patras in Greece, Zeldovich was allowed to attend a meeting outside the USSR — he was then 68. His energy and imagination were undimmed and he presented one of the invited discourses entitled 'Remarks on the Structure of the Universe.' He reviewed all his work and that of his colleagues in Moscow in a survey of deep insight and vision.

I owe Zeldovich much more than can be expressed adequately in a few words. My contacts with him and his colleagues in Moscow revealed to me completely new ways of thinking about science. His clarity of thought made even the most difficult problems tractable. His thinking was always based upon a very sure physical intuition, formed by complete absorption in physics and theoretical physics for a lifetime. To this was added his instinctive abilities as an inspired teacher, who could look at even the simplest problem in an enlightening way. His example and inspiration have remained with me all these years.

A man of unlimited energy

Keith Moffat

I met Zeldovich for the first time in 1975 in Poland during one of the biennial conferences on fluid mechanics. That year, it took place in Bielovezha, very close to the border with the USSR. At that time, such conferences provided an important opportunity for the meeting of scientists from East and West. This conference was no exception. In the West the name of Zeldovich was already legendary. It was well known that he had an unusually broad field of scientific interests. In particular, I knew that he had worked on dynamo theory and had proved a famous antidynamo theorem in 1956. I had also studied his more recent review paper written together with Samuel Vainstein (*Soviet Physics Uspekhi* 1972).

I remember discussing with him, during that conference, the problem of the 'fast' dynamo as a sequence of stretching, twisting, and folding operations which leads to doubling of the strength of the magnetic field. I asked whether he knew how to prove that many repetitions of such a sequence would in fact act as a fast dynamo. Showing with his hands the stretch-twist-fold sequence, he asked, don't you think this is sufficient proof?

Over the last eight years, a lot has been written about the mechanism of the fast dynamo. Nevertheless we are still far from the mathematical proof of the result which Zeldovich considered absolutely obvious. At the same time a majority of people have little doubt that his intuition in this case (as in so many others) gives the correct result.

Zeldovich was a man of unlimited energy, irrepressible curiosity, and contagious enthusiasm in science. His famous seminars in Moscow University gave brilliant expression to this enthusiasm. In 1983 I gave a seminar in this programme on the theory of the turbulent dynamo and got a lively response from the audience, strongly intensified because Zeldovich himself simultaneously played the role of interpreter and referee. He summarised each portion of my lecture in Russian and added his own inimitable commentaries. During discussions with Zeldovich, he would penetrate immediately to the core of most difficult problems to make them more understandable and thus to find new routes to their solution. His arguments were physical but his great physical intuition would invariably lead him to the appropriate mathematics.

Yakov Borisovich Zeldovich was elected to be an Honorary Doctor of Science of Cambridge University. Unfortunately, political circumstances were such that he never had the opportunity to come to Cambridge to receive this distinction.

Yakov Borisovich Zeldovich (March 8, 1914 – December 2, 1987 Elected For. Mem. R.S. 1979)[161,162]

V.L. Ginzburg

Yakov Borisovich Zeldovich was an exceedingly bright star in the firmament of Soviet and world physics and astrophysics. To these fields can be added physical chemistry or, as is sometimes said, chemical physics. More specifically, mention may be made of researches in catalysis, phase transitions, hydrodynamics, combustion and detonation theory, nuclear chain reactions, nuclear physics, the theory of elementary particles and, finally, the general theory of relativity and cosmology.

To mark YaB's 70th birthday, two volumes of his selected works were published under the titles: *Vol. I Chemical Physics and Hydrodynamics* and *Vol. II Particles, Nuclei and the Universe*. These volumes contain a complete list of YaB's works. As far as I know, both volumes are to be published in English in 1989 by Princeton University Press.[163] The second volume concludes with an autobiographical afterword dated March 3, 1984. Unfortunately, YaB died less than four years later and, although he continued to work with his former energy, these volumes can and will provide the basis for this obituary.

Access to all YaB's existing works in English translation, accompanied by expert and often highly detailed commentaries by leading specialists, together with a long editorial article of 56 pages introducing Volume I, provide abundant detail and references to many of YaB's articles.

YaB's life can to a first approximation be divided into four periods: 1914–1930, childhood and schooldays; 1931–1947, the Institute of Chemical Physics, mainly work on physico-chemical problems; 1947–1963, work on the creation of a new technology, nuclear physics, and the theory of elementary

---

[161] Reprinted with permission from The Royal Society (Memoir of Ya.B. Zeldovich, written by Vitaly Ginzburg, *Biog. Mem. Roy. Soc. Lond.*, vol. 40, pp. 429–441 (1994)). © 1994 The Royal Society

[162] This article was not included in the original Russian version of this book, published in 1993. [Translator]

[163] The English edition was published in 1992/1993: Vol. I *Chemical Physics and Hydrodynamics*, Princeton University Press, 1992; Vol. II *Particles, Nuclei and the Universe*, Princeton University Press, 1993. Editors: J. Ostriker, G. Barenblatt and R. Sunyaev.

particles; 1964–1987, mainly astronomy, with emphasis on the application of the general theory of relativity and on cosmology. This chronology, taken from Volumes I and II of his *Selected Works*, will also be used here.

CHILDHOOD AND SCHOOLDAYS (1914–1930). SOME LANDMARKS ON LIFE'S PATH

Yakov Borisovich Zeldovich was born on March 8, 1914 in Minsk (Belorussia) in his grandfather's house, but in mid-1914 his family moved to St. Petersburg (Petrograd, later Leningrad), His father was a lawyer and his mother a translator of literature (from French into Russian). In 1930 YaB completed his school education and enrolled on a course for laboratory assistants at the Institute for Mechanical Processing of Useful Minerals (*Mekhanobr*), where he examined, and possibly prepared, slices of mined rocks. In his autobiographical notes, YaB recalls a conversation with his father, who was discussing the choice of subjects with his 12-year-old son. It was generally accepted (I believe with justification) that mathematics required a certain exceptional or, at any rate, specific talent. At that time a school physics teacher solemnly read Newton's Laws first in Latin and only afterwards in Russian. In one way or another, the school physics course generated no enthusiasm; there were few popular physics texts and no suitable surroundings. It was therefore rather by chance that YaB's first inclination was for chemistry or, more precisely, physical chemistry. It was also a matter of chance that this interest should be strengthened and soon brought to fruition.

In March 1931 YaB together with colleagues from *Mekhanobr*, visited the Department of Chemical Physics of the Leningrad Physical Technical Institute (LPTI),[164] which was involved in the crystallization of nitroglycerine in two modifications. The young YaB's scientific interest was beyond his years and attracted the attention of his seniors. After completing his laboratory assistants' course, he was given the chance to work in the laboratory at LPTI led by S.Z. Roginskii. (The exact date of starting this work is known — March 15, 1931.) Soon afterwards YaB presented a report on ortho-para-hydrogen transformations at the seminar of the LPTI Chemical Physics Section (led by N.N. Semenov). Because of the evident fervour and understanding of the 17-year-old YaB, LPTI wanted to employ him and, after overcoming bureaucratic difficulties, this was achieved on May 15, 1931. YaB joined the staff of the independent Institute of Chemical Physics of the USSR Academy of Sciences, which had just split off from LPTI. (The Institute still bears this name and after the war was based in Moscow.)

YaB at once immersed himself in research. To enter university and give up this work for 4 to 5 years seemed pointless and was probably impossible for an active young man already well educated and full of energy. YaB did

[164] Now known as the Ioffe Physical Technical Institute, one of the leading physics institutes in Russia.

not, therefore, receive an official higher education. Perhaps, even given exceptional talent, such a course today would present difficulties. But at that time particularly, drastic changes were taking place in the USSR in both secondary and higher education, and diplomas were not given great attention. In one way or another, YaB was able to embark on a postgraduate career and he defended his thesis for the Candidate of Sciences Degree (roughly equivalent to a Ph.D.) in 1936. Three years later (1939) he defended his Doctor's Dissertation (equivalent to D.Sc. and generally of great importance in the USSR where it gives the right to hold a Chair). But recognition for YaB was certain and was in no way dependent on defending dissertations. By 1939 the 25-year-old YaB was the author of a number of papers on chemical physics (the difference between this and physical chemistry I have never been able to fathom). Moreover, through his exceptional talent and energy YaB had already become almost legendary. YaB himself relates some of these legends in the autobiographical notes in his *Selected Works*. I recall one story from the post-war period. When N.N. Semenov, Director of the Institute of Chemical Physics, was told of the difficulties encountered in creating a theory of nuclear forces, his reported reaction was to say 'Well, then, let's give it to Ya.B. Zeldovich and he will have it all solved in a couple of months.'

It is worth noting that during his early years YaB worked both as an experimentalist and as a theoretician. His connection with experiment always remained with him. Although the centre of gravity of his interests used to change, his earlier work was not forgotten. According to V.I. Goldanskii, literally on the day before his death, YaB, having found a new survey on combustion chemistry, expressed his great interest and intention to read it.

The discovery in 1939 of the nuclear fission of uranium immediately attracted the attention of YaB and Yu.B. Khariton as a possible way of obtaining a chain reaction — a problem close to their interests in the theory of chemical reactions. It was therefore natural that Khariton and Zeldovich should be among the first scientists called together by I.V. Kurchatov to solve the 'uranium problem' and its important applications. For many years YaB worked mainly outside Moscow, although he also concerned himself with a number of purely scientific questions. It was not until 1964 that he officially joined the Institute of Applied Mathematics of the USSR Academy of Sciences, founded and led by M.V. Keldysh. Later, in 1983, he transferred to the Institute of Physical Problems of the USSR Academy of Sciences where he led the theoretical department that had been founded by L.D. Landau.

YaB's works were widely recognized both in the USSR and abroad. In 1946 he was elected Corresponding Member, and in 1958 Full Member (Academician) of the USSR Academy of Sciences. YaB was three times awarded the title of Hero of Socialist Labour, he was a Lenin Prize winner and was four times a winner of the State Prize of the USSR. He received a number of other prizes and medals. He was elected a Foreign Member of the Royal Society of London, the US National Academy of Sciences and a number of other academies and societies.

It is time, finally, to turn to YaB's scientific contributions. It should be mentioned once again that these are described in detail in his *Selected Works* and here it is only possible (and, I feel, appropriate) to characterize briefly the main directions of his work.

CHEMICAL PHYSICS AND HYDRODYNAMICS. FIRST CONTRIBUTIONS IN THE FIELD OF NUCLEAR CHAIN REACTIONS (PRINCIPALLY 1931–1947)

The direction of YaB's early works was catalysis and adsorption (the first three papers in *Selected Works* Volume I are on these themes and his Candidate's Thesis was devoted to adsorption).[165] The main study concerned the creation of a theory of adsorption isotherms (i.e., the dependence of the quantity of an adsorbed substance on the gas pressure or on the concentration of the adsorbed substance in a solution), taking account of the inhomogeneity of the adsorbent surface.

An analysis of the flow of chemical reactions (in particular, catalytic) and of combustion led YaB to hydrodynamics, heat transfer, and turbulence. His first articles on these themes appeared in 1937 and concerned asymptotic laws of heat transfer at small velocities of a liquid and self-similarity laws of freely ascending convective flows.

It was characteristic of YaB not to overlook past experience, even when moving to completely new problems. Thus, when deeply immersed in astrophysics, he occupied himself with magnetohydrodynamics and its applications in analysing the generation of a magnetic field in a moving conducting fluid (1956, 1972, 1979, 1980 and so on). Since the chronological character of this account has already been violated (and how could it be otherwise when YaB frequently returned to problems that had excited him in the distant past?), I refer here to the solution of the problem of gas motion under the action of short-duration pressure (impulse) published in 1956 and to the theory of new phase formation published in 1942. The latter work is fundamental to the kinetics of phase transitions of the first kind (it considers the fluctuational formation and subsequent growth of vapour bubbles in a fluid at negative pressures). The theory of shock waves, and especially combustion and detonation theory, take up a very large part of YaB's legacy. Fields included here (which are difficult to separate) are ignition, the spread of flame, the combustion of powder, and nitrogen oxidation. In addition to the 15 papers on these themes included in *Selected Works* Volume I, there are several monographs.

The kinetics of chemical reactions (in particular, the theory of chain reactions), the propagation of flames as waves of combustion, the influence of different factors (the role of media boundaries, temperature, etc.) comprise an enormous field, especially if different branches and applications are taken into account. Despite my sincere wish to explain the results obtained by YaB in this field, it is clear that I cannot accomplish this in a few

[165] Zeldovich's first papers were published in 1932 when he was 18 years old.

pages. One needs, if not a monograph, a whole survey article. I must, therefore, reluctantly, limit myself to the enumeration and statement of certain assertions. On the basis of what is known to me, YaB's contribution to the theory of combustion and detonation exceeds that of any other person. But if this assessment, like many similar ones, is unavoidably subjective, there can be no doubt that all this research guaranteed YaB's success in his work on the theory of the combustion of powder (1941–1942) and on the chain fission of uranium (1939 and later). An understanding of the features of the combustion of powders served as a basis for creating the internal ballistics of solid-fuel rockets (research carried out during the war was orientated towards the 'Katyusha' rocket weapon).

The discovery of the nuclear fission of uranium (1938–1939) immediately attracted the attention of Yu.B. Khariton and Zeldovich. As YaB himself wrote: 'The discovery of uranium fission and the possibility in principle of a chain fission reaction predetermined the fate of the century — and mine as well.' The first paper by YaB (written jointly with Yu.B. Khariton) on this subject was submitted to the editors of the *Zhurnal eksperimentalnoi i teoreticheskoi fiziki* (*Journal of Experimental and Theoretical Physics*) on October 7, 1939, the second on October 22, 1939, and the third on March 7, 1940. Khariton and YaB even published two surveys devoted to splitting the nucleus and chain reaction in uranium. It is typical, and needs no further commentary, that the second part of the second of these surveys was published only in 1983. Papers of the type referred to were, as far as we know, the only publications in the world on this theme before the 1955 Geneva Conference. It is very curious that the work on uranium fission by Khariton and YaB was considered 'outside the plan' and that they worked on it in the evenings, sometimes until late.

As already noted, YaB's work on uranium fission and chain reactions in uranium, and also the theories of detonation and shock waves, predetermined YaB's attraction to the uranium problem. Unfortunately, all we know officially about this work is that he received awards (see earlier). After the lapse of about 40 years all this activity is still considered 'closed.' I would like to think that the radical changes now taking place in the USSR, in particular those connected with publicity, will bring an end to this anecdotal situation. Unfortunately, apart from what has been said, I cannot at the present time report anything concrete about YaB's work in the field of developing the new technology (as it is officially called).

NUCLEAR PHYSICS AND THE THEORY OF ELEMENTARY PARTICLES (1947–1963)

Research into uranium fission and related questions naturally drew attention to nuclear physics and the theory of elementary particles, or, as it is usually referred to today, high-energy physics. (Whether this terminology is more appropriate here is difficult to say; most particles studied are not elementary, but neither is research into all these particles and their interactions the monopoly of high-energy physics.)

In the field of nuclear physics, apart from work directly connected with nuclear fission, we recall the method of containment ('storage') of very slow-cold neutrons, suggested by YaB in 1959. It is a matter of total internal reflection of neutrons of condensed media (say, graphite blocks forming a closed cavity). YaB's idea is used today, particularly in attempts to measure the electrical moment of neutrons. In 1960 YaB examined the possibility of the existence of relatively long-living nuclei with a large isotopic spin, and of limits of stability of light nuclei relative to the nuclear emission. The possibility of observing an isotope of ^{8}He was demonstrated. (Soon after, this isotope was in fact discovered.) This research, and also the hypothesis on the possible existence of a nucleus (isomer) with a quantum vortex along the axis of the nucleus, can be found in *Selected Works* Volume II. In 1952 and 1953 Zeldovich discussed laws of conservation of baryon and lepton charges. As is clear from the commentaries to these papers, similar results were obtained at about the same time and, of course, independently by other authors. But this does not alter our conclusion that, having got involved in elementary particle physics, YaB typically succeeded in identifying straight away the pivotal problems and found himself at the centre of events. In *Selected Works* Volume II there are 16 papers devoted to the theory of elementary particles and related problems (in the sections headed 'Theory of elementary particles' and 'Atomic physics and radiation'). References list 76 papers on these subjects (some, of course, with co-authors). When the complete works of Ya.B. Zeldovich are published, they will amount to many volumes. Suffice it to say that the selected papers in Volume I have references to 19 monographs (including two dissertations) and 156 publications in journals and collections of articles, etc.; in Volume II there are references to another 14 monographs and textbooks and a further 296 papers.

All this material, of course, contains papers that are closely related in content and subject. More characteristic is the diversity of questions discussed. Undoubtedly this reflects in some measure YaB's style. Having become interested in a problem, or hearing something interesting, YaB would often make his own contribution, sometimes write a short note, leave his mark and, without stopping, move on. This kind of activity sometimes arouses criticism in scientific circles. Prolific authors are often accused of sins such as, for example, trying to maximize the number of their publications. It goes without saying that such a reproach addressed to YaB, especially in his last decade, would be simply ludicrous. By publishing short notes he could only fuel such criticism and not increase his fame. I never, unfortunately, discussed this matter with YaB, but I am convinced that he acted in accordance with a very simple motive — if something was of interest to him, even a tiny remark, he wanted to share it with others who he considered would also find it of interest. I write about this in particular because, without in any way comparing myself with YaB, in similar circumstances I would approach the matter of publication in the same way. I am convinced that requirements to publish, the ease or difficulty of deciding whether or not to publish — this is all a question of style or, if you like, the form of scientific activity. It stands to reason, that one's contribution to science is determined not by numbers of papers but by their quality and content.

I have made this digression because, once again, I have felt that I cannot simply enumerate here all YaB's results. In regard to elementary particle physics I shall confine myself to papers devoted to weak interactions (the theory of β-decay). Thus, in 1955 S.S. Gershtein and YaB for the first time formulated the important idea that a weak charged vector hadron current should be conserved, as a result of which the effective vector constant in the β-decay of a neutron does not change under the effect of virtual strong interactions. Later (in 1958) R. Feynman and M. Gell-Mann arrived at this important result in their theory of a universal weak V–A interaction. (Feynman and Gell-Mann were at first unaware of the paper by Gershtein and Zeldovich but Gell-Mann later referred to it.) In 1959 YaB expounded a very important pioneering hypothesis on the existence of neutral currents violating the conservation of spatial parity. It was also shown in this paper that the violation of parity with weak interactions must lead to the rotation of a polarization plane of light in a substance not containing optically active molecules (in the usual sense). As we know, this elegant effect was later observed. Also worthy of mention is YaB's remark of 1957 which, evidently for the first time, referred to the possibility of the existence of, as he termed it, an anapole moment. It is now customary to call this moment toroidal and in classical electrodynamics the concept is very simple. Imagine a solenoid with a constant current flowing along it and give the solenoid the shape of a torus. Then, if the solenoid is uncharged it will not have any electrical multiple moment. Further, if the azimuthal current in the torus-solenoid is zero (this is achievable with a double winding), the dipole magnetic moment will be absent. But inside the solenoid the magnetic field differs from zero and this field will interact with the current piercing the solenoid. (For this YaB suggested placing the torus in, for example, an electrolyte through which a current is passed.) Such a torus also has a toroidal moment. In the case of a small torus it will be a toroidal dipole moment.

ASTROPHYSICS, GENERAL THEORY OF RELATIVITY, COSMOLOGY (1964–1987)

Astronomy has always been closely connected with physics. At different times this connection has taken various forms. The transformation from optical to all-wave astronomy, the concept of neutrino astronomy and gravitational wave astronomy — all this is essentially the outcome of the post-war period, the second half of the twentieth century. There is often talk of a second astronomical revolution (the first being associated with the names of Copernicus and Galileo). It would be superfluous here to discuss the basis of this terminology and, more particularly, the content of the truly remarkable changes taking place in astronomy (including, of course, astrophysics and cosmology) resulting from the discovery of quasars, pulsars, relic (thermal) microwave radiation, 'X-ray stars,' cosmic masers, etc.

It is important that, at the beginning of the 1960s, Ya.B. Zeldovich, with his keen reaction to what was new and important, evidently felt he was ideally prepared for work in the field of the new astronomy. In fact, YaB's experience of research in hydrodynamics and explosions, the theory of elementary particles and much else was clearly useful. He made a decisive choice to give fundamental attention in the last quarter century of his life to astronomy. In attempting to characterize his research in these fields, we again encounter the oft-mentioned difficulty: there is so much material that this memoir cannot possibly embrace it all. An impression may be gained, albeit superficial, from the content of sections in the second part of the *Selected Works* Volume II under the general heading 'Astrophysics and cosmology,' where the bibliography lists 17 surveys. There are sections on elementary particles and cosmology (25 papers); general theory of relativity and astrophysics (38 papers); neutron stars and black holes, accretion (30 papers); interaction of matter and radiation in the universe (16 papers); formation of the large-scale structure of the universe (44 papers); observational effects in cosmology (15 papers). It is clear that we can dwell only on a few results and particular examples.

In 1966 YaB and S.S. Gershtein considered the question of limits on the rest mass of a neutrino on the basis of cosmological considerations (this work in concept was preceded by a paper by YaB and Ya.A. Smorodinskii in 1961). The limit obtained for the mass of a muonic neutrino, $m_0(\nu_\mu) < 400$ eV, was several orders of magnitude below the limit resulting from laboratory data. The same limit was obtained for the electron neutrino, which was at that time somewhat higher than the laboratory value. It was important that cosmological considerations imposed an upper limit on the sum of the rest masses of neutrinos of all possible kinds and, especially, on the sum of masses of weakly interacting particles of all possible types. This important concept is now an organic part of physics and astronomy.

In 1967 YaB turned his attention to estimates of the cosmological constant Λ, first introduced by Einstein in 1917 in the general theory of relativity. As we know, the term Λ now plays an exceptionally important role in cosmology (and is applied to the earliest inflationary stages of evolution of cosmological models). YaB already realized (this was not widely known) that the introduction of Λ was equivalent to assuming the existence of vacuum energy with density

$$\varepsilon_0 = \frac{\Lambda_c^4}{8\pi G}$$

and negative pressure $p_0 = -\varepsilon_0$. In this work YaB pointed out that estimates of the magnitude of the constant Λ resulting from the theory of elementary particles exceeded the value of Λ obtained from observations by many orders of magnitude. (As far as I know, this had been done earlier, but YaB was not aware of it.) The problem of Λ and the reasons for its smallness still remain a focus of attention.

In connection with this paper I venture to make the following remark. After A.A. Friedmann in 1922 and 1924 had discovered non-stationary solutions of the equations of the general theory of relativity for isotropic and homogeneous cosmological models, the term Λ fell into disfavour. Einstein himself declared that his introduction of Λ was almost a mistake. Pauli, Landau and probably many others spoke out against the possibility or, more precisely, the relevance of introducing Λ. True, all the arguments on this question known to me can be said to be aesthetic in nature — the introduction of Λ into Einstein's gravitational equations was not obligatory and seemed an unnecessary complication. One can understand such a position. If the analysis of fundamental physical problems in fields where there is very little or no experimental data is not governed by principles of simplicity and a minimum of propositions it is extremely difficult to move forward. But for me personally the introduction of Λ into the equations of the general theory of relativity always seemed so natural and essentially unambiguous (there is no place here to amplify this remark) that I can be said to have been a 'supporter' of Λ. I remember how we argued this point with YaB knowing that he was sometimes rather aggressive in his rejection of some hypotheses (but this was more his style, in the spirit of Landau). But when the first indication appeared that Λ had a real, albeit hypothetical, role (it was a question of observations of a red shift of the absorption lines in the quasar spectrum), YaB's instantly applied himself to the question. This showed his flexibility, lack of prejudice and ability to change his opinions rapidly when faced by facts or even by a hint of new facts from preliminary observational data. It is curious that such a trait — a rapid change of opinion — is frequently held against a person. Sometimes, of course, such reproaches are deserved, but this does not, on the whole, apply to physics nor, I believe, to science in general.

In his afterword in Volume II of his *Selected Works* YaB himself states that in the field of astrophysics his most important individual work was the non-linear theory of formation of the structure of the universe or, as it is now called, the 'pancake theory.' In fact, YaB devoted a number of papers to this great problem, some jointly with colleagues; YaB's first paper on this subject was published in 1970.

A few words should be added here about other important directions of YaB's work in the field of astronomy. He paid great attention to the diffusion of radiation in a hot interstellar and intergalactic gas. One of the 'products' of this research is the Zeldovich–Sunyaev effect (1970 and later) which consists of a temperature reduction of the relic radio-frequency radiation passing through a gas in clusters of galaxies. Further, one cannot overlook YaB's elegant remark made in 1962 concerning the possibility of transforming a body of any mass, given sufficient compression, into a black hole. Such an assertion now seems obvious, but I clearly remember that YaB's conclusion was for many of us at the time unexpected. Other important, and in practice incomparably more important, remarks and results by YaB are included in his papers, some jointly written with I.D. Novikov, O.Kh. Guseinov, and others, on the accretion of matter on white dwarfs, neutron stars, and black holes. The importance of this group of questions for X-ray astronomy, especially in the case of double stars, is now generally known. It is also impossible to overlook YaB's joint paper with O.Kh. Guseinov in 1966, in which it is proposed (as far as I know for the first time) to detect collapsed stars, particularly black holes, by observations on spectrally double stars.

Finally, the last of YaB's directions on which we shall dwell is the cosmology of the early universe, as it is now sometimes called, bearing in mind the early phases of evolution in cosmological models with singularities. At the beginning of his work on cosmology, before relic radiation (with temperature $T = 3$ K) was discovered in 1965, YaB made an erroneous conclusion that hot models of the universe contradicted observations. In 1962 he suggested a 'cold' model variant. It is characteristic that this paper, obviously with YaB's agreement, is included in his selected works and that in the afterword, YaB self-critically speaks of this and of certain other errors, stressing, however, that he 'did not insist on his mistakes.' The significance of this has already been noted. Immediately after the discovery of relic radiation YaB not only 'acknowledged' the hot model but, more importantly, opened up a broad front of research into its development and verification. The beginning of this can be said to be his programme paper published in 1966. YaB and his co-workers, and sometimes his co-workers alone, did much in the ensuing years in this direction — I have in mind different aspects of the hot model considered not too close to a singularity after, say, some hundredths of a second and subsequently. But long ago it was clear that special mystery attached to the singularity itself and its 'neighbourhoods,' in particular the 'nearest' neighbourhoods, when the general theory of relativity, as the classical theory, failed and it was necessary to turn to the quantum region of gravitation and to quantum cosmology respectively. It is natural that the quantum region, and also the later inflation region, should attract particularly close attention from YaB in the final years of his life. He writes about this at the end of his afterword, and I know of two articles published posthumously on the subject. It would be inappropriate for me to attempt here to reflect YaB's views on the contemporary state of cosmology of the early universe — everything that he wished to say on the subject is contained in easily accessible articles written only a few months before his death. Moreover, as far as I can judge, it is a question not of completely new ideas from YaB but of his positive assessment of the direction that cosmology is now taking (inflation, the role of the scalar field, multidimensional generalizations, connection with high-energy physics, etc.). As already stated, YaB worked to the very end of his life and was always enthused, in the first place by cosmology. Besides, his interests were always very broad and his very last thoughts on science we shall never know. But perhaps this is all to the good, since the life of such a person as Yakov Borisovich Zeldovich does not end with his last thought but continues in his works and in the works of those who follow him.

CONCLUDING REMARKS

Ya.B. Zeldovich can most accurately be described as a theoretical physicist of broad, one might say universal, profile. If we speak of the recent past, R. Feynman, L.D. Landau, and a few others also belong to this category, but their numbers are decreasing. The evident cause of this is the colossal expansion of the forefront of physics. As a rule, experimentalists have long lost the chance to work simultaneously, or even consecutively, in different fields of physics. The unity of methods and forms of theoretical physics makes the situation of theoretical physicists in this respect more favourable. But the breadth of the range of investigations accomplished by YaB is unusual even for versatile theoreticians.

YaB's connection with experimental work has already been noted as characteristic. At the beginning of his career he himself did a lot of experimentation. Another strong trait was his mastery of mathematics. Mathe- matics is, of course, the language of theoretical physics, but the degree of mastery of this language fluctuates enormously, even among fully qualified theoretical physicists. Some, using only standard methods, work entirely successfully because the physics extends beyond the mathematics. But in many fields, particularly the most modern (quantum field theory, string theory), the role of mathematics and of its most recent advances becomes dominant. YaB occupies here a special place. Apart from his mastery of standard mathematical methods he solved a number of problems not by using refined results, but by blazing new trails. The mathematician V.I. Arnold, in a special section of the introductory article in *Selected Works* Volume I, entitled 'Mathematics in the work of Ya.B. Zeldovich,' considers that some of YaB's achievements 'are essentially mathematical discoveries and rank with the most modern researches by mathematicians.' Since I myself belong to those theoretical physicists far removed from contemporary mathematics, I shall not interpret this statement (all the more because such an interpretation would reduce to rewriting V.I. Arnold's remarks). I would only add that YaB wrote a textbook, *Higher Mathematics for Beginners and its Applications to Physics*, that has been reprinted many times and translated into several languages. This book is in some measure the fruit of YaB's adverse reaction to conventional university mathematics courses, which are overloaded with formalism and are ineffective for a rapid mastery of the mathematics used in physics.

If YaB's scientific career is reckoned from the date of his joining the Institute of Chemical Physics (May 15, 1931), and here there is no uncertainty, he worked for more than 56 years. A massive heart attack struck him down suddenly and he died without regaining consciousness on December 2, 1987. All those 56 years were years of uncommonly intensive work. But at the same time, YaB was no workaholic. He appreciated sport (swimming and skiing), entertainment and fiction, and gave much attention to his large family — one might say to his physics family, because the majority are physicists. One family member well known to me is YaB's son — Boris Yakovlevich Zeldovich — who, like his father, is a first-class theoretical physicist.

In conclusion, I recognize clearly that I have not sufficiently reflected the scientific merits of Yakov Borisovich Zeldovich, nor have I communicated much about him as a man. However, I believe the introductory article to the selected works, the autobiographical afterword and the readily accessible articles by V.I. Goldanskii and A.D. Sakharov, who were very close to YaB, will in some measure help to fill out the picture. I would like to end by using the final words from Sakharov's obituary for Zeldovich: 'Now, when Yakov Borisovich Zeldovich has departed from us, we, his friends and colleagues in science, understand how much he himself did, and how much he gave to those who had the chance to share his life and work.'

At the source of graviphysics[166]

A.M. Fridman

My memories of Yakov Borisovich Zeldovich (Ya.B. as his friends called him) go back to the moment when I first became aware of myself. When I was about three, he gave me a basket on wheels, at the bottom of which I found a glove puppet that bore a great resemblance to a fair-haired chap in blue dungarees with his trousers tucked into shiny black boots. Using three fingers of one hand to operate the puppet's head and arms and two fingers of the other to work its legs, Ya.B. performed a sketch in which the doll's legs took on a life of their own, vying to see which could kick the highest while the top half of the puppet fell into a slumber. One booted leg tried to get the better of the other, which in turn came out on top, and finally they started kicking each other. At that instant the puppet woke up, irritated, and quieted its pugnacious legs. I loved that little play and performed it successfully in front of my friends afterwards.

Ten years later, in the spring of 1953, Ya.B. and my uncle, D.A. Frank-Kamenetskii, flew to the town of Frunze (Central Asia), to which my family had moved from Moscow not long before. Besides holding top governmental awards, Ya.B. and D.A. had special certificates that gave the holder the right to travel free of charge by plane, in the first-class train car and in the first-class cabin of a ship for all routes within the USSR. (Similar certificates were given to their wives and children under 16; however, the privilege was withdrawn a few years later.)

After spending the night at our house, the two physicists and my parents went around Lake Issyk-Kul in a taxi. They came back two days later, delighted and bearing gifts. The physicists brought optical devices — my uncle gave me a pair of binoculars with 6× magnification and Ya.B. presented me with a camera, which allowed me to take pictures in 2.4″×2.4″ format

[166] This article was not included in the original Russian version of this book, published in 1993. [Translator]

without an enlarger. Ya.B. gave me a short lecture on the schematic movement of the light beam in a camera and the principles of developing black and white and colour films (amateur colour photography was a novelty in the USSR in 1953), and this stuck in my mind for ever. One's first impressions are the strongest, which is why that lecture in physics and chemistry imprinted itself on my mind and possibly influenced my turning into a scholar. The principles of colour vision were also touched upon. The substance of his lecture amounted to far more than just the facts (though they were precious too); I valued much more the discovery that explanations concerning the essence of the many objects and phenomena which surround us were simple and achievable not only by reading books but mainly through thinking and analysing. I was surprised to discover that images appear to be upside down on the eye's retina just as they do on the focal plate of a camera lens, and that they become 'normal' in vision only after certain further developments.

I didn't meet Ya.B. again for another 15 years, although having kept in touch with Frank-Kamenetskii I was aware of the main turns in his life, for example his move from the 'hush-hush' institute to Moscow, where he founded a department at the Institute of Applied Mathematics (IAM). At that time (since 1962) I worked at the Institute of Nuclear Physics in the Academy town in Novosibirsk. Once, when I happened to be in Moscow, I called on Ya.B. and suggested reporting on the stability of compressible gravitating rotation figures at his seminar. Fridman is a common name in Russia, and as Novosibirsk was a long way from Moscow and Leningrad, where my relatives lived, and from my parents' home in Central Asia, Ya.B. had not associated it with the urchin he once met.

He seemed to be interested in my report, probably because of certain advances achieved in the theory of rotation figure stability. I used the methods developed in plasma physics, which I had already been dealing with for a few years at that time. The theory of gravitating systems' instability appeared to be much more complicated than the theory of plasma instability. The latter developed at the time from problems of homogeneous plasma to more complicated problems of inhomogeneous plasma. While very important physical mechanisms have been investigated in homogeneous plasma, equilibrium homogeneous rotating gravitating systems did not exist at all (except in the form of a cylinder with an unlimited constituent, which is inapplicable for metagalactic bodies directly.) Thus, even to define 'equilibrium' demanded the solution of serious problems, not to mention the stability of complex inhomogeneous systems. No wonder that the first tentative attempts to investigate the stability of compressible gravitating systems in the scientific literature turned out to be not quite satisfactory. Two types of errors occurred: either due to the fact that the equilibrium was incorrectly accounted for, or due to an incorrect analyses of the stability of inhomogeneous gravitating systems.

At the seminar, I spoke of the results I achieved in the course of solving some problems of the stability of the main rotating figures of equilibrium (sphere, cylinder, disk) applicable for star systems (namely strongly compressible — collisionless). They are widely known nowadays and are often quoted. After the seminar Ya.B. asked me to call him rather early the next morning. I did. When the scientific part of our conversation was nearly over, Ya.B. suddenly asked me:

'Where are you staying?'

'At the Frank-Kamenetskiis'.'

'Then you'll surely whisk away their younger daughter Masha, just like Roald Sagdeev whisked away Tema, their elder, won't you?'

At breakfast my relatives burst out laughing when I told them about our conversation. Later that day, when I met Ya.B. at the IAM, he looked confused. I had rarely seen him like that before. 'So, you're *that* Alex Fridman! I didn't recognise you ...'

Since then, every time I met him on my trips to Moscow or at conferences, Ya.B. was very warm towards me. In March, 1987, at a non-linear physics seminar in Nizhnii Novgorod, Ya.B. said to his last wife, Inna (who was a close friend of my cousin, Masha Frank-Kamenetskii): 'Alex is like family to us.' ...

... The birth of that awful morning of December 2, 1987, came nine months later when I learned that Ya.B. had a heart attack during the night and was at the Academy hospital. I called Rashid Sunyaev at the Institute of Space Research and his secretary Lora rushed to find him, realising from my voice that he was needed urgently. A moment later, I broke the news to a panting Rashid.

'I can stay by his side as long as necessary,' Rashid said, shocked and distressed. 'How can I help?'

I'll never forget the heart-breaking pain and hope voiced in his question. We could hardly believe at the time that nothing would help the man we loved so much. An hour and a half later, when I dialed Ya.B.'s home phone number and heard the voice of Academician V.I. Goldanskii, his neighbour, I realised that was the end. Having promised to take Masha Frank-Kamenetskii to Inna's as soon as possible, I was hardly aware when I was rudely stopped by a traffic patrol demanding to see my driving licence. Their radar screen had indicated that I was doing 75 mph along a narrow lane near Masha's place. The patrolman declared that I would never ever have my licence back. And indeed I still have no licence. When necessary, my wife drives the car for me.

The part Ya.B. played in guiding me towards my future occupation as a specialist in the physics of gravitating systems is hard to overestimate. However, while the latter concerns the biography of a scholar, and can hardly be of interest to anybody, the outstanding contribution YaB made to the foundation of the modern physics of gravitating systems as a branch of science somewhere between physics and astrophysics is of much greater interest to the history of science as well as to the many enthusiastic admirers of one of the last brilliant encyclopedists in the field of physics. I would like to illustrate this with a few examples.

In the autumn of 1964, Ya.B. drew the attention of his students to an article by Alar Toomre which described the stability criterion for a galactic disk with regard to the short-wave radial perturbations splitting the disk into narrow rings. Radial perturbations were investigated as monochromatic waves. That raised a natural question: how does the stability criterion change when a perturbation has an arbitrary form? The variational methods worked out by Lindquist for general hydrodynamics and by Kadomtsev for magnetic hydrodynamics and plasma physics were chosen to solve the problem. The variational approach had only been adopted for the investigation of stable systems. Therefore we needed to generalise it for a rotating self-gravitating disk. Within a few months we succeeded.

Then, together with Ya.B. and his student Bisnovatyi-Kogan, I managed to determine the stability of two further types of star systems with regard to radial perturbations, namely, the axial and spherical symmetric rotating systems such as a cylinder and a sphere, respectively.

As a result, the following statement concerning the stability of collisionless stellar systems possessing axial or spherical symmetry with regard to the radial perturbations was formulated: a system is locally stable for a given radius if the gravitational force acting on that radius depends upon the 'inner' gravitational mass only — i.e., the mass concentrated inside of the sphere restricted by that radius.

In accordance with the criterion concerned, spherical and cylindrical collisionless gravitating systems, unlike disks, are stable with respect to the radial perturbations.

Among various rotating figures Zeldovich singled out spherical systems. Most of all, he was interested in a spherical system with gravitating masses suggested by Einstein (1939) as the potential carrier of a great red shift. Thereafter, we started to call that system the Einstein model of a spherical star cluster, though this is not an accurate term, because in contrast to the Einstein model, the orbits of real spherical star clusters are far from circular. The Einstein model represents a star cluster as a number of spheres nested one inside the other (like a Russian doll) with the stars moving along ideally circular paths and the distribution of velocity vectors in any plane tangent to any sphere in any arbitrary point being isotropic at the angles. Consequently, the angular momentum of any single sphere is equal to zero. This means that the general angular momentum of the whole spherical star system is zero too. A system of this type can be arbitrarily massive, with the gravitation force being equal to the centrifugal force arising in the course of the star's rotation. Consequently, the idea that a spherical star cluster could obtain a great gravitational shift was most inviting for Zeldovich, who had in mind the recent discovery of quasars.

The stability of a sphere as well as a of rotating cylinder with respect to the radial perturbations on condition that the layers are not intersecting is obvious, because each particle (a star) moves in a gravitational field caused by a constant attracting mass. In the case of radial perturbations, the angular momentum of a particle is preserved and its steady state corresponds to the minimum energy (the Keplerian problem), causing its stability.

But this simultaneously proves the stability of a system as a whole, and is also true for the perturbations under consideration, because a single layer oscillates independently of the others. When the layers intersect, the spherical stability with regard to radial perturbations is, *a priori*, uncertain. The position was aggravated by the inevitable intersection of the layers in the course of time, whatever small perturbation existed initially, as was shown by Zeldovich. In Zeldovich's opinion, this caused the system's instability.

However, my analytical calculations proved the opposite: it was true that the intersection of the layers in the course of time was inevitable; it was not instability, but additional stability that the system acquired instead. In other words, when the radial perturbations exist in a system whose layers intersect each other, it turns out to be more stable compared with a similar system without intersections.

A bottle of mineral water was the token prize in a bet made during our scholarly dispute. Ya.B. suggested that I come to his place the next morning at 6 A.M. to finally clarify the instability dependence on intersecting layers — to be or not to be. At six sharp he answered the door and greeted me with the words, 'I am tearing my hair out. A bottle will be placed in the cabinet at the IAM today, with an inscription explaining why I lost the bet.' The bottle bearing such an inscription really appeared there.

A much more complicated problem appeared to be how to clarify the question of the stability of the Einstein model presented as a spherical star cluster with respect to arbitrary perturbations, including radial ones. There, the stability criterion alone, as predicted in a stroke of genius by Zeldovich, played a decisive role. I am conscious of the fact that the reader will be shocked by my use of such an emphatic word as 'genius' to describe Zeldovich's prediction. Nevertheless, I believe, as is shown below, that the word is quite appropriate to reveal the essence of the events that followed. First of all, it is necessary to note that there was no strong evidence proving the stability of the Einstein model with respect to the arbitrary perturbations, which is why the application of it as a model for investigating any real star system with a great red shift was impossible. Einstein clearly understood this, paying attention to the fact that his model of stability was called into question.

It was only in the summer of 1968, when I was working together with R. Sagdeev and G. Bisnovatyi-Kogan, that I realised the reason why Einstein was unable to solve the problem of the stability of his model. Then we came to understand that the eigenfunctions of the system are generalised functions never dealt with by Einstein. Several years later, when a precise solution of the Einstein problem representing an arbitrary perturbation as a series of δ-function and its derivative and using some group properties of differential equations had been obtained by me jointly with my co-authors A. Mikhailovsky and Ya. Apelbaum, I began to understand the actual complexity of the task. In any case, it was this work of which M. Leontovich said to A. Mikhailovsky, 'You did it to the best of your ability, Anatolii Borisovich!'

However, let us go back to the summer of 1968. At that time, none of us, including Sagdeev and Bisnovatyi-Kogan, had the answer to the problem yet and were probably quite unaware of its complexity. Nevertheless, we obtained the 'key' with which to define a precise stability criterion applicable to the Einstein model with respect to the arbitrary perturbations. This 'magic key' was nothing but following 'Zeldovich invocation' (in those days we could not call anything his unique prediction): a criterion of stability for the Einstein model is equal to a criterion of stability met in two nested, collisionless, self-gravitating cylinders rotating in opposite directions around aligned axes.

In August 1968, we solved the final problem, which was much easier than the initial task. The answer was that the system is stable if the angular rotation velocity is constant while the radius is increasing. With reference to the stability condition, this meant that a star system whose density is constant within the radius's growth must be stable. In so far as the frequency of all the star systems known to us at that time declined toward the boundary, the criterion obtained proved the stability of the Einstein model applied to a spherical star cluster with respect to the arbitrary perturbations.

A little later, while working with V. Polyachenko, I realised that the star clusters whose density increases towards the edge within the radial segment could exist in reality. These are star clusters surrounding massive black holes in the centres of the galaxies, whose density increases toward the centre. However, in the Roche region of a black hole stars are bound to be destroyed by tidal forces, their plasma added to the black hole by means of accretion. Thus, according to our criterion, the inverse star density distribution causing instability should exist in the vicinity of a massive black hole (see Fridman, Polyachenko, *Physics of Gravitating Systems*, vol. 2, Springer Verlag, 1984).

However, at that time the criterion of stability of two rotating cylinders in itself seemed rather strange to us — Sagdeev, Bisnovatyi-Kogan, and myself who were co-authors to Zeldovich's article, published in 1969. In fact, according to the criterion two self-gravitating cylinders which are homogeneous in density and rotating in opposite directions had to be stable. But the system under consideration is similar to two opposite flows of charged particles which should be unstable, as is well known from plasma physics. However, we found that in the case when azimuthal mode $m = 2$ the system was on the stability boundary as one of its eigenfrequencies turned to zero, with $m = 2$. This made it possible to explain the reason (as proved later) why the majority of spiral galaxies have two arms.

The paradox on the stability of the system consisting of two self-gravitating homogeneous cylinders rotating in opposite directions was also solved by Zeldovich. Without going into detail, it has to be noted that his solution of the paradox concerns only a hydrodynamical beam instability. The kinetic instability turned out to be possible even for homogeneous densities, as is observed in plasma.

Thus, a beam-gradient hydrodynamical instability in the theory of gravitating systems was discovered. The criterion of stability of the Einstein model obtained later from the exact solution turned out to be identical to the criterion on the stability of the system of two cylinders, inserted into each other, which rotate in opposite directions. Thus, Zeldovich's prediction was confirmed.

****

Thirty-five years have passed since then. For such a fast-developing branch of science as graviphysics — the term used to describe the physics of gravitating systems, which was accepted by Zeldovich after a few disputes — that is a tremendously long time. Thousands of papers and dozens of monographs, some of them over 700 pages long, have been written during that period.

Modern graviphysics is like a wide and mighty river. Peering towards the opposite bank, one can hardly discern what is on the other side or reach the ears of those standing there. Yet this powerful stream goes back to just a few tiny springs. Mine was a story of one of them, which I have attempted to recount, in memory of a man who stood at the beginning of this river created by human hands. The very fact of knowing him was a gift of life to me. His name is Yakov Borisovich Zeldovich.

An outsider's view of a great scientist-human being[167]

G. J. Wasserburg

I had been a frequent visitor to the Soviet Union for several years following the Apollo 11 mission and then served as a member of the U.S. delegation working on the U.S.–Soviet Union bilateral cooperation in space. There was also an opportunity in an exchange visit by an arrangement between Mstislav Keldish and Harold Brown after the first visit by a President of the Soviet Academy to the United States.

During that period the relationships were distant as there was an enormous gap between our countries with great tensions and apprehensions even in personal contacts and relationships. I met a large number of scientific leaders, managers, and scientists at all levels, and although there were almost always large formal barriers, there were still occasions to break through and to speak to each other as real human beings and fellow scientists.

[167] This article was not included in the original Russian version of this book, published in 1993. [Translator]

Usually this occurred when we would go out for a walk, away from it all. Many of the meetings were rather formal, sitting about long, green baize tables arranged with water bottles and ash trays. One clear impression that I got was that the Russian scientists I met often thought more deeply about the theoretical implications of their work than did many from the US. The limitations on experimental equipment, data, and computers seemed to enhance the depth of understanding which they would achieve. Maybe too much of a good thing is not always good! This was the era of the Sakharov affair and many other very difficult issues reflecting the Cold War but with a warming up in the areas of some science where it seemed of interest to both nations to begin 'defrosting' the relationships. It had even become possible for Piotr Kapitsa to come for a visit to the U.S. He spent some time in the Lunatic Asylum laboratory clean room, looking at lunar samples and talking. He later invited me to his Institute and his home with the family. We often spoke about Don Quixote. We were both fans of Don Quixote in some ways.

My most frequent meetings as a member of the bilateral working group were through the Space Institute, under the direction of Roald Sagdeev, and the Vernadsky Institute, under the direction of Aleksandr Vinogradov and then Valeriy Barsukov.

Roald Sagdeev was a brilliant scientist and a true intellectual leader and manager of the space science enterprise in the Soviet Union. He was trying to bridge over raging torrents that appeared to be subsiding but were always threatening. These meetings led to contacts with a large number of scientists at all levels. They were a very impressive group of highly capable and knowledgeable people. There were efforts on the part of Roald Sagdeev, the director of ISI, to introduce the stellar members of the Soviet Science establishment, one of whom was Joseph Shklovsky. When I last saw Shklovsky he had just gotten a fancy new leather jacket purchased when he had finally been allowed to go out of the country. Circumstances arose where at special meetings I was privileged to be introduced to Yakov Zeldovich. These meetings clearly reflected an effort to present the finest intellects in the Soviet establishment, the intellectual crown jewels of the Soviet Union, individuals of the highest level of scientific capability and achievement.

I had, of course, heard of the great Zeldovich from Kip Thorne and other colleagues. At some COSPAR meeting Zeldovich made an impressive presentation on cosmologic problems. The quality and depth of his presentation was awesome. The physical impression on meeting him at a later time, at a sort of science–politics meeting arranged by Roald Sagdeev, was of a physically solid, physically trim and tough person, an individual devoted to science and to the deep understanding of nature. There was no sense of a manager, or a manipulator. No sense of a showman except in demonstrating great science.

Zeldovich was a scholar of great fame, but the impression of intellectual greatness came from his ability to perceive scientific problems of major importance, and from the powers with which he would address these problems. This was done, of course, with the associated great confidence provided by continued success. References to him by ex-students (including expatriates in the U.S.) showed an aura of quality and integrity that emanated from him. Even students with limited contact with him would still glow from the experience. He was THE MAN.

I do not know of any individual in my society who would be held up as THE MAN, enormously respected, possibly admired. Dick Feynman was a very different kind of person. He was a 'THE MAN' kind of person at Caltech. When Dick was very ill, I would visit him rather often. One day the phone rang as we were sitting in the garden. I went to answer it and found Zeldovich on the line. He wanted to talk to Dick and to express his respects. The phone was near the dining room where there was a wonderful large hand crafted model of a troika drawn by horses, a gift from Russia many years before. The two chatted for a while and then, of course, it was soon over for Dick.

In the diverse group of scientists, leaders of science, and science managers that I have met and worked with, there is one small group that appeals to me. It is composed of individuals who really want to do science and cannot live without it. They are sustained by doing creative science. No other puzzles or games can divert them from trying to really understand something about nature. Yakov Zeldovich was one of these rare and gifted individuals.

The eye of the Universe[168]

B.P. Zakharchenya

When I attended his lectures on astrophysics or gravitational phenomena, it was clear to me as a specialist in solid-state physics that Yakov Borisovich Zeldovich was as great in physics as Izyk Stern was in playing his violin. I did not understand the details but listened to him with admiration mixed with excitement one feels when hearing a jubilant triumphant belle canto. Self-confidence and creative power are imparted to a listener. I remember Evgeny Fedorovich Gross, my tutor in crystal optics (who discovered Brillouin scattering in quartz in pre-laser era and then a well-known spectrum of exciton quasi-particle in semiconductors) telling me, when listening to Yakov Borisovich:

'Boris, can you imagine how difficult it is to take the floor after such an ace!'

Gross was to make a report at the same session at the Physical Technical Institute. If I am not wrong, the session was dedicated to the memory of

[168] This article was not included in the original Russian version of this book, published in 1993. [Translator]

Abram Borisovich Ioffe, the founder of our Institute. Yakov Borisovich's scientific youth was at the PTI. He liked to come to us.

Having started my recollections, I took a high note at once. But one can't live in major all the time. Toning it down a bit, I would say that there were blunders in his reports, too. I would never forget how in the early 1960s, when opposing in the defence of Dmitrii Varshalovich's DSc thesis (also at the PTI) and commenting the candidate's assumptions on the re-distribution of atomic levels' populations due to the anisotropy of radiation in the Universe, which inevitably led to the maser effect, Zeldovich said that it would be good to simulate this situation in Earth's conditions. He didn't know that a dynamic overpopulation of levels had already been carried out by a Frenchman, Kastler, in mid-1950s, using the method of optical pumping; later, that work was even awarded a Nobel prize. This was explained to Yakov Borisovich in a little while, but I don't think he got confused. He was one of those who abide by the German saying '*Wer zufiel liest, der wird nie gelesen werden.*'[169] This man was never embarrassed to ask a lecturer the questions which might show his poor knowledge of scientific literature. Creation not erudition was his nature. When such a man is possessed by an idea, he can work under any conditions. In this connection, I can recall something.

In the 1960s, an unusual session of the Department of General Physics and Astronomy was held in Middle Asia. I was not a member of the USSR Academy of Sciences at that time, but found myself among the participants of that interesting event. First we flew to Tashkent, where the first part of the session, with scientific presentations, was held. Then we travelled through Bukhara and Samarkand to Ashkhabad in two sleeping cars coupled to a common passenger train. We enjoyed entertaining tours around the first two towns and came back again to science in Ashkhabad. It was spring. Hot. Fields still rich in green dotted with scarlet poppies and tulips were passing by the window, with herds of camels running across that fairy carpet. What a fantastic scenery it was! Zeldovich and his colleague and pupil Igor Novikov, who shares his life and scientific work now between Russia and Denmark, were in one compartment with us. The carriage was full of celebrities. Vladimir Aleksandrovich Fock was still alive. Zavoisky, who discovered the paramagnetic resonance, Lev Arzymovich. Gaponov-Grekhov, the creator of gyrotron, corresponding member of the Academy of Sciences, looked very young at that time. He wore sports costume and shoes to do jogging when the train stopped. The stops of the train in Bukhara and Samarkand were badly organized. The train arrived there very early. The carriages in which we travelled were uncoupled and driven to the dead-end to wait for us to return after excursions and banquets. Certainly, train conductors immediately locked the toilets. I remember the old Fock tugging the knob of the locked door in despair. Yakov Borisovich and I ran to a nearest

[169] 'He who reads much will never be read.'

'duval' (a deserted enclosure made of clay) overgrown with grass and bushes. There was nothing to do! Once we were met by a middle-aged Uzbek, a rifle in his hand, who was the owner of the cherry orchard behind the 'duval.' We had it out peacefully... In Bukhara (or Samarkand) I didn't see Zeldovich during the excursion, he appeared late at the restaurant where we enjoyed Uzbek cuisine. He confessed that he stayed at our compartment in a train heated by the southern sun the whole day long. Was busy with science. He told me with excitement:

'Boris Petrovich, you cannot imagine how happy I am! I made good to work and even solved the task which was not easy for so long!'

Really I could not imagine! He was able to work for hours, naked, sitting in a red-hot metal box of a carriage.

Yakov Borisovich generated joy. He loved life in a way a man who found his place in it could do. In that very 'Oriental express' he and the wife of Gaponov-Grekhov put on Uzbek clothes and went dancing Uzbek dances (it was mostly shaking of hands raised above their heads and stamping of feet) along the carriages of the train in which local people travelled. There was a bet taken up, I don't remember with whom. Zeldovich disguised himself in an Uzbek gown 'chapan,' his dame wore a silk dress ornamented in a national pattern. They had tyubeteikas on their heads. Yakov Borisovich put on a traditional dark-coloured tyubeteika embroidered in white flowers of cotton plant. It resembled a yarmulke and matched his lively and curious face perfectly well. A bet was won, Zeldovich came back, his tyubeteika filled with money. A tyubeteika is a good deal bigger than a yarmulke! Certainly, I felt like saying 'Begging is your second nature!' — the words put into the mouth of Ostap Bender — the character of a popular Soviet satire by Ilf and Petrov. But I held my tongue.

Academician Zeldovich wasn't ashamed to dance in a train, but he felt awkward to bear his rare medals of honour, including three Stars of Hero of Socialist Labour he had been awarded for the development of the bomb. I remember one time, on his way to the Presidium of some noble meeting which I was also present at, he asked me:

'Boris Petrovich, please help me to pin my Stars.'

We went to hide at the restroom and there, standing in the 'pissoir room,' I was pinning a medal holder with the stars taken out of his trouser pocket to his jacket, with an effort. The clasp was rather inconvenient; in addition, Yakov Borisovich was much smaller than me, a lanky fellow of six feet four. A man entered the restroom and immediately rushed out frightened. He might have thought that a perplexed professor was being robbed of his gold by a plunderer.

Yakov Borisovich had many students, most of whom sponged not only his ideas and methods of scientific research, but also his creative zest.

Becoming a member of the USSR Academy of Sciences comparatively early, I took part in the elections and my vote meant something. Till now, I keep as a token Zeldovich's letter written in a decisive and steady hand requesting to cast my vote in favour of Rashid Sunyaev. Yakov Borisovich convincingly described the outstanding abilities of one of his favourite pupils who soon became an academician and then a laureate of several top international awards. Unfortunately, Yakov Borisovich didn't live to see the glory and wide recognition of Rashid's merits. I keep interesting hand-written letters and delete e-mails because there's no live human soul in e-mail messages. I do not value them. Who only invented e-mail!

Yakov Borisovich invited me to his 'family seminar' organized at his place on Vorobiev Road. I don't know why, but I never made any report at the seminar. And he invited me because my paper on optical orientation of electron and nuclear spins in semiconductors aroused his interest, especially the part in which I explained the effect of deep cooling of a nuclear spin system in semiconductors obtained with the help of light. The effect is complicated, but its physical nature is obvious. Electron spins oriented by the circular-polarized light orient the spins of the atomic nuclei which form a crystal lattice. It happens due to the so-called superfine interaction which shows up most strongly in semiconductors. After having completed its task, an electron polarized along the spin gets out of the process because it annihilates with a hole, another electron comes out, etc. The electron oriented along the spin plays the part of a 'Maxwellian demon.' A well-known 'gedanken experiment' conducted by Maxwell to prove that the second law of thermodynamics could be broken in principle, as entropy reduces. An atom capable of distinguishing a cold molecule from a hot one and separating one from another acted as a 'demon' while entropy simultaneously reduced. But the 'demon' got overheated and lost its ability to distinguish between molecules. Leo Szillard managed to solve the Maxwell paradox, but only a few knew about it. It is impossible for 'a spin demon' to be overheated in semiconductors as it gets out of the process because of annihilation. Vladimir Fleisher and Roslan Dzhioev from my laboratory showed that the spin temperature of nuclei thus achieved (their spin relaxation was one order of magnitude lower than in electrons) reached 10^{-5}–10^{-6} K.

Yakov Borisovich came late to my lecture in the large assembly hall at the Leningrad Scientific Centre, but in an instant he understood all and greatly appreciated it. He asked a few questions, one of them having no reference to physics at all. As a matter of fact, to enliven the lecture I had read a piece of poetry:

> 'Professor Umov crammed my mind
> with formal physics. Through cosmic dark,
> wild hair flying, his neck arched,
> he sang that Maxwell with his paradox annihilates entropy.'

Zeldovich inquired who the poet was and added that Landau liked to recite it though he didn't know whose poetry it was. I explained that it was an extract from 'The First Encounter' by Andrew Bely, a well-known poet and writer, recalling his young years when he studied at the Moscow University to become a physicist. Landau died in 1968. It was difficult to buy Bely's poems at that time as practically they were not published. Landau preferred to read poems by Soviet authors, such as Simonov's poetry. I am not aware of Zeldovich's opinion on art, but I know that he had a love for literature. It can be said that his mother Anna Petrovna was a writer and a talented translator, many of her translations from French are brilliant. I happened to be acquainted with her. She lived in Leningrad.

She had a heart attack and after intensive care therapy she found herself in the same ward with my mother, Nina Petrovna. I visited both of them bringing food and fruit, it is usual for free but poor public hospitals in this country. Anna Petrovna told me a great many stories about Yakov Borisovich's childhood and youth. Her recollection of evacuation from Leningrad on the eve of the siege was strange to listen to. All the time she recalled a chest with family belongings which had been stolen by some crooked organizer, one of those who evacuated the personnel of the Physical Technical Institute. The episode is mentioned here to show the poor conditions in which the Russian scientists lived. Young Zeldovich was a star in science of that time, in 1939 at the Leningrad Physical Institute he and his co-author Yu.B. Khariton published a famous article on the chain reaction of uranium fission. The mother of such a man recalling a thing an American considers 'garbage!' It's very likely they will never realize that a brilliant like Yasha Zeldovich is incomparably dearer than a million barrels of oil.

I remember Anna Petrovna calling me up once:

'Boris Petrovich?'

'Speaking, Anna Petrovna.'

'Oh, you already recognize my unpleasant voice! Want to know if you've read the article on the 60th birthday of Yasha in *Advances in Physical Sciences*? He is compared with Enrico Fermi there. What do you think of it?'

'Certainly, Anna Petrovna. And why not? Don't you think Yakov Borisovich is as good as Fermi?'

'Believe me, Boris Petrovich, I am happy to hear your words. But I'd like to tell you that Yasha is pretty dumb compared with his father, Boris Naumovich. What a clever man he was! Yasha doesn't hold a stick to him.'

It seems to me that his father was a lawyer. She had a great many of similar stories in store for me, it's impossible to put down all of them, more than that, it's impossible to describe her provincial Jewish accent exaggerated on purpose and emphasized in joke, I think, to imitate a well-known Soviet actress Ranevskaya. Anna Petrovna always played a part. She lived in France for a long time and worked at Sorbonne. Her translations were published in the journal *Foreign Literature* loved by the intelligentsia of the former USSR. She made acquaintance of the outstanding writers in this country and abroad. She was a person of 'refinement,' so to say, but loved acting. She died in Moscow where Yakov Borisovich had moved her to. Old people feel bad about being moved.

I think her great son, who produced an abundance of ideas in nuclear physics, physics of combustion and explosion, gravitational physics, relativity astrophysics, had suffered the 'sin of histrionics' too, to some extent. Just remember the oriental dancing in the train. He loved somewhat to show off.

When he was called to report or be rewarded on the stage he always dashed omitting steps to the platform from the hall. It was an acting trick! Still, acting was not for him, though he couldn't help it. He was captured by science, which was Alpha and Omega of his life.

Assumption, experimental fact and its theoretical explanation attracted his attention most of all. It was true not only of physics. I remember how thrilled he was discussing the works by Alexandr Mikhailovich Ugolev, a brilliant physiologist, his works being far from star worlds. He created fresh notions regarding a digestion system as one of the main functions of the body. Academician Ugolev called himself a 'boweler' (one who studies bowels). His stories of evolution, of the wonder world of the small intestine, of the generation of opiates — narcotic substances — in the human organism were simply fascinating. By the way, he was involved in space programmes working out the diets for cosmonauts based on his digestion physiology. Zeldovich reproached me and Zhores Alferov for being his mates and not inviting that brilliant scientist to lecture at the PTI.

Bright and clever, his mind alert, Zeldovich was agile and quick-motioned in appearance. I never saw him look melancholy, with lacklustre eyes or a stooping walk of a dull-looking middle-aged man. His slightly protruding Pinocchio's eyes, fast-moving out of natural curiosity, always made me think of an aphorism by a wise khasid. That philosopher said: 'The man is the eye of the Universe.'

Zeldovich predicts that black holes radiate[170]

Kip S. Thorne

The year was 1971; the month, June; the place, Moscow: an elegant private apartment on Vavilov Street near October Square, where Yakov Borisovich Zeldovich had ensconced me for six weeks. At 7:00 A.M. I was roused from my sleep by a phone call from Zeldovich. 'Come to my flat, Kip! I have a new idea about spinning black holes!' Knowing that coffee, tea, and peryozhki (pastries with spicy ground beef inside) would be waiting, I sloshed cold water on my face, threw on my clothes, grabbed my attache case, dashed down five flights of stairs into the street, grabbed a crowded trolley, transfered to a trolley bus, and alighted at Number 2B Vorobyevskoye Shosse, in the Lenin Hills, 15 kilometers south of the Kremlin. Number 4, next door, was the residence of Alexei Kosygin, the premier of the USSR; after Brezhnev, the second most powerful person in the Soviet Union.

I walked through the open gate in the eight-foot high fence and entered the 5-acre, forested yard surrounding the massive, squat apartment house

[170] Adapted from a preliminary (1989) version of Chapter 12 of K.S. Thorne, *Black Holes and Time Warps* (Norton, New York, 1994).

Number 2B and its twin Number 2A, with their peeling yellow paint. Zeldovich lived in one of eight apartments: the southwest quarter of the bottom floor. In a city where, typically two, three, or even more families shared a 700-square foot apartment, with one kitchen, one bathroom, a living room and a bedroom, Zeldovich and his wife had 1500 square feet to themselves — one reward for his monumental contributions to Soviet nuclear might.

Zeldovich met me at the door, a warm grin on his face. I removed my shoes, put on slippers from the pile beside the door, and followed him into the shabby but comfortable living/dining room, with its overstuffed couch and chairs and its tiny, two-foot square blackboard on one wall. Across from the blackboard was a map of the world, with colored pins identifying all the places to which Zeldovich had been invited (London, Princeton, Beijing, Bombay ...), and to which the Soviet State, in its paranoid fear of losing nuclear secrets, had forbade him to go.

Zeldovich, eyes dancing, sat me down at the long dining table dominating the room's center, and began to draw a sketch on the blackboard.

'Suppose we have a rapidly spinning, metal cylinder, a few centimeters in size,' he said.[171] 'And suppose that incoming electromagnetic waves, with wavelengths a centimeter or so long, graze the cylinder's surface. The oscillating electric field that the waves carry will drive electric currents to flow in the metal surface, and those currents will emit waves of their own.' This is called *stimulated emission* because the incoming waves stimulate the cylinder to emit the new waves. 'If the cylinder spins rapidly enough and the incoming waves are properly configured,' Zeldovich claimed, 'the stimulated waves will emerge from the cylinder in phase with the incoming waves, and the two will reinforce each other. More wave energy will come out than was sent in; the incoming waves will have been amplified by the rotating cylinder. This is obvious.' (It was far from obvious to me, but I had developed an awe for Zeldovich's intuition; presumably I would be able to verify his claim in the privacy of my own apartment that evening.)

'Where does the added energy come from?' Zeldovich asked. 'From the spin of the cylinder. The currents, interacting with the incoming waves that produce them, slow the cylinder's spin a bit; kinetic energy is thereby released from the spinning cylinder's mass; some of the energy heats the cylinder, and some goes into the emitted waves.' (This sounded reasonable, but was far from obvious; I would have to check it that evening, after Zeldovich released me.)

'This is a very general phenomenon,' Zeldovich asserted. 'Any kind of waves impinging on any kind of spinning object can produce stimulated emission, and resulting wave amplification. This should be true even for a black hole: Send electromagnetic waves, or gravitational waves, or any other

[171] Here and below I have reconstructed the gist of what we said. The ideas, as I describe them are faithful to the conversation, but the words are not. I have translated the conversation into less technical language than we actually used.

kind of wave toward a black hole. If you configure the waves just right, they should stimulate the hole to emit additional waves that amplify the original ones. And the hole, to provide the added energy, should slow its spin ever so slightly.'

This was too much for me. Zeldovich's intuition about other kinds of objects was far greater than mine, but not about black holes — I thought. Four weeks of intense discussion had shown me the weakness of Zeldovich's mastery of general relativity. For technical details of relativity he relied heavily on Igor Novikov, and now, during my weeks in Moscow, on me. Zeldovich, evidently, had called me in this morning to act as a skilled opponent against whom to test his idea.

I counterattacked: 'A black hole is not an electrical conductor. Incoming electromagnetic waves cannot generate electric currents around the hole because electric currents consist of moving electric charges, and the curved space around the hole contains no electric charges. Without current to radiate, there can be no stimulated emission. And because the hole's horizon will absorb some of the incoming waves, there actually will be a net loss of wave energy, not a gain.'

'No, Kip,' Zeldovich insisted. 'If configured carefully, incoming electromagnetic waves *must* find a way to spin down the hole, and must use the liberated energy to amplify themselves. This *always* happens when an object is spinning.' Zeldovich was not sure just how the hole would amplify the waves, but his intuition made him *feel* that it must happen. (He was, in fact, right: calculations by Charles Misner a few months later would reveal that the waves are amplified by interacting with the hole's spacetime curvature, a process to which Misner gave the name 'superradiance'; and this revelation would trigger two of my students, William Press and Saul Teukolsky, to invent an [impractical] 'black hole bomb').

'Let me go on,' Zeldovich said. 'Wherever there is stimulated emission, spontaneous emission must also be possible. If our spinning metal cylinder is completely alone in space, with no incoming electromagnetic waves whatsoever, it will spontaneously emit electromagnetic waves — though rather more weakly than when stimulated. The spontaneous waves will get their energy from the cylinder's spin; as they emerge from the cylinder, the spin will slow a bit. Similarly, if a spinning black hole is capable of stimulated emission, then it must also be able to emit waves spontaneously. A spinning black hole, all alone in space, with no incoming waves whatsoever, should spontaneously emit all forms of waves: electromagnetic, gravitational, ... and as the waves flow out, the hole's spin should slow.'

'That's crazy!' I told him. 'It violates everything we know about black holes. Waves and particles can go down a hole, but nothing at all can ever come out.'

Zeldovich, though unsure, was insistent. 'Let's look closely at how the spinning metal cylinder spontaneously emits,' he said. He then reminded me that vacuum fluctuations are a consequence of quantum mechanics' uncertainty principle. The uncertainty principle insists that nobody can ever know simultaneously, and with complete precision, both the position of an electron and its velocity (i.e., its rate of change of position). The uncertainty in the rate of change of position is inversely proportional to the uncertainty in the position.

The uncertainty principle also applies to an electromagnetic wave. There the analog of an electron's position is the strength of the wave's electric field. Thus, the field strength and its rate of change cannot both be known simultaneously and with complete precision. The uncertainty in the rate of change of field strength is inversely proportional to the uncertainty in the field.

This uncertainty principle forces electromagnetic waves, in each cycle of their oscillation, to always contain a certain minimum total energy. To see why, assume that you have found a wave that contains precisely zero energy in a cycle of its oscillation. The wave must have vanishing electric field strength throughout the cycle, since energy density is proportional to the square of the electric field. This, however, implies that throughout the cycle both the field and its rate of change vanish; they are both known to be precisely zero. But this simultaneous precise knowledge violates the uncertainty principle. The wave energy thus cannot vanish.

Suppose we extract from a wave all the energy that the uncertainty principle allows. What will be the form of the remaining, irremovable energy? It will consist of a randomly fluctuating electric field (and also a randomly fluctuating magnetic field, which is governed by the same arguments as we are exploring for the electric field). At any moment of time the electric field will be close to zero but will differ from zero by a small unknown amount. The field's uncertainty, while small, must be large enough to permit the uncertainty in its time rate of change also to be small, thereby permitting the field uncertainty to remain small as time passes.

If we remove all extractable energy from every wave that can exist in spacetime, we will leave spacetime as empty as possible. This state of most extreme emptiness is called the *vacuum state* and the fluctuating electric fields and magnetic fields that all the waves in this vacuum state must have are called *vacuum fluctuations*.

Because the energy of the vacuum fluctuations is not extractable, the fluctuations are difficult to observe. They are all around us, and inside us; they exist everywhere in the universe; but they are nearly invisible. The only way they ever become visible is by interacting with other objects.

'Suppose,' Zeldovich said to me on that June morning in 1971, 'that we have a spinning metallic cylinder surrounded by a perfect vacuum. Examine one of the waves that, when in a non-vacuum state, stimulates the cylinder to emit as it skims over the cylinder's surface. Although the incoming wave is now in its vacuum state, the vacuum fluctuations of its electric field will still create fluctuating electric currents in the cylinder's surface, and those fluctuating currents will still emit outgoing waves. Thus, the wave will arrive at the cylinder in its vacuum state, but will leave with excess, non-vacuum energy.' 'The excess energy,' Zeldovich reminded me, 'is said to have been *spontaneously emitted*, even though the emission was really stimulated by the incoming vacuum fluctuations. The energy for the spontaneous emission comes, not from the vacuum fluctuations (one can never extract their energy), but rather from the cylinder's spin. As the outgoing waves carry off energy, the cylinder gradually slows its spin. When, ultimately, the cylinder stops spinning, no more energy can be extracted from it, and the spontaneous emission will shut off.'

'In a similar way,' Zeldovich asserted, 'a spinning black hole should spontaneously emit waves. Vacuum fluctuations impinging on the hole should stimulate the emission. As the waves carry off energy, the hole should gradually slow its spin, until finally the spin stops and the spontaneous emission stops.'

'Yes,' I agreed, '*if* a black hole can amplify incoming waves, then it will probably emit waves spontaneously, in response to vacuum fluctuations. However, it is outrageous to think that the hole can amplify incoming waves! And it is equally outrageous to think that the hole can spontaneously emit. Black holes are fundamentally different from spinning cylinders and other spinning objects,' I said. 'They have horizons; other objects do not; waves can go down the horizons but cannot come out, and thus a black hole cannot emit.'

Zeldovich offered me a wager. In the novels of Ernest Hemingway (which he enormously admired), Zeldovich had read of White Horse scotch, an elegant and esoteric brand of whisky. If detailed calculations with the mathematical laws of physics show that a spinning black hole spontaneously emits waves, then I was to bring to Zeldovich from America a bottle of White Horse scotch. If the calculations show that there is no such spontaneous emission, Zeldovich would give me a bottle of fine Georgian cognac.

I accepted the wager; but I knew it would not be settled quickly. To settle it would require understanding, clearly and unequivocally, the laws of *quantum fields* (the laws governing quantum mechanical features of electric, magnetic, and other fields) in curved spacetime. The laws of quantum fields in flat spacetime were well developed, but not in curved spacetime. To combine quantum fields with spacetime curvature was a difficult first step toward the marriage of quantum mechanics and general relativity, a first step on which Leonard Parker, Bryce DeWitt, and a few others had begun to work in the 1960s, but which was far from complete.

Two years later, in September 1973, I made my fourth visit to the Moscow, this time accompanying Stephen Hawking and his wife Jane. This was Stephen's first Moscow trip. I accompanied him because of his severe physical disability (amiotrophic lateral sclerosis or ALS, a motor neuron disease that by 1973 had confined him to a wheelchair and made his speech intelligible only to his family and others who knew him well). Stephen, Jane, and our Soviet hosts, uneasy about how to cope in Moscow with Stephen's special needs, thought it best that I, being both a close friend of Stephen's and familiar with everyday Moscow life, act as companion, translator for physics conversations, and guide.

We stayed at the Hotel Rossiya, just off Red Square near the Kremlin. Although we ventured out nearly every day to give a lecture at one institute or another, or to visit a museum or the opera or ballet, our interactions with Soviet physicists occurred for the most part in the Hawkings' two-room hotel suite, with its view of St. Basil's Cathedral. One after another the Soviet Union's leading theoretical physicists came to the hotel, to pay homage to Hawking and to converse.

[Hawking already in 1973 was a towering figure in theoretical physics. Though only seven years past his PhD, he had already deduced from general relativity the second law of black hole mechanics (areas of horizons can only increase) and other laws about the dynamical evolution of holes. With Roger Penrose he had proved, relying on no idealizing assumptions, that the interiors of black holes and the beginning of the universe must contain *singularities;* and with George Ellis he had published a monumental treatise on the 'global methods' of calculation used to deduce these remarkable results. The golden age of black hole theory was in full swing, and by 1973 Hawking had become its dominant figure.]

Among the physicists who made repeated trips to Hawking's hotel room were Zeldovich and his PhD student Alexei Starobinsky. Hawking found Zeldovich and Starobinsky as fascinating as they did him. On one visit, Zeldovich described to Hawking his conjecture that a spinning black hole should spontaneously emit all types of waves: gravitational, electromagnetic ... and then he described a partial formulation of the laws of quantum fields in curved spacetime that Starobinsky and he had developed, and a tentative proof, using those laws, that a spinning hole does, indeed, emit waves spontaneously. Zeldovich was well on his way toward winning his bet with me.

Of all the things Hawking learned from his conversations in Moscow, this one intrigued him most. However, he was skeptical of the manner in which Zeldovich and Starobinsky had formulated the laws of quantum fields in curved spacetime; so, after returning to Cambridge, he began to develop his own formulation and use it to test their claim that spinning holes should spontaneously emit waves. In the meantime, several very young physicists in America were doing the same thing; among them William Unruh at Berkeley, Larry Ford at Princeton, and Don Page, a student of mine at Caltech.

By early 1974, Unruh, Ford, and Page, each in his own way, had tentatively confirmed Zeldovich's prediction: A spinning hole should spontaneously emit waves, until all of its spin energy has been used up and the emission turns off. Then came a bombshell. Stephen Hawking, first at a conference in England and then in a brief technical article in the journal *Nature*, announced an outrageous conclusion that conflicted with everyone else. His calculations predicted that, as the hole spontaneously emits waves and gradually stops spinning, its spontaneous emission does *not* turn off. With no spin left, and no spin energy left, the hole keeps on emitting waves; and as it emits, it keeps on losing energy, or equivalently mass. Since the spherical horizon of a nonrotating hole has a circumference proportional to its mass, its surface area is proportional to its mass squared. Thus, as the nonspinning hole emits waves and loses mass, its surface area must shrink, in violation of Hawking's own second law of black hole mechanics. Ultimately, if one waits long enough, the hole's mass will shrink to zero, the area of its horizon will shrink to zero, and the hole will be gone. It will have evaporated away to nothing.

The world's half dozen experts on quantum fields in curved spacetime were quite sure that Hawking had made a mistake. His conclusion violated everything then known about black holes. Perhaps his formulation of the laws of quantum fields in curved spacetime, which differed from other peoples' formulations, was wrong; or perhaps he had the right laws, but had made a mistake in his calculations.

For the next few years, the experts examined minutely Hawking's formulation of the laws and their own formulations, Hawking's calculations of the waves from black holes, and their own calculations. Gradually one expert after another came to agree with Hawking.

How was it possible to reach agreement on the fundamental laws of quantum fields in curved spacetime, without any experiments to guide the choice of the laws? How could the experts claim near certainty that Hawking was right, without experiments to check their claims? Their near certainty came from a demand that the laws of quantum fields and the laws of curved spacetime be meshed in a totally consistent way. The new, meshed laws had to be consistent with general relativity's laws of curved spacetime in the absence of quantum fields, and with the laws of quantum fields in the absence of spacetime curvature. This demand for a perfect mesh, analogous to the demand that the rows and columns of a crossword puzzle mesh perfectly, turned out to constrain the new laws almost completely. While the demand for consistency among laws is powerful, rarely has it been so powerful as this. For example, perfect meshing was not enough, by itself, in 1915 to guarantee that gravity is described by Einstein's curved spacetime laws.

In September 1975 I returned to Moscow for my fifth visit, bearing a bottle of White Horse scotch for Zeldovich. To my surprise, I discovered that although all the Western experts by now had agreed that Hawking was right and black holes can evaporate, nobody in Moscow believed it. Although several confirmations of Hawking's claims, derived by new, completely different methods, had been published during 1974 and 1975, those confirmations had had little impact in the USSR. Why? Because Zeldovich and Starobinsky, the greatest Soviet experts, were disbelievers. I argued with them, to no avail; they knew so much more about quantum fields in curved spacetime than I that although I was sure I had truth on my side, I could not counter their arguments.

My return flight to America was scheduled for early Tuesday morning, September 23. On Monday evening, as I was packing my bags in my tiny room at the University Hotel, the telephone rang. It was Zeldovich: 'Come to my flat, Kip! I want to talk about black hole evaporation!' Tight for time, I sought a taxi in front of the hotel. None were in sight, so in standard Muscovite fashion I flagged down a passing motorist and offered him five rubles to take me to Number 2B Vorobyevskoe Shosse. He nodded agreement and we were off, down back streets I had never travelled. My unease abated when we swang onto Vorobyevskoe Shosse. With a grateful 'Spasibo!' I alighted in front of number 2B, jogged through the gate and forested grounds, into the building, and up the stairs to the second floor, southwest corner.

Zeldovich and Starobinsky greeted me at the door, grins on their faces and their hands above their heads. 'We give up. Hawking is right; we were wrong!' For the next hour they described to me how their version of the laws of quantum fields in a black hole's curved spacetime, while seemingly different from Hawking's, was really completely equivalent. They had concluded black holes cannot evaporate because of an error in their calculations, not because of wrong laws. With the error corrected, they now agreed. There is no escape. The laws insist that black holes must evaporate.

****

In 1986, about a year before his death, Zeldovich and I spent an evening together in has flat, eating zakuski, drinking wine, Georgian brandy, and vodka, and talking. In a somewhat black mood, he recounted to me a long list of major discoveries in physics and astrophysics that he had failed to make, including the evaporation of black holes. I was amazed by his mood and told him so. It seemed to me then, and now, that no theorist in the second half of the 20th century had a greater impact on our understanding of the astrophysical universe than Zeldovich. Much of that impact emerged from his own research, but even more from the research of others — research stimulated in seminal ways by conversations with Zeldovich (as in Hawking's case) or by his writings. Yakov Boris'ch (as I learned to call him, with a mixture of affection and awe) was a profoundly insightful man.

Part VII

From the Popular Science Legacy

In memory of a friend[172]

Ya.B. Zeldovich

Boris Pavlovich Konstantinov. Academician, director of the Leningrad Physical Technical Institute (1955–1963), Hero of Socialist Labour ... A bronze bust in front of the entrance to the Institute ...

To me, he will always be a friend, a relative, just Boris, my first wife's brother, and to my children — kind 'Uncle Borya.' Anyone who thinks that people should write impartially or passionlessly — should not read this essay. I loved Boris, and I'm now sad that he is not with us, that he left us so early. He remained sharp-witted, inventive, deep and merry, kind and active up to his last months, when his heart tragically grew tired of beating. Boris Pavlovich didn't see his sixtieth birthday. These notes are dedicated to the 75th anniversary of his birth. And can it really be that we can't dedicate them to a living person? How much Boris Pavlovich could have done with his mind and talent over these fifteen years.

We met in 1932 in the Konstantinovs' large flat on Malaya Podyacheskaya Street near Griboedov Canal in Leningrad. We were introduced by Varvara Pavlovna, Varya.

I learned that Boris was a physicist, that he worked in an acoustic laboratory, and that he — like me, at the Institute of Chemical Physics — didn't have a higher education. I learned that he had recently been seriously ill — at that time, it was called 'heart disease,' and that Nina Nikolaevna Ryabinina put him back on his feet and became his wife. Of the large family — six brothers and three sisters — Boris was the most talented, song-loving, and lively. But the entire family was splendid. In difficult years, the brothers and sisters remained solid, helped each other live and obtain educations. It is likely that, in Boris Pavlovich's ability to work with people, understand them, and help them, a large role was played by his childhood and youth living in this large family. By the time I arrived in the family, the parents had passed

---

[172] *Nauka v SSSR*, 1984, no. 5, pp. 109–12.

away. Their father died in 1918. Recently, in the thick book *All of Petrograd, 1916*, I found the line 'Konstantinov Pavel Fedorovich, hereditary honourable citizen, contract construction, Malaya Podyacheskaya, building 10.'

He was a naturally gifted person, who began as a poor peasant son under Galich, was sent away to study and, through untiring honest labour, rose to an independent status. To this day, I remember the story of how he sat to work with his documents and accounts every day before dawn at five.

The house in which the Konstantinovs lived and the six-storey building opposite were constructed by him. We have now moved away from the primitive idea that the one who builds is he who lays the bricks with his hands (Pavel Konstantinov began from this); we understand the importance of the organiser of the work. However, in the 1920s and 1930s, when the Konstantinovs were studying and applying for work, they wrote on their application forms 'father — peasant' — the record in the book *All of Petrograd* could have shattered their fates.

Many years later, Varya told us about how lists of expelled students were hung in the Polytechnical Institute, and each time she fearfully searched for her name.

In 1929, their mother, Agrippina Fedoseevna, died. The eldest sister, Ekaterina Pavlovna Termen (her husband's name) was married, went abroad with her husband, and settled in the USA. The eldest brother — the talented radio engineer Aleksander Pavlovich Konstantinov — was married and somewhat distanced himself from his family. He has been described as an inventor in the field of television. He worked at the Pulkovo Observatory and was connected with the geophysicists. I remember his instrument for recording earthquakes from changes in the capacity of a condensor at an exhibition of the Academy of Sciences on the embankment of the River Neva.

In 1937, Aleksander Pavlovich was arrested; in 1946, I received reliable information that he had died, and ten years later, in 1956, there followed a complete exoneration. Aleksander Pavlovich very much valued the mind, talent, and inventiveness of Boris Pavlovich, helped him apply for a position in the Acoustic Laboratory, and gave him the courage to study on his own. The two of them looked similar in many ways — tall, fair-haired, with large facial features. However, Boris' childhood was spent under more difficult conditions than the healthy Aleksander's, and this also made its mark on their appearances. It may be that Aleksander was more self-assured, while Boris was kinder and more attentive.

In 1932, the younger brothers and sisters were very close; they helped each other, didn't have enough to eat, and tried to save money by not taking the tram — but by fair means or foul they managed to get into the gallery of the Mariinskii Opera Theatre, sometimes appearing on stage as extra, and went on ski trips.

Involuntarily, I think: should I write about all this? That which is dear and touching to me for understandable reasons — does the reader want or need to hear about this? Is it relevant to someone who, for example, works at the Physical Technical Institute or studies at the Physics–Mechanics

Department of the Polytechnical Institute and has only seen the bronze bust, but never met B.P. Konstantinov while he was alive?

The scientific works, inventions, and results of Konstantinov have been published or exist in reports or in various plants and laboratories; they have become impersonal. But no! It only *seems* that they are impersonal. If we try to understand his creativity more deeply, the question arises of why this person took on a specific problem, why his colleagues came to him and followed him, why he became a leader. And then, to understand the person, we must return to his childhood, his family, the source of his personality.

During the war, Boris and his family, like my family and I, were in Kazan'. I knew about Boris' serious defence work at the Physical Technical Institute, about the brilliant defence of his Doctor's dissertation in 1943, about his very good relations with senior scientists — Abram Fedorovich Ioffe, Petr Leonidovich Kapitsa, and Nikolai Nikolaevich Andreev.

Even before the end of the war, I ended up in Moscow with the Institute of Chemical Physics. In parallel with my work on the internal ballistics of solid-fuel rockets, I was attracted by Kurchatov and Khariton to atomic issues. For many years, this became the main business of my life.

Together with the Physical Technical Institute, Boris Pavlovich returned to Leningrad. For several years, we met only rarely.

In the course of our work, there arose a complex problem for which there were several very different possible technical solutions. This problem was also put to the Leningrad Physical Technical Institute, and in particular, to Boris Pavlovich.

He exhibited enormous good sense in his fundamental choice of solution, and equally enormous inventiveness in its constructional implementation. This began a new period in his life: work in the laboratory in Leningrad; discussion of the work; agreement on plans; material support from Moscow, from government departments and the Academy of Sciences; construction and operation of the process far from Moscow.

In most cases, Boris preferred to stay in our flat when in Moscow. He would sleep on the divan in our room, and sat with us in the evenings, especially with his sister. In the mornings (I'm ashamed of this now), we would wake him up with noisy exercises and the children's racket — then a car would come to take him to the ministry or the train station.

I remember an evening when Boris was unusually pale, silent, and serious. Difficulties had been discovered, possibly associated with a mistake that had been made earlier. The fate of the project had been put under question. Boris drank coffee and didn't go to bed — and by morning, he had been able to find refinements and additional ideas that saved the work.

I don't need to say that the work was carried out successfully, that Boris Pavlovich received the Gold Star of a Hero of Socialist Labour, and was soon elected a Corresponding Member of the Academy of Sciences.

Finally — and this may be the most important — the process proposed and developed by Konstantinov and his colleagues remains unsurpassed even today, and is still applied on ever-growing scales.

In 1955, Boris Pavlovich was named the director of the Leningrad Physical Technical Institute. I have already referred to the bronze bust in front of the entry to the institute; in fact, there are two busts on granite columns standing opposite each other, both of people dear to all of us: Abram Fedorovich Ioffe (1880–1960), the founder of the institute and its director from 1918 until 1948, and Boris Pavlovich Konstantinov (1909–1969).

The role of Abram Fedorovich Ioffe in the establishment of Soviet physics and Soviet science cannot be overestimated. We have all — including Konstantinov — acknowledged in the past and acknowledge now that we are indebted to him, consider him our teacher in the highest sense. Therefore, the question of whether the bronze Konstantinov has the right to stand together with the bronze Ioffe is a difficult one. I prefer to present my defence of Konstantinov openly, leaving no room for reservations.

First, the main thing to note is that Ioffe did not cease to be the director of the PTI because Konstantinov became the director. This is already clear from a comparison of dates (1948 and 1955). After Ioffe, a person from outside the Physical Technical Institute was named to replace him. This person believed it to be his task to change the style of the institute and to eradicate love for Ioffe.

Konstantinov is not the second, but the third director of the PTI. His arrival marked the beginning of a rebirth of the institute, which had somewhat lost its authority and traditions in the unpleasant intervening period. As well as providing energetic support for the old, classical research directions of the institute, Konstantinov very actively:

1. supported the development of processes for separating substances (chromatography, technical electrolysis) in the best style of the Physical Technical Institute)
2. developed astrophysical research (I'll talk about this in more detail later)
3. began a large scale work on thermonuclear-plasma diagnostics and found his place in many years of research on atomic collisions
4. strengthened theoretical departments and theoretical research (I don't mean to say that Konstantinov supervised theoreticians, such as Gribov — but Gribov, Shekhter, and many others understood that there was no need for them to find other places of work, that they were valued, and that they had opportunities to gather young researchers to them and teach them to work on their own)
5. advanced nuclear physics (a cyclotron was constructed in Gatchina, and the Nuclear Institute bearing the name of Konstantinov arose and then acquired an independent existence)

It is likely that this is not a complete list, and that I have failed to refer to dozens of individual affairs brought about by Konstantinov that were beneficial to the institute — in connection with people and equipment. For this he sacrificed his time, strength, and health and died before reaching the age of sixty. Ask any researcher at the PTI who has been there for 15 to 20 years, and he will tell you what Konstantinov meant for the institute.

One more important detail: Ioffe supported Konstantinov over the course of many years. I have already noted how Konstantinov defended his doctoral dissertation while the Physical Technological Institute was in evacuation in Kazan'.

Being a researcher under Ioffe, with his blessing, Konstantinov took on work of state importance, about which I've written above. Ioffe put forth Konstantinov as a candidate for the Academy. And Ioffe had no need to regret this.

Konstantinov became director of the Physical Technical Institute at a time when Ioffe already had the Semi-conductor Institute, and no longer wanted to return to the PTI. In turn, Konstantinov went to visit Ioffe, consulted with him, and supported him.

Let us recall the one hundredth birthday of Ioffe in Leningrad. The words of A.P. Aleksandrov and many others still resound, describing what the institute founded by Ioffe was and remains for this country, for science, for physics and physicists. But do we understand sufficiently well, do we comprehend, that all this would have been different if Konstantinov and then Tuchkevich had not preserved the institute!

I will not dwell here on Konstantinov as Vice President of the Academy of Sciences. No doubt others will write about this, and write well. I greatly value his activities as Vice President, activities in which the knowledge and talent of the physicist and his human qualities — his intellect and kindness — were united. At the same time, I very clearly saw that this responsibility was an extreme physical load for Boris Pavlovich. This load, taken on voluntarily, together with his role as director, sapped his health over the course of several years. His heart weakened. I will never forget how Boris came to our dacha to rest one Saturday evening. It was difficult for him to climb the single flight of stairs. We carried an armchair into the garden, sat around Boris Pavlovich, laughed and joked. And in our hearts there arose an enormous anxiety for him, for his life. In that period, he didn't leave his responsibilities. Attempts to help him in devious ways didn't work. Should a doctor or someone close to him have told him about the deathly danger? He didn't leave any final or parting letter. He was probably optimistic and hoped to regain his health — this is a necessary component of a healthy psyche, even if one's heart is ill.

I would like to talk in more detail about only one work of Boris, which greatly occupied him and was very controversial. Again, I warn you — don't expect impartiality from me!

I'm referring to Konstantinov's astrophysical idea of 'antimatter at home,' in the solar system, in meteors. This idea is unquestionably incorrect. It contradicted experiments conducted in Konstantinov's laboratory. Moreover, one could show from a careful analysis of indirect data that the idea of antimatter in the solar system (or even in our galaxy) was highly improbable.

Now, after 20 to 30 years, the question is sometimes raised on a personal level: is it appropriate to consider Konstantinov's work on searches for antimatter an amoral act, testifying to his poor knowledge of the subject, his unwillingness to listen to the opinion of specialists, or, even worse, his desire to obtain glory, honour, and status via unjust means?

It is easier to address the last of these accusing questions. It is sufficient to point out that by the beginning of his astrophysical activities, Boris Pavlovich was already an Academician, director, and Hero of Socialist Labour. He received all these honours for undoubted service in various fields. His astrophysical activities didn't promise him anything personally except for additional work, fuss, and anxiety — and also, of course, the satisfaction of his scientific interest!

I remember well the attitude toward the idea of antimatter. I didn't believe in it, and considered it improbable. But I thought about the fact that, even if it were so improbable, the importance of such a discovery if it proved true would be enormous.

In physics and everyday life, we make the distinction between probability and mathematical expectation. One joke says that mathematical expectation is 'the product of probability and unpleasantness.' A probability of 1% = 0.01 is small — small compared to unity; probability is dimensionless, and must be compared to unity. But if this is the probability that an illness will be deadly, the disease is taken very seriously! The mathematical expectation of winning a Volga car with a probability of 1% is 100 to 150 roubles.[173] And so forth.

So here, given the small probability of discovering antimatter, I didn't consider the mathematical expectation to be that small. From this point of view, the counter-arguments weaken. Yes, it has been shown that 10 or 100 meteor showers are made of normal matter, and in the spectrum of the long-lasting glowing track there are lines of normal iron. Yes, the majority — the overwhelming majority — of meteors are 'normal.'

Does this prove — strictly logically — that the $n + 1$th, 11th, or 101st shower will not be of antimatter?

The question is moved to another plane — should one only take on work with guaranteed success, should one refuse right from the beginning work that carries a large risk? One example of work with a large risk of a negative result is searches for radio signals from extraterrestrial civilisations. However, we know that this work has continued over many years and is not considered compromising.

Let us return to the question of antimatter. Indeed, there is no antimatter in the solar system. Konstantinov's colleagues themselves came to this conclusion; for many astronomers, this conclusion was trivial right from the very start of the work.

But one can and should look more deeply, and ask the question, why is there no antimatter?!

---

173 The price of a Volga car at that time was between 10,000 and 15,000 roubles. [Translator]

Then, the pragmatic point of view 'No — and that's all there is to it' begins to resemble Chekhov's expression 'This doesn't happen because this doesn't ever happen.'

The formulation of Konstantinov's question was a naive expression of sacred anxiety. In the formation of the solar system, there is no place for antimatter. But the more general question posed by Konstantinov is: 'Why can't there be antimatter in the universe? And an especially pointed question is why can't there be antimatter in the hot universe, in which, in the distant past, during the course of several minutes, there were certainly superhigh temperatures sufficient to give rise to pairs of protons and anti-protons?

While literally incorrect, Konstantinov's idea undoubtedly stimulated an extremely important direction in modern cosmology — theoretical studies of the extremely early universe.

There is no space to discuss this research direction in more detail here. Boris Pavlovich didn't live to see the day when such questions began to be discussed widely. I refer the reader to an article by A.D. Dolgov and myself in the August 1982 issue of *Priroda*.

Let's return to the Leningrad Physical Technical Institute. Enthusiastic about his idea, Boris Pavlovich founded an astrophysics department. This department is now one of the leading centres for high-energy astrophysics research.

In 1981, Mazets and his group achieved world-wide leadership in the study of cosmic sources of gamma radiation. And to this day, Mazets is grateful to Boris Pavlovich, who attracted him, a qualified nuclear physicist, to work in astrophysics.

And so, an idea that was literally incorrect was nonetheless fruitful in connection with theoretical and experimental work. An analogy with the idea of thin-layer isolation, which played a major role in the history of Leningrad PhysTech in the pre-war period, comes to mind. There is no space to describe this theme in more detail here.

Let us glance over the entire life path of Boris Pavlovich Konstantinov; the remarkable integrity of the man and the scientist; a life which was short but vivid and fruitful; a life lived with dignity.

Autobiographical afterword[174]

Ya.B. Zeldovich

The last page of the last paper has been turned over, and naturally the question arises of the sum total of 70 years of life and 53 years of research, and of lessons for the future that can be drawn from this.

---

174 *Selected Works. Particles, Nuclei* [in Russian]. Moscow: Vselennaya, 1985, pp. 435–46.

The first question — the sum total — is the subject of the first article, the introduction composed by the editors of the collection, placed at the beginning of the first book but encompassing the contents of both volumes. In my view, the introduction presents an overestimation of my results and their influence on modern science.

It would be inappropriate, however, to argue about the greater or lesser importance of one or another piece of work. Qualitative differences between estimates of my work, and also of the general state of physics, could be interesting: external estimates by specialists, even the most well-disposed, and internal ones by me myself. For this reason, this afterword is written from a particularly subjective position, without any pretence to objectivity.

I remember well my first choice of a field of knowledge in a conversation with my father, when I was still a child (12 years old). Mathematics requires special capabilities, which I didn't feel I had. Physics seemed a complete science: this reflected the influence of my venerable school physics teacher, solemnly reading the unshakeable laws of Newton, first in Latin, then in Russian. The restless soul of the new physics had not yet penetrated to middle school in 1926. At the same time, my course in chemistry had an abundance of riddles: what is valence, catalysis? And chemists didn't hide the absence of a fundamental theory. Ya.I. Frenkel's book *The Structure of Matter* made a big impression on me, especially the first part, which is primarily dedicated to atomic theory and kinetic gas theory, the definition of Avogadro's number and Brownian motion. But atomic theory, like thermodynamics, is equally related to physics and chemistry. Later, my fate led me to the Institute of Chemical Physics (ICP).

In 1930, I was a laboratory assistant at the Institute for the Mechanical Processing of Mineral Resources (*Mekhanobr*), and was working with thin sections of mountain rock. I'll always remember the richness of the Kol'sk Peninsula, and I was impressed with respect for Academician A.E. Fersman. In March 1931, I visited the department of chemical physics at the Leningrad Physical Technical Institute on an excursion with workers from *Mekhanobr*. In S.Z. Roginskii's laboratory, I was interested by the crystallization of nitroglycerine in two forms. L.A. Sena described this to us (Roginskii was abroad).

After some discussion (in which neither I nor Sena yet knew the truth), they suggested to me that I work at the laboratory in my free time. Soon, the question of an official transfer arose. By the time I became an employee there (May 15, 1931), the department had been transformed into the independent Institute of Chemical Physics. In the interim, I remember my synopsis talk about the kinetics of the transformation of parahydrogen to orthohydrogen. Not fully understanding what it was, all the same I firmly and passionately defended the principles of detailed equilibrium. N.N. Semenov, S.Z. Roginskii, and many other future colleagues were present.

Many years later, I heard three legends. The first: *Mekhanobr* gave me to the ICP in exchange for an oil pump. The second: Academician A.F. Ioffe wrote to *Mekhanobr*, saying that I would never be any use at solving practical problems. The third: Ioffe couldn't stand 'infant prodigies,' and so gave me to the ICP.

To this day, I don't know how much truth there is in each of these. I can only testify to the fact that I didn't see Ioffe until 1932, and I saw him then under remarkable circumstances. A general seminar for PhysTech and its sister institutes had been founded. Ioffe read a telegramme from J. Chadwick about the discovery of the neutron, commented on it, and in conclusion, a resolution was adopted and a telegramme sent in response saying that we (all?!) would be working on neutron physics. For me — though not immediately — this resolution turned out to be prophetic.

In my interest in chemistry, a large role was played by the purely visual impression of bright colours and shapes, beginning with the 'transformation of water into blood' during the interaction of iron salts and potassium sulphocyanate, with the formation of precipitates and crystallisation. This was followed by an interest in the sharpness of the colour transitions of indicators, and further in the sharpness of phase transitions.

Atomic spectra were studied in neighbouring laboratories. I clearly remember that, in comparison with the variety of colours and shapes of macroscopic phenomena, the detailed atomic theory turned out to be boring. I'm writing about this today in order to testify to my deep lack of understanding of physical theory at that time.

Together with this, there was the correct and natural feeling that general regularities can be discerned behind this randomness of form and alternation of smoothness and roughness. Today, these regularities have acquired the name of catastrophe theory and synergetics.

In the 1930s, developing the theory of combustion, we were essentially studying specific examples of these new sciences without knowing their names. Recall Molière's muddle-headed noble, who learned in his declining years that, his whole life, he had been speaking prose.

One enormous and lasting service of Abram Fedorovich Ioffe and Nikolai Nikolaevich Semenov is the founding of institutes that could attract capable young people from all over. A 'supercritical' situation arose, with a rapid growth of promising personnel and a large output from them. For me, the possibility of studying with young theorists (who were nevertheless older than me!) played a large role. I am deeply grateful to my teachers of that time, who are now my friends, L.E. Gurevich, V.S. Sorokin, O.M. Todes, and S.V. Izmailov. I studied at (but didn't graduate from) the correspondence division of the university for about two years. I went to splendid lectures on electrodynamics by the late M.P. Bronshtein. I remember the words 'gradient invariance,' which I didn't understand at the time ...

The combination of experimental and theoretical research on a single question was a great joy. I first observed the Freundlich adsorption isotherm experimentally while studying a MnO_2–CO–O_2–CO_2 system. Only after this was the corresponding theory developed (see Paper I in my book *Chemical Physics and Hydrodynamics*). Without delay, I verified experimentally the temperature dependence of the exponent n in the formula $q = cP^n$.

There was nothing fundamentally new in the experiment; the Freundlich isotherm, as shown by its name, had been discovered by Freundlich and not by me. However, the experiment awoke my desire to understand the phenomenon and to construct a theory for it. I think that this is a general phenomenon. I insistently advise theoreticians working in macroscopic physics to participate in experiments!

One particular cycle of work on adsorption and catalysis made up my Candidate of Science dissertation. Those blessed times when the VAK[175] would allow people without a higher education to defend such dissertations! The defence was in September 1936.

Even earlier, I began to swim independently and decided to study fuel elements. My interest in electrochemistry was warmed by respect for Academician A.N. Frumkin, who was kindly disposed toward my research on adsorption, which was to a substantial degree done in parallel with work by him and M.I. Temkin. Reflections about paths for the transformation of fuel energy into electricity naturally arose under the influence of A.F. Ioffe.

However, I essentially ended up alone in Leningrad at the ICP in connection with the question of fuel elements. The work proceeded very slowly.

In 1935, the extremely energetic and go-getting A.A. Rudoi from Odessa came to — or rather burst into — the institute. He was inspired by the chain theory of chemical reactions. What was preventing us from finding a means to transform combustion energy into the energy of active centres, and use it for endothermic nitrogen oxidation reactions? Why not obtain several litres of nitric acid from a single kilogram of fuel and the free air? Behind a misty fog, a very idyllic picture was painted: a tractor ploughing a field simultaneously supplies it with nitrous fertilizer, and the classical installations for the synthesis of ammonia lie in desolation. Semenov took Rudoi into the institute, and simultaneously created a serious group to investigate the question. This group included the late P.Ya. Sadovnikov, D.A. Frank-Kamenetskii, and A.A. Koval'skii. It also included me. It turned out that the formation of oxides of nitrogen during the burning of hydrogen in air had been observed by Cavendish, before the composition of the air was established.

I won't describe here the results of this large collective work — it is laid out in my book *Chemical Physics and Hydrodynamics*.

I again worked as both an experimentalist and a theoretician. The work forced me to study and apply the theory of dimensionality, similarity and self-similarity, and expanded my horizons. It led me to the problems of turbulence, convection and thermal technology. A.A. Gukhman's book *Similarity Theory* was inspiring. A strong and fruitful friendship with David Albertovich Frank-Kamenetskii was forged. An engineer by education, he sent a letter to the ICP, in which N.N. Semenov recognized his talent. He

---

175 *Vyshchaya Attestatsionnaya Komissiya*, or Supreme Attestation Committee, a governmental body responsible for awarding the highest educational degrees, such as Candidate and Doctor of Science. [Translator]

called David Albertovich from Siberia to Leningrad, and soon drew him into the work on nitrogen oxidation. It was from Frank-Kamenetskii, with his engineering education, that I learned about the Reynolds number, supersonic flow, the Lavalle nozzle, and many other things.

Much later, also in connection with nitrogen oxidation, I met with Ramzin, who was at that time, having already received a State Prize, still active but terminally ill. Working at home in the evenings, he carried out, in two weeks, work that a scientific and design institute would have stretched out for years. However, the qualitative answer was elucidated earlier. In the best case, with the heating of air and fuel, a relatively low concentration of nitrous oxides was obtained, even with the addition of oxygen. The limiting process was the transformation of NO into NO_2 via the classical three-molecule reaction $2NO + O_2 = 2NO_2$. Only NO_2 can be absorbed and used, but the technological volumes necessary for its formation are excessively large. The dream was not realized, and it is only in recent decades that the theory of nitrogen oxidation has acquired new, ecological meaning. The theory of nitrogen oxidation was the theme for my Doctor of Science dissertation, which I defended at the end of 1939. It is pleasant for me to note that the opponents included Aleksander Naumovich Frumkin. A natural continuation of this work, in which combustion was a source of high temperatures, was investigation of the process of combustion itself.

Combustion appears in many forms: the combustion of explosive mixtures, the combustion of unagitated gases, detonation, and so forth. These processes were all studied earlier, but without penetrating into the chemical kinetics of the reactions. The preceding generation of researchers based their work on thermal technology and gas dynamics. One brilliant exception was the Frenchman Taffanel, who in 1913 and 1914 published works anticipating many later developments. In 1914, he fell silent. It was only in April 1985 that I learned that Taffanel lived until 1946, successfully working on engineering problems.

Before us stood a wide field of activity, and the period from 1938 through 1941 was fruitful. The lively interest of N.N. Semenov was expressed here. As a rule, ten minutes after I got home in the evenings, Nikolai Nikolaevich would telephone, and dinner would be delayed for an hour. There were discussions of individual parts of Semenov's well-known review article in *Uspekhi Fizicheskikh Nauk* (1940, vol. 23, p. 251; vol. 24, p. 433).

A combustion laboratory was organised in the institute, where we systematically investigated the kinetics of the reaction $2CO + O_2 = 2CO_2$, right up to the highest temperatures. It may be that it was more important that a laboratory of internal combustion engines had already existed for a long time in the institute, right nearby, where K.I. Shchelkin studied detonation. I felt the most influence from the neighbouring laboratory of explosive materials. My colleagues A.F. Belyaev and A.A. Appin — who were the same age as me — worked there. This laboratory was organised and supervised by Yulii Borisovich Khariton, who remains my friend and teacher to this day. Much will be written below about my current work with Yulii Borisovich.

As a theoretical physicist, I consider myself a student of Lev Davidovich Landau. Here, there is no need to explain the role of Landau in laying the basis for and developing theoretical physics. At the same time, without belittling his role, I would like to note that, with years of maturity — I am aging, alas! — I began to understand and value more the role of other schools and people as well. Most of all, I am thinking of Ya.I. Frenkel, with his enormous intuition, optimism, and breadth; V.A. Fok, with the depth and brilliance of his mathematical technique; I.E. Tamm and his students; and the school of oscillation theory that came from L.I. Mandel'shtam. There are also many others, including thriving mathematicians who are now successfully working in theoretical physics.

Please do not read the above paragraph in the wrong way. If I write that Frenkel had intuition, and Fok was a good mathematician, you shouldn't conclude that Landau had neither intuition nor knowledge of mathematics — that isn't what I meant! Landau's talent was harmonious, and his judgement was harsh, but almost always fair. These words about schools of theoretical physics can be applied to physical schools as a whole.

In my youth, my horizon was bounded by the ICP and PhysTech. There is no doubt that PhysTech has given us a brilliant group of physicists, providing Igor Vasil'evich Kurchatov and his comrades-in-arms, who carried out state work of the utmost importance. This is excellently described in many articles and books. However, in the pre-war years, and also in the first years of the war, it seemed to me, for example, that optics was a science in which fundamental questions had already been exhausted. Today, it is sufficient to name Cherenkov radiation and lasers in order to refute that incorrect superficial judgement. The line going from Lebedev through Rozhdestvenskii and Vavilov, Mandel'shtam and Tamm, Cherenkov, Frank, Ginzburg, Prokhorov and Basov proved to be infinitely more fruitful than seemed to me in the 1930s.

It is now difficult for me to establish whether this was my own personal colour-blindness or in some measure an underestimation of another school (other schools) by some of my colleagues. In any case, from the very open reminiscences of Gamow and some remarks of Skobel'tsyn, I can now judge confidently the views of representatives of other directions in physics. The school of Lebedev very definitely felt its existence to be separate from the school of Ioffe. But let us present this theme to historians of science. Currently, there is happily no such opposition, and there has been a rather close mixing of schools which were quite separate in earlier times.

Returning to my work at the end of the 1930s, I see one significant defect: insufficient attention to publishing my results abroad. I was well acquainted with foreign research, and had published several papers in Soviet journals in English. However, it never occurred to me to send my reprints to foreign scientists. There was never any talk of trips abroad. The time was to blame, but, perhaps to some extent also to blame were senior colleagues who should have worried more about maintaining connections.

Let us continue. The discovery of uranium fission and the fundamental possibility of fission chain reactions predetermined the fate of the century — and of me. The associated papers by Yu.B. Khariton and myself are published at the beginning of this book, and I have nothing to add to commentaries on their scientific importance (and there is no reason to add to them). I would like only to note the guiding role of my teacher — Khariton — in the understanding of the overall human significance of the task. I was perhaps more interested in specific questions in the calculation methods and so forth. It was not by chance that Yulii Borisovich became a member of the Uranium Committee in 1940 (see *Uspekhi Fizicheskikh Nauk* from March 1983). The further development of the work is well known from the many reminiscences of participants.

Yulii Borisovich notes a curious detail: we considered research on the theory of uranium fission to be beyond our normal work plan, and worked on it in the evenings, sometimes very late ... By the way, apparently the institute administration also maintained that same point of view — a capable, but more practical, colleague requested 500 roubles to perform a survey of the theory of isotope fission, but the sum wasn't granted ...

Speaking of my subsequent work, I would like to emphasize the role of the theory of detonation and explosions.

The surprise of American scientists when tests of air samples showed that their nuclear monopoly had ended in August of 1949 is well documented. August 1949 — the first test of a Soviet atomic weapon — was the result of an enormous goal-oriented effort by all the people; the scientific potential of the country, accumulated back in the pre-war years, also played a role. The surprise in the USA would have been less if they had read our papers from the pre-war years, published in Russian. I am speaking here not only of work on the chain fission of uranium. Work on explosions and detonation theory is also a necessary part of such research, without which the problem cannot be solved. Recall that Khariton formulated the condition for a detonation limit as early as 1938. I formulated the final one-dimensional detonation theory in 1940. In the USA, the same problem was solved by John van Neumann — an outstanding mathematician — only in 1943. Note that van Neumann worked on this task precisely in connection with the problem.[176]

Soon after the beginning of the war, the Institute was evacuated to Kazan'. The task arose of performing a detailed analysis of processes associated with our rockets — 'katyushas.' The theory of combustion of powders, which was sufficient for the internal ballistics of barrel artillery, required refinement. A delicate balance between the arrival of powdered gases during combustion and their departure through the nozzle must be maintained in the combustion chamber for a reactive discharge. New

---

[176] For more detail about the history of the development of detonation theory, see the book *Chemical Physics and Hydrodynamics*, in both the chapters and accompanying commentaries.

concepts about the combustion of powders, the phenomenon of blowing (discovered in our laboratory by O.I. Leipunskii), the role of the heated layer of powder were all unfamiliar to artillery workers, and obtained varying evaluations from powder workers and specialists in internal ballistics.

I would like to note the interest in and support for the work shown by General Professor I.P. Grave, the well-known rocket constructor Yu.A. Pobedonostsev (both now departed ...), and the still thriving G.K. Klimenko. However, we did not always encounter such support, and there were also keen arguments and attempts at exerting administrative influence, as well as shouted exchanges of scientific arguments.

In connection with our work on powder combustion, our group was transferred to Moscow. We turned out to be the advance party, and the entire Institute of Chemical Physics was directed to Moscow (and not back to Leningrad) at the end of the war. Work on combustion and detonation, as well as on powder combustion, continued at the ICP even after the group of theoreticians (including me) had moved to work on new themes. I would like here to express my deep gratitude for this to A.G. Merzhanov and his group, B.V. Novozhilov, G.B. Manelis, A.I. Dremin, and many others (the Institute of Chemical Physics of the USSR Academy of Sciences). In the course of their research, they did not forget my work — and did not allow others to forget. Without this continuity, there is no doubt that very much would be discovered anew abroad. There is no task more thankless than a belated battle for priority ...

One's first love cannot be forgotten — and so in 1977, a science council on the theoretical foundations of combustion processes was organised. I am still working in the field of combustion up to the present day, although not at full steam. In connection with combustion problems, with the close collaboration with G.I. Barenblatt in the 1950s, the concept of an intermediate asymptote was formulated, which has general value for mathematical physics. Also together with him, a very general solution was found in the theory of perturbations of self-wave processes (for example, the propagation of flames), corresponding to a shift and having an increment that is identically equal to zero. Physicists working in field theory will see here an analogy with so-called Goldstone particles.

Together with A.P. Aldoshin and S.I. Khudyaev (ICP), I studied the transition from the theory of Kolmogorov, Petrovskii, Piskunov, and the Englishman Fisher to the theory of Frank-Kamenetskii and myself. In the most general case of reaction kinetics and arbitrary initial conditions, the correct approach to the problem of propagation again proves to be linked with the idea of intermediate asymptotes.

The question of the hydrodynamic instability of flames, discovered by L.D. Landau, has proven to be far from simple: here, after the very fundamental work of A.G. Istratov and V.B. Librovich, it was only in the 1980s that it became possible to progress, together with V.B. Librovich and N.I. Kidin.

Ideas taken over from field theory enable us to approach the non-linear theory of spin combustion in a new way. Recently, in the framework of the

Council, much attention has had to be paid to organisational work associated with the large energetics of coal combustion.

Let us return to the atomic problem and to the 1940s and 1950s. A huge collective was led by Igor Vasil'evich Kurchatov. A very important part of the work was supervised by Yulii Borisovich Khariton. Soon, this problem fully encompassed my work as well. In those difficult years for the country, no expense was spared to create the optimum conditions for this work.

For me, those were happy years. Large-scale new technology was created in the best traditions of great science. Attention to new proposals and to criticism, completely independent of the ranks and titles of the authors; the absence of concealment and suspiciousness — that was the style of our work.

The country survived the difficult post-war years. However, the enormous authority of Kurchatov created a healthy atmosphere. Moreover, our work exerted a favourable influence on Soviet physics as a whole. Once, when I was in Kurchatov's office, a telephone call came in from Moscow: 'So should we publish an article in *Pravda* by a philosopher that refutes the theory of relativity?' Without reflecting a minute, Igor Vasil'evich responded: 'In that case, you can close our entire establishment.' The article was not printed.

By the middle of the 1950s, some of the most urgent problems had already been solved. New trends appeared, and the Geneva conference on the peaceful use of atomic energy and Kurchatov's famous presentation in Harwell (England) on thermonuclear reactions became landmarks of *détente*.

Some works associated with applied themes presented more general scientific interest and were published. These included work on strong shock waves, their structure, and their optical properties.

Interest in phenomena occurring at high temperatures also led to the fundamental question of the establishment of thermodynamical equilibrium between photons and electrons. The specific nature of this problem was that, at sufficiently high temperatures, scattering becomes predominant over emission and absorption. A.S. Kompaneets carried out splendid work in this area. This was published in 1965, and proved to be extremely important for cosmology and astrophysics, for plasma in the hot universe, and for the emission of material falling in the gravitational field of a black hole.

Work on the theory of explosions prepared me psychologically for investigations of explosions of stars, and of the largest explosion — that of the universe as a whole.

Simultaneously, my industrial work stimulated interest in nuclear physics and the physics of neutrons. In the 1950s, from here it was possible to reach out and touch the physics of elementary particles. The thin book *Theory of Elementary Particles* by Enrico Fermi provided an enormous stimulation for me. In the English edition, which I used, but not in the Russian translation, the publisher (not Fermi) gave the following warning on the cover:

'This book is published using the means of a certain wealthy lady, bequeathed for proof of the existence of God. Revealing the laws of nature and their harmony proves the existence of God more than theological texts.'

If by the existence of God we mean the objectivity of the laws of nature, which exist independent of our knowledge and desires, then any Marxist philosopher could put his signature under this thesis.

As part of my self-education, I worked through the best description of the general theory of relativity — the second part of *Field Theory*, the second volume in the course in theoretical physics by Landau and Lifshitz.

I would like once more to emphasize the enormous role played by my association with Lev Davidovich Landau. In Kazan', and then in Moscow, we lived near each other and were in close contact in connection with our work. The possibility to visit him, ask for advice, bring my proposals, ideas and work for his judgement — I feel all this was a huge blessing. I learned about the tragedy of January 1962, when Landau ceased to be a theoretical physicist (although he remained among the living) when I was far from Moscow. The unforgettable anxious days, weeks, months of fighting to save his life, the solidarity of physicists, overstepping state boundaries. The school founded by Landau was preserved! It lives in those who continue the monumental course of theoretical physics — E.M. Lifshitz, L.P. Pitaevskii. It lives in the L.D. Landau Institute of Theoretical Physics of the USSR Academy of Sciences. Its organisation, the selection of people, the maintenance of the highest professional level of theoreticians is a huge service provided by I.M. Khalatnikov and his colleagues. In a narrow sense, the theoretical department of the Institute of Theoretical and Experimental Physics of the USSR Academy of Sciences can also be considered part of the school of Landau — the offspring of I.Ya. Pomeranchuk, currently headed by L.B. Okun'. In a wider sense, the ideas and methods of Landau, together with those of other eminent Soviet theorists (I briefly mentioned them above), became an organic part of all Soviet theoretical physics.

Returning to the memoir genre, I would like to say that my work with Kurchatov and Khariton gave me very much. The main thing was and remains the inner sense that we have carried out our duty to the country and people. This gave me a certain moral right to study in the following period such questions as particles and astronomy, without worrying about their practical value. Above, I wrote about how scientific interest in these questions ripened. Together with this, I should talk self-critically about my own weaknesses and the difficulties I encountered during the new turn in my scientific activity. I remember that in 1964, I officially transferred to the Institute of Applied Mathematics of the USSR Academy of Sciences (IAM), founded by M.V. Keldysh in 1953. A.N. Tikhonov has headed this institute since his death. I worked in this institute for 19 years (until my transfer to the Institute of Physical Problems at the beginning of 1983).

Before my transfer to the IAM, my research on particles and astronomy was done out-of-hours, and was to some extent optional — and now I see that this was reflected in its quality. Until recently, I was proud that I had obtained the maximum physical results using a rather elementary reserve of mathematical knowledge; but now, and especially in connection with elementary particle theory, the opposite side of this assertion stands in front of me. And why should one essentially be limited to some modest volume of mathematical knowledge? However, I am thinking about this now as applied to a professional theoretical physicist.

There is the completely different question of how the teaching of mathematics begins in middle school. When my children were growing up, I examined their school textbooks and decided to write a new one. This was the origin of my book *Higher Mathematics for Beginning Physicists and Technicians*.

I'll present part of my letter published in the American journal *Physics Today* in connection with a discussion in that journal about the reasons for the decline in the level of physics teaching in the USA.

'In connection with the discussion of how to teach the younger generation physics, I would like to recall one general difficulty.

The laws of physics are formulated in the form of differential equations: such, for example, are the Newtonian laws of motion of a material point, a solid body, or a gyroscope. Maxwell's laws for the electromagnetic field are equations in partial derivatives, and the laws of gas dynamics are also described in this way.

Schoolchildren are capable of understanding all this material.

However, it would be more precise to assert that they are not capable of deeply understanding and loving physics if the necessary reserve of mathematical terminology is lacking. Here is my main comment: in most cases, the teaching of mathematical analysis begins only with some delay, and includes the difficult elements of the set and limit theory.

So-called 'rigorous' proofs and existence theorems are much more complicated than an intuitive approach to derivatives and integrals.

As a result, the mathematical ideas that are necessary for understanding physics reach schoolchildren too late. This is like serving salt and pepper not at lunch, but a bit later — with your mid-afternoon cup of tea.'

But let us return to the mathematics that is used and operates in modern theoretical physics.

To an enormous degree, particle theory is developing under the influence of supporting mathematical ideas and along directions indicated by mathematical elegance. I won't discuss the classical textbook example of the Dirac theory of relativistic electrons, which led to the concept of antiparticles. Let us turn to isotopic invariance. A discrete symmetry is observed experimentally: the exchange of a proton with a neutron (or vice versa) in the same quantum state does not change the energy of the nucleus. However, Heisenberg considered it necessary to introduce a continuous rotation group in isotope space, which smoothly transformed a neutron into a proton via a rotation of 180° through mystical intermediate states! Not the simplest, but the most complex and refined formulation proved to be the most fruitful. The depth of Heisenberg's formulation was revealed in the transition from nuclei to mesons. Concepts constructed by analogy with isotopic rotation shone especially brightly in connection with the theory of quark colour, gradient invariance, and Young–Mills theory.

I won't describe in detail my papers on particles — they are presented in this book and are very competently commented on. It is clear from the commentaries, wiped of their jubilee politeness, how many mistakes I've made. There are even more mistakes in papers which have been published, but which are not included in this collection.

In this book are my papers in the field of astrophysics and associated commentaries. It is not expedient to dispute these commentaries. Today, my most important separate piece of research is my non-linear theory for the formation of structure in the universe, or, as it is now called, the 'pancake' theory. The structure of the universe, its evolution, and the properties of the matter making up the hidden mass have not yet been resolved conclusively. A large role in this work was played by A.G. Doroshkevich, R.A. Sunyaev, S.F. Shandarin, and Ya.E. Einasto. The work continues. However, the 'pancake' theory is beautiful even by itself; if the initial assumptions are valid, the theory gives a correct and non-trivial answer. The 'pancake' theory is a contribution to synergetics. It was especially pleasant for me to learn that this work to some extent incited mathematical investigations by V.I. Arnol'd and others. A large volume of work on the spectrum of the relict radiation in the presence of perturbations has been hanging in the air — the universe turned out to be very smooth, and the perturbations too small.

The diagnostic proposed by myself and R.A. Sunyaev for hot plasma based on scattering of relict radiation with spectral distortion has survived and remains of great interest.

To a significant extent, my work (together with my closest co-workers, above all R.A. Sunyaev, A.G. Doroshkevich, S.F. Shandarin, and — until 1978 — I.D. Novikov) in the field of astrophysics proved to lend itself to promotion, popularisation, and pedagogy. All this is necessary and useful, however, it is evaluated according to criteria other than the acquisition of the original results.

At the beginning of my astrophysical activity, I was bothered by habits acquired in the course of my applied activities. An astrophysicist should pose the questions: how is nature constructed? What observations provide the possibility of elucidating this? However, I formulated the problem more like this: how would it be best to construct the universe, or a pulsar, in order to satisfy given technical conditions — forgive me, I meant to say direct observations? This is how the idea of a cold universe arose, and my idea of a pulsar as a white dwarf in a state of strong radial oscillations. As justification for these ideas, I can only say that I was never stubborn about my errors. Apparently, on the whole, my activity — scientific and propagandistic — has been useful. Astronomers took me into their ranks. My election to the National Academy of the USA and to the Royal Society, as well as my gold medals from the Astronomical Society of the Pacific and the Royal Astronomical Society are associated with my astronomical research. It was a great honour for me to be invited to give a talk about modern cosmology at the XIII General Assembly of the International Astronomical Union. Greece, the columns of an ancient theatre, a black starry sky above me, the audience on marble benches, my anxiety before the talk, as well as during the talk and its favourable conclusion. Life goes on, and cosmology goes deeper into the area where physics is far removed from experimental verification. The new generation of theorists talk not about the first three minutes or seconds, nor about nuclear reactions in plasma. Processes on the Planck scale of 10^{-33} cm, over the Planck time of 10^{-43} s, with the Planck energy of 10^{19} GeV are discussed. Stephen Hawking, A.D. Linde, A.A. Starobinskii, A. Guth, and others lead the way. Field theory considers 5-, 11-, and 26-dimensional spaces. Under laboratory conditions, these will necessarily imitate our usual (3+1)-dimensional space-time, and the extra dimensions are hidden, rolled up, leaving traces only in the systematics of particles and fields. Twenty-year-old kids come and immediately, without the burden of previous work and traditions, sink their teeth into the new topics. Do I look like a mastodon or archaeopteryx among them?

I am comforted by the reconstruction of my psyche with age. Currently (a few days before my seventieth birthday), I am less interested by competitive motives, whether precisely I will utter that 'eh,' about which Bobchinskii and Dobchinskii argued.[177] The final result, the physical truth, interests me, almost regardless of who finds it first. It would be enough if only I had the strength to understand it!

Humanity, like never before, is on the threshold of amazing discoveries. The idea of all-unifying theories presents itself more and more vividly, and an ever larger role is played by geometry. It could be that in the higher, rather than the literal, sense, Einstein is right, and his theory, which reduces the gravitational force to geometry, will provide a model for comprehensive theories.

It is probable that cosmology will provide a testing chamber for the verification of the new theories. In this connection, I recall the work of S.S. Gershtein, V.F. Shvartsman, S.B. Pikelner, L.B. Okun', I.Yu. Kovzarev, M.Yu. Khlopov, and myself as the first timid applications of cosmological reasoning to the solution of questions in particle theory that are not currently accessible experimentally. Together with L.P. Grishchuk and A.A. Starobinskii, I am attempting to progress in the analysis of the birth of the universe. In the middle of the 1980s, the most difficult and fundamental questions in natural science were braided together into a tight knot. There is no desire in me stronger than to live to see their solution and to understand it.

Moscow
March 3, 1984

---

[177] The heroes of Gogol's comedy *Revizor*. Their silly argument about who first pronounced the sound 'eh' triggered a whole chain of events. [Translator]

The social and human importance of fundamental science[178]

Ya.B. Zeldovich

In solving the problem of how to maximally satisfy the material and spiritual needs of humankind, a large role is assigned to science. The thesis that science has become an industrial force is well known. When characterising the economy of one or another country or region, people talk about scientifically charged technologies (i.e., when production and competitiveness is directly connected to the level of science). Scientifically charged technologies include the production of microelectronic circuits and their application in computational and information technology, or the production of pharmaceutical preparations using genetic engineering. The list could be continued endlessly.

Developed countries spend some percentage of their gross national product on science. This fact also demonstrates the importance of science in modern manufacturing.

Together with practical goals, science also pursues fundamental tasks, such as elucidating the structure of the microworld and investigating the universe, understanding life and its origin.

Elementary-particle accelerators are built — they are technically complex and very elaborate installations occupying many square kilometres, with a cost of several billion dollars. Gigantic telescopes are constructed. Radio astronomy is being developed. Astronomical instruments are carried outside the atmosphere into space. Theoretical groups consisting of extremely capable people work intensively on the development of fundamental theories using, in particular, modern computational means. Specialists from other fields could probably also present a long list of problems, instruments and methods.

There is also a general pattern to scientific investigations: in many fields, there is a more or less well-defined division of research into two types. First, there is work where a practical goal is envisioned ahead of time — so-called applied science. Second, there is work that sets the goal of acquiring knowledge, of understanding the micro- and macroworlds, without orienting toward specific practical tasks ahead of time. Without belittling the importance of work of applied science, we call this second type of research fundamental science.

This terminology is not very good. If we were to understand the word fundamental to mean basic, firm, solid, we would need to change the terminology. When solving practical problems, we use firm and solidly established laws. And the opposite is true: when investigating unsolved questions about the microworld and the universe, we must to a large degree use bold hypotheses. Some of these must then be thrown out due to the

[178] *Voprosy filosofii*, 1985, no. 6, p. 57.

weight of new experimental data or as a consequence of internal contradictions that emerge as the ideas are developed. The degree to which a task is fundamental matches the boldness of various theories.

All the same, in the absence of a better generally accepted term, we will talk about 'fundamental' science. What is its role in society? Why do societies spend quite large amounts on the development of fundamental science?

I will begin with the arguments of sceptics, with which I disagree. A major Soviet scientist, now deceased, once said: 'Science is the means of satisfying one's personal curiosity at the state's expense.' I won't name him; I'll only comment that this scepticism most likely interfered with, rather than aided, the scientific activity of the author of this biting, but fundamentally wrong aphorism.

In this same vein, the sceptics tell a historical anecdote. Laplace was interested in the shape of the Earth, that is, the ratio of the length of the equator to the length of a line passing through the pole. However, in the stormy and difficult years of the French Revolution, it was extremely difficult to obtain grants for the precise geodesic measurements required. I note in passing that now, such measurements are conducted not only for peaceful but also for military purposes. But let us return to the times when intercontinental ballistic missiles didn't exist, and to Laplace. Laplace proposed to the Convent a new revolutionary unit of length — the metre, defined to be 1/40,000,000 of the Earth's equator. A decree was adopted, a grant was given, and the measurements were conducted. Together with determining the unit of measure — the metre — Laplace also learned how the Earth is flattened at the poles and bulging at the equator due to the action of the centrifugal force of its rotation. But did the great French scientist abuse the trust of the members of the Convent?

Thus, one view of fundamental science is the object of the personal curiosity of one or more scientists who obtain state money in any way. This attitude is not new; recall the satire of Jonathan Swift in *Gulliver's Travels*, the description of the island of Laputa, the scientists living there and the themes of their investigations.

However, this view is superficial and quite incorrect. I hope that I will be able to demonstrate this to even the critical reader.

Another view of fundamental science is based on still more convincing historical examples.

Newtonian mechanics is necessary for the engineering industry — from weaving looms to aviation, for our entry into space and its applications in communication technology.

The electrodynamics of Faraday and Maxwell became the basis of the electrification of industry, and led Popov to the development of radio communications.

The theory of the atom and quantum mechanics led Einstein to predict the existence of stimulated emission. Based to a large extent on these predictions, Basov and Prokhorov created lasers, with all their applications in technology and medicine.

The development of the theory of the atomic nucleus begun by Rutherford led to the creation of nuclear energetics via the ignition of uranium, followed by the thermonuclear ignition of deuterium and tritium (heavy isotopes of hydrogen). The tragic side effect of this was the creation of nuclear and thermonuclear weapons.

This list could be continued almost infinitely, and could fill an encyclopaedia of many volumes.

In the examples I have presented, it is characteristic that in most cases, the original discoverers did not foresee the far-reaching consequences of their discoveries. Sometimes they were able to make guesses about these consequences: when asked about the importance of the discovery of electromagnetic induction, Faraday answered with the question, 'Maybe you, looking at a new-born baby, can say what he'll accomplish in his lifetime.' Without being specific, Faraday anticipated a great future for his discovery.

But Einstein didn't suspect that lasers would be discovered. Rutherford, who found schemers tedious, to the end of his life (1937) firmly denied the possibility of energetic applications of nuclear physics.

We come to a concept that can be briefly expressed as follows: 'Every good piece of fundamental science brings practical results.' However, we will maintain an unbiased attitude toward this thesis. We will not rely on authorities, and will not make reference to historical experience. Its strength is that it is experience, and its weakness is that it is historical; in other words, it relates to the past, to the situation in science in the past, which differs from the situation in the present time.

Now, we not only know more individual scientific facts — we understand much deeper the internal relations between various fields of science.

One main achievement can be called the correspondence principle. This is a very broad generalisation of the specific 'correspondence principle' used by Bohr in the development of his theory of the atom. Briefly, this principle is characterised as follows: theories exist, and become established forever in their field; new theories must correspond to these established theories, developing and changing them only in new fields of application.

Let me present just two widely known examples:

1. Newtonian classical mechanics has been established forever for large bodies moving with small velocities.

2. Science penetrates the depths of the proton and neutron — the bricks from which the nucleus is constructed. It turned out that these, in turn, are composed of so-called quarks — particles with fractional charge, which it is fundamentally impossible to extract from the nucleus. However, the general picture of an atom consisting of a nucleus and electrons has not changed, and remains in perpetuity. A new step in our understanding of nuclear physics does not change chemistry and atomic physics, and does not promise new sources of energy!

The conclusion following from these examples and the generalised correspondence principle is that an understanding of general laws for the development of science has emerged. This knowledge does not allow us to make vague promises. Together with examples of fruitful practical applications of the achievements of fundamental science (aviation, electronics, radio and television, atomic energy, informatics), we can also present counter examples.

The general theory of relativity — the geometrical theory of gravitation created by Einstein in 1916 — is undoubtedly a splendid achievement. This theory has greatly advanced our concepts of the forces of nature, and fully explained the essence of one of these forces — gravitation reducing it to geometry. At the end of our century, the general theory of relativity is becoming a model for the further development of fundamental physics. Enormous progress in astronomy, and particularly in cosmology — the science of the universe as a whole — has been associated with this field. At the same time, the general theory of relativity does not have practical (i.e., energetic, informational, medical) applications. This means that we cannot (and should not) say that each good theory must yield practical fruits.

Everything I've said above must be considered a brief foreword to the third view of fundamental science, which is the main thesis of this article.

Fundamental science is also necessary because it satisfies the spiritual needs of humankind.

These spiritual needs cannot be reduced to the appreciation of art, music, and the beauty of nature. Knowledge and understanding of the structure of nature is also an extremely important need.

I especially want to emphasize that this is a need in most people, and not only scientists. Here, it is appropriate to draw a comparison between those who listen to music (and there are many such people) and composers.

Appreciation of the beauty of science must be taught. It may be that the specialists (I don't exclude myself) are guilty of insufficiently promoting understanding of the essence of science. Middle school could to a large extent teach pupils the most general concepts about the problems and methods of science, even without the specific details required by specialists.

In the 1950s, the dramatic applications of nuclear physics greatly increased the prestige of this science.

The competition for entrance to physics departments increased. Grants were given to increase the size of accelerators. I remember the comment of a major foreign physicist: 'Large accelerators have become an object of governmental prestige, as in the Middle Ages it was prestigious to construct huge cathedrals.' Here, there is a challenge — an allusion to the similarity between the functions of fundamental science and religion. I accept this challenge, and take up the gauntlet. In certain historical periods, religion played a progressive role, united nations, and ordered the life of society. Currently, the role of religion is waning. I am an atheist; I do not believe in God, and I hope that a rational world-view will become common. At the same time, the existence of religion is an objective historical fact. The socio-psychological conclusion following from this fact, in essence, supports the main thesis of my article: humankind objectively has spiritual needs that are deeply lodged in its consciousness. It can only be a good thing if interest in science occupies the place in the spiritual life of humankind that was previously occupied by religion.

But I also emphasize the difference between science and religion. There is no single religion; there are many different religions, created by different social and historical (or pre-historical) conditions.

Battles between different religions have taken the most unnatural, brutal, and bloody forms. In the name of God, those with other beliefs have been burned, executed, and driven out (you only have to look at any text on the history of religion).

Science differs from religion in that it studies objectively existing laws of nature. There is only one science, and its conclusions, verified by experiments, are the same, whatever the country in which the experiment is conducted, and whatever the colour of the experimenter's skin.

This simple and obvious fact links scientists all around the world. With properly formulated propaganda of scientific knowledge, the international character of fundamental science calls forth mutual respect between the peoples of various countries. The importance of this factor in the current time of dangerous growth in international tensions cannot be overestimated.

During international competition in technology and applied science, it is possible to place an artificial restriction on the dissemination of new information so that it cannot be used by one's competitors.

In fundamental science, withholding information about new results and printing envelopes with the inscription 'Open in 10 or 20 years' became a thing of the past long, long ago. The modern scientist rushes to publish his results, or even his hypotheses and guesses. Together with the journals, preprints are issued within two or three weeks. The role of personal contact at conferences and symposia has grown immensely. Fundamental science plays an increasingly beneficial role in strengthening international relations that are impervious to local or temporal fluctuations.

The ancient Greeks valued science highly, but looked down on its applications. Currently, applied science has won a position in society where it does not need to be defended. But let's not forget about fundamental science.

Let's not forget the role of fundamental science in the creation of applied science, and also preserve our respect and delight in fundamental science in its own right.

It is a splendid creation of the human mind, and, in turn, it perfects the human mind and soul.

In order for this not to be considered unsubstantiated, I'll present two examples of the achievements of the 1980s in fields that are closest to me.

One example is related to the theory of elementary particles.

The theory of electromagnetism has been developed for more than 100 years. Radioactivity has been studied for nearly 100 years. But until recently, it seemed that these were fundamentally different phenomena.

Electromagnetism is connected with the motion of charged particles — electrons. The constructor of an electrical generator, electrical motor, or radio apparatus imagines electrons as tiny, very light, point-like charges moving along wires, creating currents and magnetic fields, collecting on capacitor plates, and giving rise to discharges in an electric field. Electrons can only move from place to place, but cannot be created or destroyed.

In radioactivity phenomena, an atomic nucleus can emit, that is, create and eject, an electron.

It is important to emphasize that the electron was not previously present in the nucleus — it was created. There are also nuclei that absorb and destroy electrons that were previously abiding peacefully among the atoms.

Thus, it seemed that electromagnetism and radioactivity were in no way connected. However, over the past 20 years, a theory unifying these two fields of physics has emerged.

The motion of the electron is imagined as its disappearance in one place and appearance in another. An electromagnetic field can be viewed as an ensemble of special particles — photons, or quanta (bundles) of electromagnetic field, flying at the speed of light. Analysing both the similarities and differences between radioactivity and electromagnetism, theoretical physicists came to the conclusion that other particles should exist as well, which resemble photons in many ways, but differ from them in their large mass. New predictions of the existence of three sorts of particles — positive, neutral, and negative — that are nearly one hundred times heavier than a hydrogen atom! Accordingly, their velocities are always less than the speed of light. These particles are unstable and decay, and so cannot be used in an electrical motor or for radio communications. In 1983, these particles were discovered as a result of long, difficult, but well-directed research. The experimental verification of theoretical predictions can always be considered an example of the strength of a theory, and also of science as a whole. This was the case when Mendeleev predicted the existence of elements unknown at the time, and Leverrier predicted the existence of a previously unknown planet. The time of great predictions in science has not passed.

The second example of achievements in fundamental science is modern progress in cosmology — the science of the entire universe. The general picture of an expanding universe long ago — several decades ago — passed into the ranks of firmly established facts. In the new formulation characteristic of the past five or ten years, answers are emerging to the questions of why the universe is expanding, and how the initial conditions whose evolution we observe could have arisen. Here, there is no space to explain in any detail these difficult theoretical questions. I'll only note that: (1) the gravitational interaction with existing matter can compensate the loss of energy in the creation of new matter; (2) under certain conditions, gravitation leads to separate parts of the system pushing off from each other, giving the system as a whole the motion characteristic of the expanding universe; (3) the emergence of ordinary matter in a hot plasma in which matter and antimatter were initially present in equal amounts is possible. None of these points have been proven by direct experiments (and won't be proven any time soon), but they are in agreement with modern theories, and do not contradict general principles, such as the law of conservation of energy. Together, these theoretical ideas provide the opportunity to understand the origin and properties of the current universe.

There is no doubt that a biologist, geologist, or solar-system researcher would have presented another set of achievements of fundamental science in his own field of knowledge. And all the same, the world of elementary particles and the universe as a whole are two extremes, between which lie everything else.

One other extraordinarily important function of fundamental science is its role in educating a qualified workforce.

I remember a discussion in 1938, when Academician Abram Fedorovich Ioffe, the director of the Leningrad Physical Technical Institute (LPTI) was charged with conducting active work in the field of nuclear physics at the institute. They said that studying nuclei was not in agreement with the word 'technical' in the name of the LPTI.

Several years went by, nuclear physics became the most important problem on a national scale, and an alumnus of LPTI, Igor' Vasil'evich Kurchatov was in charge of this most important work, attracting to the field first and foremost his colleagues from the institute. But even when such a dramatic turnaround doesn't occur, scientists who have studied fundamental science and displayed their abilities in this field provide an extremely valuable workforce for the solution of applied problems. They bring the methods and working style of fundamental science to these problems: courage, collectivism, the highest qualifications.

The exchange of personnel, expertise, and equipment between various branches of science is an extremely important condition for the flowering of all fields of science.

In conclusion, I would like to recall the words of Tyutchev:

> Blessed is he who visited this world
> In its fateful minutes:
> He was called by the All-holies,
> Like a companion, to their feast.

The All-holies are gods, let's forgive Tyutchev his manner of expression. Let our time not be fateful in the sense of horrors and catastrophe. The fateful minutes in this context are the years and decades when the fate of our understanding of Nature — infinite, yet knowable — is decided.

With children and son-in-law (1960).

With wife, V.P. Konstantinova (Warsaw, 1970).

With A.D. Sakharov at a conference (1987).

With N. Manson and Li (Poland, Karpach, 1977).

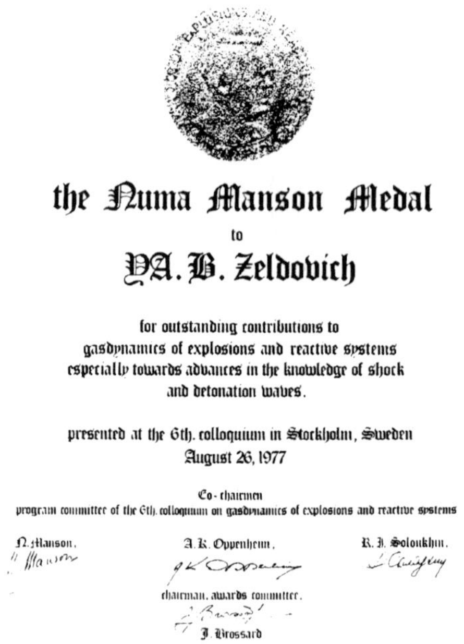

the Numa Manson Medal

to

YA. B. Zeldovich

for outstanding contributions to
gasdynamics of explosions and reactive systems
especially towards advances in the knowledge of shock
and detonation waves.

presented at the 6th. colloquium in Stockholm, Sweden
August 26, 1977

Co-chairmen
program committee of the 6th. colloquium on gasdynamics of explosions and reactive systems

N. Manson. A. K. Oppenheim. R. I. Soloukhin.

chairman, awards committee.

J. Brossard

Diploma to the Manson medal.

Far from all organisational problems solved: Ya.B. Zeldovich, A.D. Sakharov, and R.A. Sunyaev at the international workshop on cosmology and X-ray astronomy within the framework of the Conference on the 30th anniversary of the launch of the first satellite (Moscow, 1987).

With I.S. Shklovskii.

Ya.B. Zeldovich and Prof. Valentin Telegdi during the High Energy Physics Conference in Kiev, 1959.

At a lecture in Sternberg Astronomical Institute.

In the department of relativistic astrophysics of Sternberg Astronomical Institute. From left to right: Rashid Sunyaev, Lev Rozhansky, Nikolai Shakura, Valentin Rudenko, Yakov B. Zeldovich, Vladimir Lipunov, Leonid Grishchuk, unidentified student, Alexei Starobinsky.

With Academicians I. Khalatnikov and G. Flerov (1984).

Ya.B. Zeldovich in the day of his 70th anniversary.

Ya.B. Zeldovich, his wife Angelika Yakovlevna Vasilyeva, the wife of Ed Salpeter Mika, Gyusal Sunyaeva, and Rashid Sunyaev. (Courtesy of Ed Salpeter.)

John Archibald Wheeler, S. Chandrasekhar, Ya.B. Zeldovich, and I.D. Novikov at the IAU symposia in Warsaw and Krakow in 1973. (Courtesy of Bill Press.)